METHODS IN MOLECULAR BIOLOGY

Series Editor
John M. Walker
School of Life and Medical Sciences
University of Hertfordshire
Hatfield, Hertfordshire, AL10 9AB, UK

For further volumes:
http://www.springer.com/series/7651

Morphogen Gradients

Methods and Protocols

Edited by

Julien Dubrulle

Integrated Microscopy Core, Baylor College of Medicine, Houston, TX, USA

✲ Humana Press

Editor
Julien Dubrulle
Integrated Microscopy Core
Baylor College of Medicine
Houston, TX, USA

ISSN 1064-3745   ISSN 1940-6029   (electronic)
Methods in Molecular Biology
ISBN 978-1-4939-8771-9   ISBN 978-1-4939-8772-6   (eBook)
https://doi.org/10.1007/978-1-4939-8772-6

Library of Congress Control Number: 2018956811

Preface

During embryonic development and tissue growth, progenitor cells receive cues from their environment that instruct them to differentiate according to their spatial location. This positional information is conveyed by molecules called morphogens that form a concentration gradient across the tissue. These morphogens, whether they are secreted peptides, metabolites, or diffusible transcription factors, are produced from a source and travel through the responsive field by passive diffusion, advection, or active cell-to-cell transport. Different concentrations of the morphogen trigger different responses in the receiving cells, leading to a spatial arrangement of cells with specific identities that depend on their distance from the source.

The study of morphogen gradients combines biophysics, cell and developmental biology, and applied mathematics approaches to understand how they form, what their biological functions are, and to predict their behavior in space and time. This book presents methods focusing on the visualization of morphogen gradients, the analysis of their biophysical and biological properties, and the theoretical aspects underlying their functions.

Houston, TX, USA   *Julien Dubrulle*

Contents

Contributors

DIANA BARAC • *Department of Biosystems Science and Engineering, ETH Zurich, Basel, Switzerland*

DAVID CHEUNG • *Division of Developmental Biology, Cincinnati Children's Research Foundation, Cincinnati, OH, USA; Division of Biomedical Informatics, Cincinnati Children's Research Foundation, Cincinnati, OH, USA*

MATHIEU COPPEY • *Institut Curie, PSL Research University, CNRS, Sorbonne Université, Physico Chimie, Paris, France*

MARAYSA DE OLIVERA-MELO • *Department of Cell Biology, University of Virginia School of Medicine, Charlottesville, VA, USA*

NATHALIE DOSTATNI • *Institut Curie, PSL Research University, CNRS, Sorbonne Université, Nuclear Dynamics, Paris, France*

LIJUAN DU • *Department of Cell Biology and Molecular Genetics, University of Maryland, College Park, MD, USA*

MARÍA FLORENCIA ERCOLI • *IBR (Instituto de Biología Molecular y Celular de Rosario), CONICET, Universidad Nacional de Rosario, Rosario, Argentina*

TERESA FERRARO • *Ecole Normale Supérieure, PSL Research University, CNRS, Sorbonne Université, Laboratoire de Physique Théorique, Paris, France; Institut de Biologie Paris-Seine, Sorbonne Université, CNRS, Developmental Biology, Paris, France*

CÉCILE FRADIN • *Institut Curie, PSL Research University, CNRS, Sorbonne Université, Nuclear Dynamics, Paris, France; McMaster University, Hamilton, ON, Canada*

PAUL FRANÇOIS • *McGill University, Montreal, QC, Canada*

CAMILA GOLDY • *IBR (Instituto de Biología Molecular y Celular de Rosario), CONICET, Universidad Nacional de Rosario, Rosario, Argentina*

FENG HE • *Division of Medical Genetics and Genomics, Children's Hospital, Zhejiang University School of Medicine, Hangzhou, Zhejiang, China; Institute of Genetics, Zhejiang University School of Medicine, Hangzhou, Zhejiang, China; Key Laboratory of Pollinating Insect Biology of the Ministry of Agriculture, Institute of Apicultural Research, Chinese Academy of Agricultural Sciences, Beijing, China*

ADRIEN HENRY • *McGill University, Montreal, QC, Canada*

NATHALIE HOUSSIN • *Department of Cell Biology, University of Virginia School of Medicine, Charlottesville, VA, USA*

YAN HUANG • *Agricultural and Biological Engineering, Purdue University, West Lafayette, IN, USA*

DAGMAR IBER • *Department of Biosystems Science and Engineering, ETH Zurich, Basel, Switzerland*

LAURENT JUTRAS-DUBÉ • *McGill University, Montreal, QC, Canada*

ANNA KICHEVA • *Institute of Science and Technology Austria (IST Austria), Klosterneuburg, Austria*

ANATOLY B. KOLOMEISKY • *Department of Chemistry and Center for Theoretical Biological Physics, Rice University, Houston, Texas, USA*

CHRISTINE LANG • *Department of Biosystems Science and Engineering, ETH Zurich, Basel, Switzerland*

JUN MA • *Division of Medical Genetics and Genomics, Children's Hospital, Zhejiang University School of Medicine, Hangzhou, Zhejiang, China; Institute of Genetics, Zhejiang University School of Medicine, Hangzhou, Zhejiang, China; Division of Developmental Biology, Cincinnati Children's Research Foundation, Cincinnati, OH, USA; Division of Biomedical Informatics, Cincinnati Children's Research Foundation, Cincinnati, OH, USA; Laboratory of Systems Developmental Biology, Institute of Genetics, Zhejiang University School of Medicine, Hangzhou, Zhejiang, China*

YOSHIKATSU MATSUBAYASHI • *Division of Biological Science, Graduate School of Science, Nagoya University, Nagoya, Japan*

PATRICK MÜLLER • *Friedrich Miescher Laboratory of the Max Planck Society, Tübingen, Germany*

MICHAEL D. MULTERER • *Department of Biosystems Science and Engineering, ETH Zurich, Basel, Switzerland*

XUE WEN NG • *Department of Chemistry and Centre for Bioimaging Sciences, National University of Singapore, Singapore, Singapore*

JAVIER F. PALATNIK • *IBR (Instituto de Biología Molecular y Celular de Rosario), CONICET, Universidad Nacional de Rosario, Rosario, Argentina; Centro de Estudios Interdisciplinarios, Universidad Nacional de Rosario, Rosario, Argentina*

CARMINA ANGELICA PEREZ-ROMERO • *Institut Curie, PSL Research University, CNRS, Sorbonne Université, Nuclear Dynamics, Paris, France; McMaster University, Hamilton, ON, Canada*

MANAN'IARIVO RASOLONJANAHARY • *Department of Mathematical Sciences, University of Liverpool, Liverpool, UK*

RAMIRO E. RODRÍGUEZ • *IBR (Instituto de Biología Molecular y Celular de Rosario), CONICET, Universidad Nacional de Rosario, Rosario, Argentina; Centro de Estudios Interdisciplinarios, Universidad Nacional de Rosario, Rosario, Argentina*

SOUGATA ROY • *Department of Cell Biology and Molecular Genetics, University of Maryland, College Park, MD, USA*

KARUNA SAMPATH • *Division of Biomedical Sciences, Warwick Medical School, University of Warwick, Coventry, UK*

HIDEFUMI SHINOHARA • *Division of Biological Science, Graduate School of Science, Nagoya University, Nagoya, Japan*

GARY H. SOH • *Friedrich Miescher Laboratory of the Max Planck Society, Tübingen, Germany*

HOLLY STAINTON • *Department of Biomedical Science, University of Sheffield, Sheffield, UK*

ANNA STOPKA • *Department of Biosystems Science and Engineering, ETH Zurich, Basel, Switzerland*

HAMID TEIMOURI • *Department of Physics and FAS Center for Systems Biology, Harvard University, Cambridge, MA, USA*

BERNARD THISSE • *Department of Cell Biology, University of Virginia School of Medicine, Charlottesville, VA, USA*

CHRISTINE THISSE • *Department of Cell Biology, University of Virginia School of Medicine, Charlottesville, VA, USA*

MATTHEW TOWERS • *Department of Biomedical Science, University of Sheffield, Sheffield, UK*

HUY TRAN • *Institut Curie, PSL Research University, CNRS, Sorbonne Université, Nuclear Dynamics, Paris, France; Ecole Normale Supérieure, PSL Research University, CNRS, Sorbonne Université, Physique Théorique, Paris, France*

DAVID UMULIS • *Agricultural and Biological Engineering, Purdue University, West Lafayette, IN, USA; Weldon School of Biomedical Engineering, Purdue University, West Lafayette, IN, USA*

BAKHTIER VASIEV • *Department of Mathematical Sciences, University of Liverpool, Liverpool, UK*

RODRIGO VENA • *IBR (Instituto de Biología Molecular y Celular de Rosario), CONICET, Universidad Nacional de Rosario, Rosario, Argentina*

ALEKSANDRA M. WALCZAK • *Ecole Normale Supérieure, PSL Research University, CNRS, Sorbonne Université, Laboratoire de Physique Théorique, Paris, France*

LUCAS D. WITTWER • *Department of Biosystems Science and Engineering, ETH Zurich, Basel, Switzerland*

THORSTEN WOHLAND • *Department of Chemistry and Centre for Bioimaging Sciences, National University of Singapore, Singapore, Singapore; Department of Biological Sciences, National University of Singapore, Singapore, Singapore*

HONGGANG WU • *State Key Laboratory of Brain and Cognitive Sciences, Institute of Biophysics, Chinese Academy of Sciences, Beijing, China; University of Chinese Academy of Sciences, Beijing, China*

PENG-FEI XU • *Department of Cell Biology, University of Virginia School of Medicine, Charlottesville, VA, USA*

MARCIN ZAGORSKI • *Institute of Science and Technology Austria (IST Austria), Klosterneuburg, Austria*

Part I

Visualization of Morphogen Gradients

Chapter 1

Analysis of Expression Gradients of Developmental Regulators in *Arabidopsis thaliana* Roots

María Florencia Ercoli, Rodrigo Vena, Camila Goldy, Javier F. Palatnik, and Ramiro E. Rodríguez

Abstract

The regulatory mechanisms involved in plant development include many signals, some of them acting as graded positional cues regulating gene expression in a concentration-dependent manner. These regulatory molecules, that can be considered similar to animal morphogens, control cell behavior in developing organs. A suitable experimental approach to study expression gradients in plants is quantitative laser scanning confocal microscopy (LSCM) using *Arabidopsis thaliana* root tips as a model system. In this chapter, we outline a detailed method for image acquisition using LSCM, including detailed microscope settings and image analysis using FIJI as software platform.

Key words miR396, *PLETHORA*, Laser scanning confocal microscopy, Root, *Arabidopsis thaliana*

1 Introduction

Both plants and animals rely on stem cells to generate the different cell types that shape their bodies, although plants, unlike animals, generate new organs throughout their life cycle. Plant growth and development is supported by meristems, which are collections of proliferating cells. The shoot and root apical meristems are responsible for the aboveground and belowground organs of the plant, respectively. They both harbor small collections of stem cells (SCs), which generate by asymmetric formative cell divisions the various cell types that constitute plant organs [1]. The daughters of the SCs usually engage in several rounds of fast cell division cycles combined with limited cell elongation to provide enough cells for organ growth. Later on, cell proliferation ceases and cells enlarge to reach their final size, shape and function.

The *Arabidopsis thaliana* root is an excellent model to study plant development [2] as the meristem located at the tip of the root supports the undetermined growth of the organ. It has a simple,

Julien Dubrulle (ed.), *Morphogen Gradients: Methods and Protocols*, Methods in Molecular Biology, vol. 1863, https://doi.org/10.1007/978-1-4939-8772-6_1, © Springer Science+Business Media, LLC, part of Springer Nature 2018

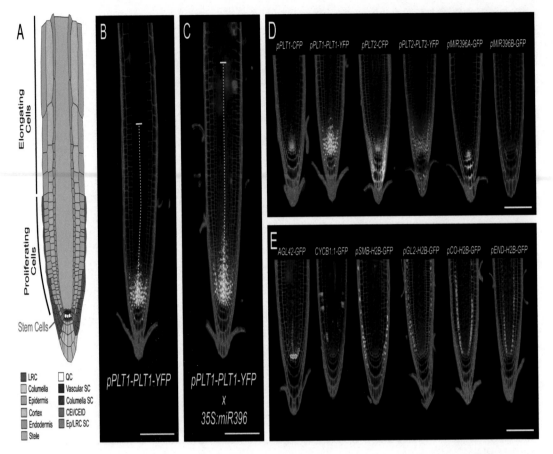

Fig. 1 Laser scanning confocal microscopy to study gradients of gene expression in *Arabidopsis thaliana* roots. (**a**) Scheme of a medial longitudinal section of an *Arabidopsis thaliana* root tip. Different colors indicate different cell types. The developmental zones are indicated at the left of the scheme. *QC* quiescent center, *SC* stem cells, *CEI/CEID* cortex–endodermis initial and cortex–endodermis initial daughter, *Ep/LRC SC* epidermis and lateral root cap stem cells. (**b**) Expression gradient of *Arabidopsis thaliana PLT1* determined by laser scanning confocal microscopy of propidium iodide-stained roots from a *PLT1* YFP reporter (*pPLT1-PLT1-YFP*). The reporter is a C-terminal translational fusion of YFP to the complete *PLT1* gene including the upstream regulatory sequences. The dotted line indicates the trace used to measure the size of the meristem. (**c**) Modifications in the *PLT1* expression gradient by miR396 overexpression (*35S:miR396*). The dotted line indicates the trace used to measure the size of the meristem. (**d**) LSCM images of *Arabidopsis thaliana* transgenic lines expressing fluorescent protein reporters of *PLT1*, *PLT2*, *MIR396A*, and *MIR396B*. *pPLT1-CFP*, *pPLT2-CFP*, *pMIR396A-GFP*, and *pMIR396B-GFP* are transcriptional reporters of *PLT1*, *PLT2*, and miR396-coding genes. *pPLT1-PLT1-YFP* and *pPLT2-PLT2-YFP* are translational reporters of *PLT1* and *PLT2*. (**e**) LSCM images of selected *Arabidopsis thaliana* transgenic lines expressing fluorescent reporters suitable for analyzing the developmental consequences that result from the modification of the expression gradients of developmental regulators (*see* Table 1 for a full description). Scale Bar: 100 μm

stereotyped, and stable cellular organization, with a limited number of cell types and tissues organized in a radial pattern that is easily accessible experimentally (Fig. 1a). In the longitudinal axis, SCs are located in the tip of the meristem in a region named the stem cell

Table 1
Description of selected *Arabidopsis thaliana* transgenic lines expressing fluorescent reporters suitable for monitoring the *PLT1* expression gradient and for analyzing the developmental consequences that result from its modification

Line	Description	Expression pattern	References[a]
pPLT1-CFP	Transcriptional reporter line of *PLT1*, a transcription factor of the *PLETHORA* family that functions as master regulators of root development	Expressed in a gradient with the highest levels in the stem cell niche	[3, 4]
pAGL42-GFP	GFP reporter of *AGAMOUS-LIKE42* (*AGL42*), a MADS box transcription factor	Quiescent center plus adjacent stele and ground tissue stem cells	[5]
pCYCB1;1-GFP	Fusion of GFP to the promoter of the mitotic cyclin CYCB1;1. Includes also the destruction box located in the amino terminal region of the cyclin	Cells in the G2/M phases of the cell cycle. Can be used to monitor the cell proliferation activity of the different developmental regions of the root meristem	[6–8]
pSMB-H2B-GFP	A root cap-specific NAC domain transcription factor, is expressed just after the asymmetric cell division that generates the lateral root cap cells	Expressed in lateral root cap cells. Additional layers of LRC can result from changes in the activity of the SCN niche	[7, 9]
pGL2:H2B-YFP	Contains a YFP fusion to histone H2B. Expressed only in root epidermal cells	Expressed in non-hair epidermal cells	[10]
pCO:H2B-YFP	Contains a YFP fusion to histone H2B. Expressed only in root cortex cells	Expressed in cortex cells	[10]
pEND:H2B-YFP	Contains a YFP fusion to histone H2B. Expressed only in root endodermal cells	Expressed in endodermal cells	[10]

[a]The references indicated here describe the preparation of the reporter line and/or its utilization to analyze the developmental consequences of modifying the expression of *PLT* expression gradients

niche (SCN), proliferating cells can be found in the next 300 μm of the meristem, and elongating cells are located further above (Fig. 1a).

These spatially separated regions are determined, at least partially, by the gradient-shaped expression of regulatory molecules. Both axes, radial and longitudinal, have been shown to be controlled by the expression gradient of transcription factors and small noncoding regulatory RNAs [6, 11]. These regulatory molecules act in a similar way to animal morphogens defining positional developmental outcomes according to their expression levels.

PLETHORA1 (*PLT1*) and *PLETHORA2* (*PLT2*) are transcription factors whose expression forms a gradient along the longitudinal axis of the root (Fig. 1b, d). The highest level of PLT1 protein are found in the SCN and specifies SC identity, a reduction of PLT1 levels is required for cells to start the fast proliferation phase, while a further decrease is required for cells to elongate [4, 6, 12, 13]. These expression gradients are established from a narrow transcriptional domain located in the SCN region from which PLT protein spreads thanks to cell-to-cell movement and mitotic protein segregation [6]. Different regulatory networks have been linked to the establishment of PLT expression gradients. For example, transcription factors of the *GROWTH-REGULATING FACTOR* class have been shown to repress *PLT* expression in proliferating cells. In turn, *GRFs* are regulated negatively by the small regulatory RNA microRNA (miRNA) miR396 [14, 15]. As a consequence, miR396 overexpression down-regulates the GRFs and leads to changes in PLT expression gradients [7] that result in modifications in root development (Fig. 1c). Interestingly, miR396 is expressed differentially between the developmental zones of the root (Fig. 1d) and is induced by *PLT* genes [7]. These regulatory interactions define a gene regulatory network that shapes the *PLT* expression gradient and helps to delineate the different developmental zones found in the root tip.

To study the expression gradients of plant regulators and their biological functions, it is essential to determine how their gradients are shaped and the developmental consequences when they are modified. Here, we describe how to estimate the expression gradients of *PLETHORA* genes in *Arabidopsis thaliana* roots using quantitative laser scanning confocal microscopy (LSCM). We also provide indications on how to evaluate the changes that occur in root patterning and development when the expression gradients of *PLT* genes are perturbed.

2 Materials

2.1 Material for Plant Growth

1. *Arabidopsis thaliana* seeds from transgenic lines expressing selected fluorescent reporters (*see* Table 1 for a list of selected reporter lines and their description).

2. 1.5 mL reaction tubes.

3. Seed sterilization solution (70% V/V ethanol +0.05% V/V Tween 20).

4. 96% V/V ethanol.

5. Square culture plates (125 × 125 × 15 mm) with plant growth media: 1% W/V plant cell culture tested agar, 1× Murashige–Skoog basal salts (MS), 1% W/V sucrose and 2.3 mM

2-(N-morpholino) ethanesulfonic acid (MES), adjusted to pH 5.6–5.8 with KOH.

6. Sterile nylon mesh squares (10 × 10 cm, Nitex Cat. 03-100/44, Sefar) (*see* **Note 1**).

7. Tweezers.

8. Alcohol burner.

9. Sterile wooden toothpicks.

10. Micro pipette.

11. Sterile pipette tips.

12. 3 M™ micropore™ surgical tape.

13. Plant growth chamber.

2.2 Laser Scanning Confocal Microscopy

1. A laser scanning confocal microscope Zeiss LSM880, Axio Observer Z.1, configured with a Quasar detector (2 multialkali photomultiplier (PMT) plus 1 array of 32 Gallium arsenide phosphide (GaAsP) detectors. Objective: 20× plan apochromat 0.8 numerical aperture (NA).

2. Software: ZEN BLACK edition 2.1 2012 SP2.

3. Microscope slides and 20 × 20 mm #1.5 cover slips.

4. Style 5 metal tweezers.

5. Micro pipette and tips.

6. Ultrapure water.

7. Propidium iodide (PI) (stock solution: 1 mg/mL in water, working solution: 0.015 mg/mL in water) (*see* **Note 2**).

2.3 Image Analysis

1. Windows-, Macintosh-, or Linux-enabled computer.

2. Software for image processing. A suitable option is FIJI. FIJI stands for "*Fiji Is Just ImageJ*" and is a distribution of ImageJ that contains numerous plugins for scientific image analysis. FIJI is freeware that can be downloaded from *https://fiji.sc/*. The site also contains a comprehensive description of the various plugins available and several tutorials for image analysis.

3 Methods

3.1 Seed Sterilization and Plating

1. Use 20 mg of seeds of the selected marker line for sterilization in a 1.5 mL microcentrifuge tube. Table 1 contains a list several root marker lines that can be used to analyze the expression gradient of *PLETHORA* genes or the developmental consequences that occurs after its modification.

2. Add 1 mL of seed sterilization solution to the tube containing the seeds. Incubate for 5–10 min on a rocking platform to ensure thorough sterilization of the complete seed surface. In a laminar flow hood, remove the sterilization solution, then rinse the seeds three times with 1 mL of 96% ethanol. Allow the seeds to settle between washes.

3. After the final wash, remove completely the ethanol. Use a different tip for each of the genotypes to avoid cross-contamination of the seeds. Leave the tubes open in the laminar flow hood until the seeds are completely dry.

4. While seeds are drying up, prepare the plates. In a laminar flow hood, add 50 mL of melted plant growth media in each square plate and let solidify for 10 min.

5. When the medium is solid, add a piece of sterile nylon mesh to each plate (*see* **Note 3**). To transfer the mesh a pair of sterilized tweezers will be necessary. An easy way to sterilize the steel tweezers is to flame them using an alcohol burner. Once the tweezers are sterilized, carefully remove a piece of mesh from the autoclaved package and place it on top of the solidified medium. Add a piece of mesh to each plate, sterilizing the tweezers between every plate to avoid contamination.

6. Using a sterile wooden toothpick, carefully place 30 seeds in a line on top of the mesh. Wet the toothpick with the media, stick a dry seed from the tube to the tip of the toothpick and deposit the seed on the mesh by just touching the mesh with the seed in the desired position.

7. Seal the plates with surgical tape. Keep the seeds at 4 °C in the dark for 2 days to stratify.

8. Transfer plates to a growth chamber set at 21 °C and continuous light conditions for 5–6 days. Place plates vertically in the growth chamber so that roots grow along the surface of the mesh (Fig. 2a).

3.2 Image Acquisition Under the LSCM

We describe here how to acquire images under the LSCM similar to those of Fig. 1d that can reveal the expression gradient of a gene of interest, using *pPLT1-CFP* as an a example. This construct express the Cyan Fluorescent Protein (CFP) under the control of the upstream regulatory sequences of the *PLT1* gene (*pPLT1*). Simultaneously, staining of the cell walls with PI allows for the observation of the cellular structure of the root tip.

A typical configuration for the acquisition of CFP and PI fluorescence would employ two channels in the sequential scanning mode, with different excitation and emission LSCM tracks for each fluorophore. We propose a simpler and faster configuration in which both channels are collected simultaneously using a single excitation laser. This configuration takes advantage of the wide

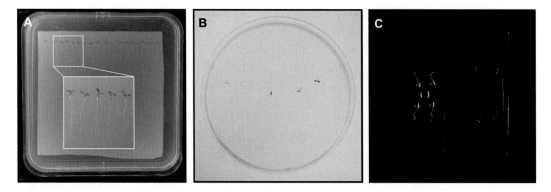

Fig. 2 Plant growing conditions, staining, and mounting for LSCM imaging. (**a**) Image of a square plate with 6-day-old seedlings directly germinated on a sterile nylon mesh on top of the culture media. (**b**) Petri dish with the PI solution were seedlings are being stained. The seedlings were removed from the plate by picking them softly from the cotyledons and then placed floating on the PI solution. (**c**) For the mounting process, it is possible to work with up to four roots per slide to speed up the analysis. On the left side of the image, a microscope slide with 40 μL of water distributed in five small drops where four stained seedlings were gently placed. To guarantee the integrity of the root during the mounting process, place first the root tip on top of one of the drops and then the rest of the plant on the dry part of the slide. The right side of the image shows the completed assembly after carefully placing a cover slip on top of the plants

excitation spectrum of PI and of the high laser power necessary for the CFP excitation. This ensures enough excitation of PI even if the laser wavelength does not match the maximum peak of the excitation spectrum for this fluorophore (*see* **Note 4**).

3.2.1 *LSCM Set Up*

This section describes the typical configuration of the microscope and it will be described following the modules of the ZEN 2.1 SP2 (Zeiss's software for confocal systems). It should be noted that the specific parameters can significantly change if another LSCM or imaging system is used for acquisition. However, the general principles of imaging remain the same, so this method can be considered as a general guide.

1. Turn on the LSCM system. Start the ZEN BLACK software and choose the "Start System" option.

   On Laser Module

2. Turn on the Argon ion laser.

   On Imaging Setup Module

3. Check box Ch1 (Channel 1) for CFP detection and Ch2 for PI detection on the "Use" column (*see* **Note 5**).

4. Select the correct emission spectrum on the "Dye" column for each channel.

5. Choose the preferred color for visualization on the "Color" column.

6. Set up the appropriated bandwidth for acquisition on the "Range" column. (*see* **Note 6**). This can be done using the scroll bar or entering the desired values directly. For CFP the bandwidth needs to be set between 465 and 505 nm, and for PI between 592 and 642 nm.

7. Select a dichroic mirror capable of reflecting the excitation wavelength used. In this case it is a combined dichroic, MBS 458/543.

8. In addition, it is desirable to acquire the transmitted light channel. To do this, check the box on "T-PMT".

In Acquisition Mode Module

9. Select the $20\times$ objective from the drop-down list (*see* **Note 7**).

10. Select the sampling resolution in pixels to 1024×1024.

11. Choose a scan speed of six which results in a pixel dwell of 2.06 μs.

12. Set the following parameters for "Averaging":
    (a) Number: 2.
    (b) Mode: Line.
    (c) Method: Mean.
    (d) Bit depth: 16bits.
    (e) Direction: ----> (mono directional scan).

The next step in the configuration is setting the range of intensities, a gain for the photomultipliers, an offset value and the laser power (*see* **Note 8**).

In Channels Module

13. Check the box corresponding to the laser line of 458 nm.

14. Set laser power at "20" (meaning 20% of the maximum power).

15. Click on "1 AU" button (*see* **Note 9**).

16. Set the value "gain (master)" for Ch1 at 800 (*see* **Note 10**).

17. Set "digital offset" at value 2 for Ch1 (*see* **Note 10**).

18. Set the value "gain (master")" for Ch2 at 750.

19. Set "digital offset" at value for Ch2.

After completing these steps the configuration for the observation of *pPLT1-CFP* expression of a root stained with PI is ready. To generate analogous setting to observe other fluorescent proteins similar criteria must be followed by changing the values indicated for a correct acquisition, i.e., excitation wavelengths, gain values, offset, power, etc. as stated previously (*see* **Note 11**).

3.2.2 Root Staining, Mounting, and Imaging.

1. Prepare the PI working solution as a 1/100 dilution of the stock solution. Pour the dilution in a recipient big enough to place roots, for example in a 90 mm diameter petri dish.

2. To stain a root take a seedling from the plate using tweezers by picking it softly from the cotyledons and place it floating in the PI solution (Fig. 2b). Incubate for 90 s (*see* **Note 12**).

3. Prepare a microscope slide with a drop of water (Fig. 2c) to a total amount of 40 µL. Take a seedling out of the PI solution by picking it from the cotyledon and place it gently on top of the drop of water. Place first the root tip on one of the drops and then the rest of the plant on the dry glass (*see* **Note 13**).

4. Take a coverslip and place it gently on top of the plants (Fig. 2c).

5. Take the whole assembly and place it on the microscope stage. Use "Live" for focusing a medial longitudinal section of the root and then "Snap" to acquire a high quality image.

3.3 Estimating the Expression Gradient of PLETHORA1 *Genes and Its Modifications in Arabidopsis thaliana Roots*

1. Start Fiji on the computer. Open the files by dragging them (e.g., from the desktop, Windows Explorer, or the Mac Finder) onto the Status Bar, or using the command File>Open in the menu bar. Image files must be in TIFF, GIF, JPEG, DICOM, BMP, PGM, or FITS format, or in a format supported by a reader plugin. Last versions of Fiji are able to open a large range of image formats including those from the major LSCM manufacturers, such as Leica, Nikon, Olympus, and ZEISS (*see* **Note 14**).

2. Rotate the image in order to orientate the root longitudinal axis in a way that the *PLT1* expression gradient, can be visualized horizontally. The function Image > Transform > Rotate will open a dialog box to rotate the active image or stack in a specified number of degrees (Fig. 3a). The superimposed Grid Lines in Preview mode can be useful to define the number of degrees to rotate.

3. Split the channels of the multichannel image using Image > Color > Split channels (*see* **Note 15**). This process will result in a number of images equal to the number of channels acquired. One of the images will contain the data for CFP, the fluorescent tag attached to PLT1, while the other will contain the data for PI, used to visualize the cellular structure (Fig. 3b).

4. Convert each image to a 32 bits image using Image > Type > 32 Bits. (*see* **Note 16**).

5. To measure the background fluorescence (*see* **Note 17**) of the CFP image, draw a region of interest (ROI) (*see* **Note 18**). The ROI does not need to be of a specific shape; however, it is important to use the same ROI (shape and size) for each image

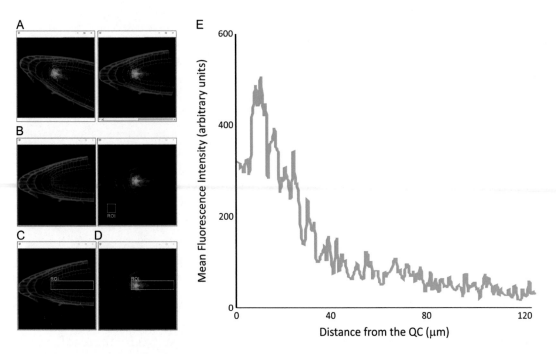

Fig. 3 Quantification of *PLT1* expression gradient in the *Arabidopsis thaliana* root meristem. (**a**) Rotation of the image so that the root longitudinal axis, and therefore the *pPLT1-CFP*, expression gradient places horizontally. (**b**) Splitting of the imaging channels into separated images and selection of a ROI (yellow) to quantify the background fluorescence. (**c**) Use of the PI image as an anatomical reference to draw and locate the ROI that will be used in the quantification of *pPLT1-CFP* expression. (**d**) Migration of the previously selected ROI to the CFP image. (**e**) Representative *pPLT-CFP* fluorescent intensity profile for one root

where the background needs to be removed. It is desirable to draw the ROI close to the defined region were measurements will be done, but it is also important that this region does not express the fluorescent protein that is being measured. Considering this, a rectangular ROI just outside the root tip is convenient in this step (*see* **Note 19**) (Fig. 3b, right panel).

6. To determine the fluorescence intensity of the background go to Analyze > Measure or directly press M, and this operation will open a new window that provides information about fluorescence intensity values for the defined ROI. Register the mean grey value (*see* **Note 20**).

7. Use Process > Math> Subtract. This will open a dialogue box where it is possible to write the value of fluorescence intensity from the background (mean grey value) to be subtracted from the complete image. Press "ok" to complete the operation.

8. Draw a rectangular ROI with a size and shape that covers the *pPLT1-CFP* expression domain. To define the ROI limits the PI image that provides the root cellular structure should be used. The suggested ROI for this gene is a rectangle starting from the base of the QC to the end of the image and as wide as

the QC (Fig. 3c). It is important to register the size of the ROI, as exactly the same ROI needs to be used for all images.

9. Copy the ROI to the *pPLT1-CFP* image (*see* **Note 21**) (Fig. 3d) and press "K" to get the plot of the fluorescence intensity profile (*see* **Note 22**). The outcome is a two-dimensional graph where the X-axis is the distance along the rectangular ROI, and the Y-axis is the pixel fluorescence intensity (Fig. 3e). Press "List" to view the profile data, and then copy the values to a spreadsheet.

10. Repeat this process at least for 15 plants of each genotype under analysis. Calculate the average of intensity and standard deviation for each position and plot the resulting data to obtain a quantification of *PLT-:CFP* expression as a function of the longitudinal position along the root tip (for an example, *see* [7].

3.4 Analysis of the Morphological Modifications Resulting from Changes in the PLT1 Expression Gradient

Modification in the expression gradients of *PLETHORA* genes are associated with morphological changes including changes in the SCN area, changes in meristem size, modified number of cell layers within each cell type, etc. [6, 7, 12]. In the following steps we specifically describe how to analyze changes in meristem size and in the number of Lateral Root Cap layers (LRC). Table 1 and Fig. 1 provide a list of fluorescent markers lines to assess the modifications in other cell types and other developmental zones of the root tip.

3.4.1 Determination of the Meristem Size in Length

This parameter is best determined by analyzing the file of cortex cells.

1. Draw a line on the PI image from the base of the QC up to the cell in the cortex file that is twice the length of the immediately preceding cell (dotted lines in Fig. 1b, c).

2. Press M and register the length of the meristem in μm.

3. Repeat this process for at least 15 plants of each genotype under analysis. Calculate the average of meristematic zone length and its standard deviation in order to generate a bar plot for this parameter.

3.4.2 Determining the Number of Lateral Root Cap Layers

Periclinal cell divisions of the epidermis–lateral root cap initials (Fig. 1a) generate the epidermis and lateral root cap cell types. Changes in the *PLT1* expression gradient result in concomitant changes in the activity of these stem cells, which in turn leads to a modification in the number of LRC layers found at the root tip [7].

1. Draw on the PI image a line 35 μm in length from the bottom of the QC up into the vascular cylinder. At this position a wild-type plant has one LRC layer in average. Count the number of LRC on both sides of the root.

2. Repeat this process for at least 15 plants of each genotype under analysis. Calculate the average of LRC cell layers and its standard deviation in order to generate a bar plot for this parameter.

4   Notes

1. The mesh is ordered by the yard. Cut to size by first drawing the mesh gently with a pencil. 20 pieces of mesh are bundled into a clean paper packet before autoclaving.

2. PI is light sensitive, so the stock and working solution need to be protected from light. Change the working solution every 2 h, as it loses staining power quite fast. The stock solution can be kept at 4 °C for 1 month in the dark.

3. The nylon mesh allows for easy transfer of the plants between plates when chemical treatments need to be performed. Also, it provides a firm support where roots can be easily dissected with a razor blade and collected in the case that samples need to be prepared for further analysis (Fig. 2a).

4. In order to use this configuration, it is important to verify that there is no bleeding between channels before you start recording images in each session.

5. The use channel 1 (Ch1) means that the detector in use will be the multialkali detector placed on the right side on the QUASAR detection unit (http://zeiss-campus.magnet.fsu.edu/tutorials/spectral imaging/quasar34ch/indexflash.html).

6. To set the collected light bandwidth place the lower limit 5 nm above the laser excitation line and the upper limit around 30–40 nm above. Previous analysis of the emission spectra from the fluorophores under study can help to ensure proper setup of the excitation wavelength and the fluorescence acquisition range. Various on lines tools provide this information (for example, https://searchlight.semrock.com/).

7. The 20× objective will provide a sufficient field of view for the visualization of a complete root meristem. The numerical aperture of 0.8 will give sufficient resolution so that both, the structure and the expression pattern are properly imaged.

8. In order to set these values, it is essential to have a previous knowledge of the experiment to execute. Considering that CFP intensity will be quantified, it is highly desirable to find no saturated and no 0 intensity values in the images. To achieve this condition, it is necessary to know which plants of the set of genotypes or conditions are expected to present the higher or lower values of fluorescence intensity when compared to control/wild-type.

9. "1 AU" sets the pinhole diameter to reach 1 Airy disc unit, which gives the best conditions for LSCM.

10. During setting up of the acquisition parameters, it is desirable to use a Look Up Table (LUT) that indicates which pixels have a saturated fluorescence intensity (in red color) and which are have an intensity equal or below 0 (in blue). In the ZEN software this LUT is named "Range Indicator". With the equipment scanning in "Continuous" mode, the offset value must be modified until only a few pixels are displayed in blue. Then, the gain value, laser power, pixel dwell, etc. must be adjusted until only a few pixels are displayed in red.

11. If low intensities of fluorescence are expected it is advisable to use the GaAsP detector since its sensitivity and signal to noise ratio are notoriously better that those of the regular multialkali PMTs detectors.

12. Up to four roots can be mounted per slide to speed up the analysis (Fig. 2c).

13. It is crucial to guarantee the integrity of the root during the mounting process. If the root suffers mechanical damage, the PI will enter the cells, dimming the fluorescence of any other fluorescent protein and making it impossible to observe the cellular structure of the root tip. Therefore, it is important to use the correct amount of water, enough to cushion the mounting process but not too much since it can leave the root to move between the glasses, which can cause it to take oblique positions and make it impossible to obtain a precise medial longitudinal section.

14. When a file is chosen to be opened the software will detect if it supports its format, and will show a dialog box called "Bio-format import options". To get information about each of the options from this window it is necessary to point over them with the mouse and a description will appear in the information box to the right.

15. The important feature of these separated images is that the actual channel information is always retained, and so the original pixel values remain available. Pixel values for each channel are in general determined from light that has been filtered according to its wavelength. In principle, any LUT (lookup table, or color table) might be applied to each channel, but it makes sense to choose LUTs that somehow relate to the wavelength ("color") of the light detected for each of the corresponding channels.

16. 32-bit images can display 2^{32} gray levels in real numbers and is the appropriate image format to perform the mathematical operations described here. Pixels are described by floating

point values and can have any intensity value including NaN (Not a Number).

17. The "Background" is the light detected that is not specific to the experiment. Background needs to be subtracted from the true signal within the region of interest; otherwise, background will contribute to the apparent "signal" and the quantifications will result in incorrect values.

18. A Region Of Interest (ROI) is an area within the image that you select to perform any of the available operations in FIJI.

19. To use a rectangular ROI of a fixed area go to Edit > Selection> Specify. This will open a dialog box that allows the user to define rectangular or elliptical selections. Width and Height are the dimensions of the selection. These allow for easy set-up of identical ROIs in different files and images.

20. The measurement "mean grey value" is defined as the sum of the gray values of all the pixels in the ROI divided by the number of pixels.

21. To transfer a ROI from one image to another, it is necessary to select the image with the desired ROI, then select the destination image, and press Ctrl > Shift > E (the keyboard shortcut for Edit > Selection > Restore Selection [E]). Alternative ways to transfer ROIs across images are the ROI Manager or the cursor synchronization features provided by Analyze > Tools > Synchronize Windows.

22. For rectangular selections or line selections wider than one pixel, the plot will display a "column average plot," where the Y-axis represents the vertically averaged pixel intensity (the function will average vertically unless "Alt+K" is pressed). The use of rectangle selections is recommended to avoid noisy plots as the pixel intensity is averaged in each position.

Acknowledgments

We thank Philip Benfey for the *AGL42-GFP* reporter, Moritz Nowack for the *pSMB-H2B-GFP* reporter, Lieven de Veylder for the *pGL2:H2B-YFP*, *pCO:H2B-YFP*, and the *pEND:H2B-YFP* reporters, and, finally, Ben Scheres for the *PLT1* and *PLT2* transcriptional and translational reporters. We thank Carla Schommer for critical reading of the manuscript. MFE and CG were supported by fellowships from CONICET. RV is a technician of CONICET. JP and RER are members of CONICET and are supported by grants from ANPCyT.

References

1. Heidstra R, Sabatini S (2014) Plant and animal stem cells: similar yet different. Nat Rev Mol Cell Biol 15(5):301–312. https://doi.org/10.1038/nrm3790

2. Petricka JJ, Winter CM, Benfey PN (2012) Control of Arabidopsis root development. Annu Rev Plant Biol 63:563–590. https://doi.org/10.1146/annurev-arplant-042811-105501

3. Ercoli MF, Ferela A, Debernardi JM, Perrone AP, Rodriguez RE, Palatnik JF (2018) GIF transcriptional coregulators control root meristem homeostasis. The Plant cell 30 (2):347–359. https://doi.org/10.1105/tpc.17.00856

4. Galinha C, Hofhuis H, Luijten M, Willemsen V, Blilou I, Heidstra R, Scheres B (2007) PLETHORA proteins as dose-dependent master regulators of Arabidopsis root development. Nature 449(7165):1053–1057. https://doi.org/10.1038/nature06206 nature06206 [pii]

5. Nawy T, Lee JY, Colinas J, Wang JY, Thongrod SC, Malamy JE, Birnbaum K, Benfey PN (2005) Transcriptional profile of the Arabidopsis root quiescent center. The Plant cell 17 (7):1908–1925

6. Mahonen AP, ten Tusscher K, Siligato R, Smetana O, Diaz-Trivino S, Salojarvi J, Wachsman G, Prasad K, Heidstra R, Scheres B (2014) PLETHORA gradient formation mechanism separates auxin responses. Nature 515 (7525):125–129. https://doi.org/10.1038/nature13663

7. Rodriguez RE, Ercoli MF, Debernardi JM, Breakfield NW, Mecchia MA, Sabatini M, Cools T, De Veylder L, Benfey PN, Palatnik JF (2015) MicroRNA miR396 regulates the switch between stem cells and transit-amplifying cells in Arabidopsis roots. Plant Cell 27 (12):3354–3366. https://doi.org/10.1105/tpc.15.00452

8. Ubeda-Tomas S, Federici F, Casimiro I, Beemster GT, Bhalerao R, Swarup R, Doerner P, Haseloff J, Bennett MJ (2009) Gibberellin signaling in the endodermis controls Arabidopsis root meristem size. Curr Biol 19 (14):1194–1199

9. Fendrych M, Van Hautegem T, Van Durme M, Olvera-Carrillo Y, Huysmans M, Karimi M, Lippens S, Guerin CJ, Krebs M, Schumacher K, Nowack MK (2014) Programmed cell death controlled by ANAC033/SOMBRERO determines root cap organ size in arabidopsis. Curr Biol. https://doi.org/10.1016/j.cub.2014.03.025

10. Dietrich D, Pang L, Kobayashi A, Fozard JA, Boudolf V, Bhosale R, Antoni R, Nguyen T, Hiratsuka S, Fujii N, Miyazawa Y, Bae TW, Wells DM, Owen MR, Band LR, Dyson RJ, Jensen OE, King JR, Tracy SR, Sturrock CJ, Mooney SJ, Roberts JA, Bhalerao RP, Dinneny JR, Rodriguez PL, Nagatani A, Hosokawa Y, Baskin TI, Pridmore TP, De Veylder L, Takahashi H, Bennett MJ (2017) Root hydrotropism is controlled via a cortex-specific growth mechanism. Nat Plants 3:17057. https://doi.org/10.1038/nplants.2017.57

11. Carlsbecker A, Lee JY, Roberts CJ, Dettmer J, Lehesranta S, Zhou J, Lindgren O, Moreno-Risueno MA, Vaten A, Thitamadee S, Campilho A, Sebastian J, Bowman JL, Helariutta Y, Benfey PN (2010) Cell signalling by micro-RNA165/6 directs gene dose-dependent root cell fate. Nature 465(7296):316–321. https://doi.org/10.1038/nature08977

12. Aida M, Beis D, Heidstra R, Willemsen V, Blilou I, Galinha C, Nussaume L, Noh YS, Amasino R, Scheres B (2004) The PLETHORA genes mediate patterning of the Arabidopsis root stem cell niche. Cell 119(1):109–120. https://doi.org/10.1016/j.cell.2004.09.018

13. Santuari L, Sanchez-Perez GF, Luijten M, Rutjens B, Terpstra I, Berke L, Gorte M, Prasad K, Bao D, Timmermans-Hereijgers JL, Maeo K, Nakamura K, Shimotohno A, Pencik A, Novak O, Ljung K, van Heesch S, de Bruijn E, Cuppen E, Willemsen V, Mahonen AP, Lukowitz W, Snel B, de Ridder D, Scheres B, Heidstra R (2016) The PLETHORA gene regulatory network guides growth and cell differentiation in Arabidopsis roots. Plant Cell 28 (12):2937–2951. https://doi.org/10.1105/tpc.16.00656

14. Rodriguez RE, Mecchia MA, Debernardi JM, Schommer C, Weigel D, Palatnik JF (2010) Control of cell proliferation in Arabidopsis thaliana by microRNA miR396. Development 137(1):103–112

15. Jones-Rhoades MW, Bartel DP (2004) Computational identification of plant micro-RNAs and their targets, including a stress-induced miRNA. Mol Cell 14(6):787–799

Chapter 2

Detection and Quantification of the Bicoid Concentration Gradient in Drosophila Embryos

Feng He, Honggang Wu, David Cheung, and Jun Ma

Abstract

We describe methods for detecting and quantifying the concentration gradient of the morphogenetic protein Bicoid through fluorescent immunostaining in fixed *Drosophila* embryos. We introduce image-processing steps using MATLAB functions, and discuss how the measured signal intensities can be analyzed to extract quantitative information. The described procedures permit robust detection of the endogenous Bicoid concentration gradient at a cellular resolution.

Key words Fluorescent immunostaining, Morphogen gradient, Bicoid, *Drosophila*

1 Introduction

The ability to quantify endogenous proteins has been facilitating our understanding of the molecular events controlling embryonic development. Antibodies against specific epitopes on target proteins are commonly used. Procedures for immunohistochemistry, imaging, and image processing have been developed and successfully applied to various biological systems. This chapter describes methods for detecting the morphogenetic protein Bicoid (Bcd) in the whole-mount *Drosophila* embryo. Bcd forms a concentration gradient along the anterior-posterior (AP) axis in the early embryo [1–4], posing both technical and analytical challenges to accurate quantification of its concentration gradient profile [5–8]. We developed tools to overcome such challenges [5, 9, 10]. The protocol that we used for embryo staining was largely based on published methods [11] with several modifications [12], including a postfixation step and the treatment of embryos with an SDS-containing hybridization buffer prior to antibody incubation. In addition, we developed analytical methods to properly measure and treat

---

Feng He and Honggang Wu contributed equally to this work.

Julien Dubrulle (ed.), *Morphogen Gradients: Methods and Protocols*, Methods in Molecular Biology, vol. 1863, https://doi.org/10.1007/978-1-4939-8772-6_2, © Springer Science+Business Media, LLC, part of Springer Nature 2018

background signals in fixed embryos [5]. This permitted accurate description of Bcd gradient properties based on measured "raw" intensity data, avoiding distortions associated with the use of normalized intensity data [5]. Applications of our developed methods have led to the discovery of, among other things [13–15], how the Bcd gradient profile is scaled with embryo size and how such scaling is controlled at a fundamental level [4, 5, 16–18].

2   Materials

2.1   Chemical and Biological Reagents

1. PEM buffer: 0.1 M PIPES, 1 mM $MgCl_2$, 1 mM EGTA, pH adjusted to 6.9 with KOH.
2. Bleach: sodium hypochlorite 50% v/v dilution with water, prepared freshly before use.
3. Fixation buffer: 2 mL PEM, 370 μL formaldehyde and 2 mL heptane in a scintillation vial, prepared freshly before use.
4. Methanol.
5. Rabbit anti-Bcd polyclonal antibody (Santa Cruz).
6. Alexa Fluor 594 goat anti-rabbit (Molecular Probes).
7. DAPI 0.1 μg/mL (Sigma).
8. ProLong Gold Antifade Mountant (Thermo Fisher).
9. 20 × SSC: 3 M NaCl, 0.3 M sodium citrate, pH adjusted to 7.0 with NaOH.
10. Hybridization buffer: 750 mM NaCl, 75 mM sodium citrate, 50% v/v formamide, 0.1% v/v Tween 20, and 0.3% SDS.
11. PBS: 137 mM NaCl, 2.7 mM KCl, 10 mM $Na_2HPO_4 \cdot 7H_2O$, 2 mM KH_2PO_4, pH adjusted to 7.4 with HCl.
12. PBST: 1 × PBS with 0.1% Tween 20.
13. Post-fixation buffer: formaldehyde 10% v/v diluted in PBST.
14. Antibody incubation buffer: 1.75× Roche blocking reagent (Roche 11921673001) diluted in PBST.
    All solutions are made with deionized water and stored at room temperature.

2.2   Equipment and Tools

1. Grape juice agar plates (*see* **Note 1**).
2. Scintillation vials.
3. Vortex shaker.
4. Tube nutator.
5. Water bath.
6. Nail polish.
7. Glass slides.

8. Coverslips.

9. Zeiss Imager Z1 ApoTome microscope [5, 16] or confocal microscopy [10, 18].

2.3 Software

1. MATLAB (R2012a or higher version). We process images semiautomatically using a custom MATLAB script [5]. The MATLAB codes for image processing and intensity measurement are available upon request.

3 Methods

3.1 Embryo Collection, Staining, and Imaging

1. 0–4 h eggs are collected on grape juice agar plates (*see* **Note 1**), washed with deionized water, and dechorionated in freshly prepared sodium hypochlorite bleach for 5 min with occasional, gentle shaking at room temperature. Wash off the bleach with deionized water.

2. Transfer embryos to a scintillation vial containing the fixation buffer; agitate at ~500 rpm on a shaker for 20 min at room temperature.

3. To devitellinize embryos, remove the aqueous phase and add 2 mL methanol. Vortex vigorously for 1 min at room temperature.

4. Transfer devitellinized embryos that are in the organic layer to a 1.5 mL tube; wash three times with methanol at room temperature (*see* **Note 2**).

5. Rehydrate embryos from methanol to PBST in a stepwise manner at room temperature, with the following methanol to PBST ratios for each of the successive steps: 4:1, 7:3, 3:2, 1:1, 2:3, 3:7, 1:4, and 0:1; let embryos equilibrate at each step for 5 min. Wash embryos three times in PBST at room temperature, 5 min each (*see* **Note 3**).

6. Post-fix embryos for 20 min with post-fixation buffer. Rinse two times with PBST; wash three times in PBST at room temperature, 5 min each.

7. Transfer embryos from PBST to hybridization buffer in a stepwise manner, with the following PBST to hybridization buffer ratios for each of the successive steps: 3:1, 1:1, 1:3, and 0:1; let embryos equilibrate at each step for 5 min. Wash embryos three times in hybridization buffer at room temperature, 5 min each (*see* **Note 4**).

8. Incubate embryos overnight in hybridization buffer at 60 °C.

9. On the following day, wash embryos three times in prewarmed (60 °C) hybridization buffer for 5 min each. Transfer embryos from hybridization buffer to PBST (at 60 °C using prewarmed

buffers) in a stepwise manner, with the following hybridization buffer to PBST ratios for each of the successive steps: 3:1, 1:1, and 1:3; let embryos equilibrate at each step for 5 min. Wash embryo three times in PBST at room temperature, 5 min each.

10. Embryos are incubated in antibody incubation buffer at room temperature for 30 min, followed by overnight incubation in the presence of primary antibody (diluted 1:400 in antibody incubation buffer) at 4 °C (*see* **Note 5**).

11. Wash embryos three times in PBST at room temperature, 5 min each. Block embryos with antibody incubation buffer for 30 min, and incubate with secondary antibody (diluted 1:400 in antibody incubation buffer) for 1 h at room temperature. Starting from this step, avoid light exposure to stained embryos. DNA is counterstained with DAPI (1:1000 dilution in PBST) for 10 min, followed by wash in PBST for three times, 5 min each.

3.2 Imaging

1. For capturing images of flattened embryos, they can be directly mounted between a coverslip and a slide. A small weight (~5–10 g) may be placed on top of the coverslip to flatten the embryos when necessary. For imaging embryos with their 3-dimentional (3D) morphology preserved, coverslip bridge is used (*see* **Notes 6** and **7**).

2. Stained embryos are imaged under either a Zeiss microscope with ApoTome or confocal microscopy. An image capturing fluorescent signals that detect Bcd and another capturing DAPI counterstain signals are taken sequentially for each embryo. To permit comparison of detected Bcd signal intensities across embryos, exposure time is set within a linear range and kept identical for all embryos in an experiment. Specifically, embryos with the strongest Bcd signals are identified through a quick scan of the entire slide(s) and used to set an exposure time (or "gain" in confocal microscopy) to avoid nonlinearity (signal saturation) during imaging.

3.3 Image Processing and Analysis Using MATLAB

1. Use the MATLAB function "imread" to load graphic (grayscale or color) files as intensity images (Fig. 1a, b).

2. A global threshold of pixel intensity is determined by applying the MATLAB function "graythresh" to each DAPI intensity image. Then, a whole-embryo mask is constructed by setting the DAPI intensities above the threshold to 1 and others to 0 (Fig. 1c; *see* **Note 8**).

3. The function "regionprops" measures many properties of each segmented object on the whole-embryo mask, such as the area size ("Area"), the smallest rectangle containing the object ("BoundingBox"), and the angle between the major axis of

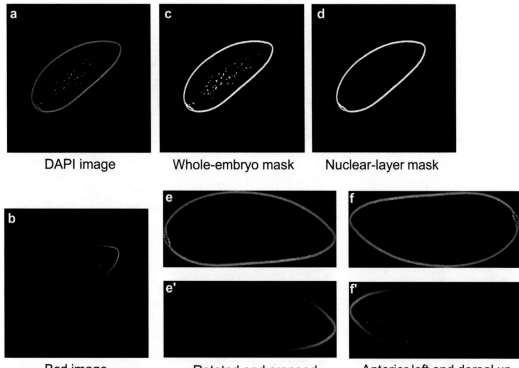

Fig. 1 Image processing. Midsagittal images of an embryo at early nuclear cycle 14 are used as an example to illustrate image processing steps (**a**: DAPI; **b**: Bcd). The whole-embryo mask (**c**) and the nuclear-mask (**d**) are constructed to calculate parameters for standardization of the images (**e–f** and **e'–f'**). For purposes of presentation, images shown have their contrast adjusted

the nuclear-layer mask and the horizontal axis ("Orientation"). The object with the largest area size is identified as the nuclear-layer mask (Fig. 1d).

4. All intensity images of a given embryo are rotated by the angle "Orientation." Then, the function "imcrop" crops away all noninformative image areas outside the box bounding the embryo region (Fig. 1e').

5. The last steps of image standardization are to set the anterior pole of the embryo to the left with the function "fliplr" and to set the dorsal side of the embryo to upside with "flipud" (Fig. 1f'). To determine the anterior side of the embryo, the pixel intensities of the rotated and cropped Bcd image are measured, and the horizontal half that has the larger median value is identified as the anterior part of the embryo. To determine the dorsal side of the embryo, the line connecting the most left pixel and the most right pixel is computed, and the vertical half in which this line lies is identified as the dorsal part of the embryo.

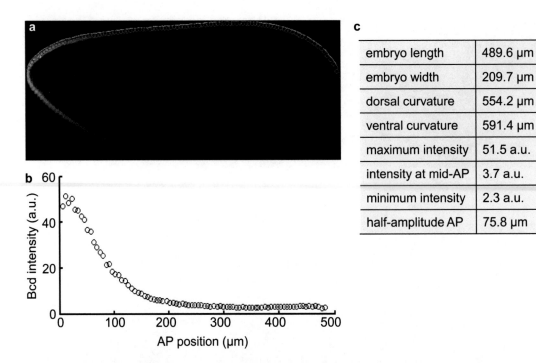

embryo length	489.6 µm
embryo width	209.7 µm
dorsal curvature	554.2 µm
ventral curvature	591.4 µm
maximum intensity	51.5 a.u.
intensity at mid-AP	3.7 a.u.
minimum intensity	2.3 a.u.
half-amplitude AP	75.8 µm

Fig. 2 Extraction of Bicoid signal intensities. Processed and illustrative image of an embryo (**a**) is shown. Scanning windows are shown as circles within the nuclear layer of the embryo. Scanning window diameter: one half of the nuclear-layer height; scanning interval: embryo length/100. Extracted Bcd intensity data plotted as a function of AP position (**b**) and other relevant information of the embryo (**c**) are shown. Bcd intensity data shown (in arbitrary units; a.u.) are "raw" and unadjusted, without background subtraction

6. Raw Bcd intensities are captured by sliding a circular window within the nuclear layer of the embryo. Both the window diameter and the scanning interval are parameters that may be optimized (*see* **Notes 9** and **10**).

7. To determine the scanning path, the contour line of the nuclear-layer mask is identified by the function "edge" (Fig. 2a, red line). Scanning is performed along the midline of the nuclear layer that is parallel to the contour line (Fig. 2a, blue circles).

8. The profile of raw Bcd intensities as a function of AP position (Fig. 2b) and other quantitative features of the embryo (Fig. 2c) are extracted and may be subjected to further analyses as desired.

3.4 Estimation of Background and Noise

Fluorescent labeling and staining can give rise to background signals. For our initial studies [5], we developed methods designed to directly and accurately quantify background signals using embryos that lack Bcd protein (derived from null mutant females). To ensure identical experimental and imaging conditions for wildtype and mutant embryos, we "mixed" them together in same tubes prior to antibody staining. The quantified signal intensities in the mutant embryos were then used in background subtraction for intensity data of wildtype embryos. An indirect, alternative method was to use the signal intensities detected in the posterior region of the embryo as a background approximation in data analysis [13, 14, 19]. In both of these methods for analyzing Bcd gradient properties, we avoided the use of normalized intensity data whenever possible.

Two types of systems errors in the procedures, i.e., imaging noise and processing noise, can be estimated by specific experiments [5]. For imaging noise, images are sequentially captured for the same embryo under identical settings within a short period, and the variance of the intensity data across different images is computed. For processing noise, the sliding path is moved by one pixel toward eight directions on the midsagittal plane, and the intensity variance among all nine datasets is computed.

4   Notes

1. The recipe for making grape juice agar plates is as follows: 550 mL dH_2O, 410 mL grape juice, 87 g sucrose, 18 g yeast, 22.5 g agar, 22 mL 1.25 N NaOH, and 11 mL Acid Mix A (to make 100 mL Acid Mix A: 50 mL dH_2O, 41.8 mL proprionic acid, and 4.15 mL phosphoric acid). NaOH and Acid Mix A are added to the rest of the melted ingredients after it is cooled down prior to pouring the plates. For egg collection, we routinely discard the first egg-laying plate before proceeding to collecting eggs of the desired ages.

2. After fixation and devitellinization, embryos can be stored at $-20\,^\circ C$ in methanol.

3. It is important that embryos are submerged in solutions at all steps during staining. Dried embryos tend to have high background signals and overly dried embryos could easily fall apart.

4. Our observations led to the suggestion that the use of SDS-containing hybridization buffer, which is commonly used for in situ hybridization assays, is beneficial to signal-to-noise ratio in our immunostaining experiments.

5. Roche blocking reagent is used to inhibit nonspecific binding of antibodies to embryos. Block time and antibody concentration may be optimized to improve signal-to-background ratio.

6. After staining, embryos should be immediately transferred into mounting medium to allow the system to equilibrate to achieve consistent imaging quality.

7. Nail polish can be used to prepare coverslip bridge, and the amount of nail polish may be tested to optimize bridge height and avoid embryo movements caused by excessively high bridge.

8. The image processing and analysis procedures described here may be applied to quantification of the expression patterns of other genes in *Drosophila* embryos at early nuclear cycle 14 before cellularization [5, 13, 20, 21]. Accurate segmentation of single nuclei is not necessary, but the contour of the nuclear layer must be well defined. For construction of the nuclear-layer mask, image improvement procedures are optional to employ, such as contrast enhancement, local and global background removal, and noise filtering.

9. The diameter of the sliding window for extracting the Bcd signals should not exceed the height of the nuclear layer. The scanning interval can be a fixed parameter (one pixel, one nuclear radius, etc.), or optimized as embryo length/number of bins (the number of bins can be arbitrary or based on the number of nuclei along the AP axis).

10. For 2D analysis of the flattened embryo, which is specifically designed to calibrate the input–output relationship at the single nucleus resolution [5, 10], individual nuclei can be identified and registered from the nuclear mask using "regionprops." Then raw Bcd intensities are captured within the nuclear boundary of each identified nucleus.

Acknowledgments

The protocol described in this chapter was developed at the Cincinnati Children's Hospital Research Foundation. We acknowledge the invaluable contributions of Ying Wen in developing the protocol. FH acknowledges support of Institute of Apicultural Research, Chinese Academy of Agricultural Sciences. HW acknowledges the generous support of his current supervisor Dr. Renjie Jiao. JM is a National "Thousand Talents" Scholar of China and acknowledges support of Zhejiang University. Feng He and Honggang Wu contributed equally to this work.

References

1. Driever W, Nusslein-Volhard C (1988) A gradient of bicoid protein in Drosophila embryos. Cell 54(1):83–93

2. Driever W, Nusslein-Volhard C (1988) The bicoid protein determines position in the Drosophila embryo in a concentration-dependent manner. Cell 54(1):95–104

3. Liu J, He F, Ma J (2011) Morphogen gradient formation and action: insights from studying Bicoid protein degradation. Fly 5(3):242–246

4. Ma J, He F, Xie G, Deng WM (2016) Maternal AP determinants in the Drosophila oocyte and embryo. Wiley Interdiscip Rev Dev Biol 5(5):562–581. https://doi.org/10.1002/wdev.235

5. He F, Wen Y, Deng J, Lin X, Lu LJ, Jiao R, Ma J (2008) Probing intrinsic properties of a robust morphogen gradient in Drosophila. Dev Cell 15(4):558–567. https://doi.org/10.1016/j.devcel.2008.09.004

6. Gregor T, Tank DW, Wieschaus EF, Bialek W (2007) Probing the limits to positional information. Cell 130(1):153–164

7. Reinitz J (2007) Developmental biology: a ten per cent solution. Nature 448(7152):420–421

8. Crauk O, Dostatni N (2005) Bicoid determines sharp and precise target gene expression in the Drosophila embryo. Curr Biol 15 (21):1888–1898

9. He F, Wen Y, Cheung D, Deng J, Lu LJ, Jiao R, Ma J (2010) Distance measurements via the morphogen gradient of Bicoid in Drosophila embryos. BMC Dev Biol 10:80. https://doi.org/10.1186/1471-213X-10-80

10. He F, Ren J, Wang W, Ma J (2011) A multi-scale investigation of bicoid-dependent transcriptional events in Drosophila embryos. PLoS One 6(4):e19122. https://doi.org/10.1371/journal.pone.0019122

11. Kosman D, Small S, Reinitz J (1998) Rapid preparation of a panel of polyclonal antibodies to Drosophila segmentation proteins. Dev Genes Evol 208(5):290–294

12. Patel NH, Hayward DC, Lall S, Pirkl NR, DiPietro D, Ball EE (2001) Grasshopper hunchback expression reveals conserved and novel aspects of axis formation and segmentation. Development 128(18):3459–3472

13. Liu J, Ma J (2011) Fates-shifted is an F-box protein that targets Bicoid for degradation and regulates developmental fate determination in Drosophila embryos. Nat Cell Biol 13 (1):22–29. https://doi.org/10.1038/ncb2141

14. Cheung D, Ma J (2015) Probing the impact of temperature on molecular events in a developmental system. Sci Rep 5:13124. https://doi.org/10.1038/srep13124

15. He F, Saunders T, Wen Y, Cheung D, Jiao R, ten Wolde P, Howard M, Ma J (2010) Shaping a morphogen gradient for positional precision. Biophys J 99:697–707 PMCID: PMC2913175

16. Cheung D, Miles C, Kreitman M, Ma J (2011) Scaling of the Bicoid morphogen gradient by a volume-dependent production rate. Development 138(13):2741–2749. https://doi.org/10.1242/dev.064402

17. Cheung D, Miles C, Kreitman M, Ma J (2014) Adaptation of the length scale and amplitude of the Bicoid gradient profile to achieve robust patterning in abnormally large Drosophila melanogaster embryos. Development 141 (1):124–135. https://doi.org/10.1242/dev.098640

18. He F, Wei C, Wu H, Cheung D, Jiao R, Ma J (2015) Fundamental origins and limits for scaling a maternal morphogen gradient. Nat Commun 6:6679. https://doi.org/10.1038/ncomms7679

19. Liu J, Ma J (2015) Modulation of temporal dynamics of gene transcription by activator potency in the Drosophila embryo. Development 142:3781–3790. PMCID: PMC4647213. https://doi.org/10.1242/dev.126946

20. Liu J, Ma J (2013) Dampened regulates the activating potency of Bicoid and the embryonic patterning outcome in Drosophila. Nat Commun 4:2968. PMCID: PMC3902774. https://doi.org/10.1038/ncomms3968

21. Wu H, Manu JR, Ma J (2015) Temporal and spatial dynamics of scaling-specific features of a gene regulatory network in Drosophila. Nat Commun 6:10031. PMCID: PMC4686680. https://doi.org/10.1038/ncomms10031

Chapter 3

Imaging Cytonemes in Drosophila Embryos

Lijuan Du and Sougata Roy

Abstract

Conserved morphogenetic signaling proteins disperse across tissues to generate signal and signaling gradients, which in turn are considered to assign positional coordinates to the recipient cells. Recent imaging studies in *Drosophila* model have provided evidence for a "direct-delivery" mechanism of signal dispersion that is mediated by specialized actin-rich signaling filopodia, named cytonemes. Cytonemes establish contact between the signal-producing and target cells to directly exchange and transport the morphogenetic proteins. Although an increasing amount of evidence supports the critical role of these specialized signaling structures, imaging these highly dynamic 200 nm-thin structures in the complex three-dimensional contour of living tissues is challenging. Here, we describe the imaging methods that we optimized for studying cytonemes in *Drosophila* embryos.

Key words Cytonemes, Cell–cell signaling, Morphogen, *Drosophila*, Embryo, Filopodia, Trachea, *Engrailed*, *Breathless*

1 Introduction

The body plan of a multicellular organism is determined by the concentration gradients of morphogens, such as Hedgehog (Hh), Wingless (Wg), Decapentaplegic (Dpp/Transforming growth factor β family protein), fibroblast growth factor (FGF), and epidermal growth factor (EGF). These proteins disperse from a restricted source and pattern a naïve field of cells into regions of distinct differentiation trajectories by inducing concentration-dependent differential gene expression [1–7]. However, the mechanisms by which signals disperse to form concentration gradients are poorly understood. Several contrasting models have been proposed to explain how morphogens disperse through tissues to establish a positional gradient. These include free-, restricted-, and facilitated-diffusion, transcytosis, localized translation of graded mRNA, and direct delivery through cytonemes [8–16]. In this chapter, we focus on imaging methods of *Drosophila* cytonemes that mediate direct delivery of signals in developing tissues. Recent research in vertebrate and invertebrate systems has revealed that many

Julien Dubrulle (ed.), *Morphogen Gradients: Methods and Protocols*, Methods in Molecular Biology, vol. 1863,
https://doi.org/10.1007/978-1-4939-8772-6_3, © Springer Science+Business Media, LLC, part of Springer Nature 2018

morphogenetic signals localize in long cellular membrane protrusions or cytonemes to travel target-specifically from the source to the recipient cells (reviewed in [17, 18]).

Cytonemes were first described in the *Drosophila* wing imaginal discs as thin actin-based filopodia that originate from the apical side of the GFP-marked wing disc cells and orient toward the Dpp morphogen source at the center of the disc [19]. These cytonemes were found to be responsive to Dpp and contained the Dpp receptor Thickvein [20]. Several subtypes of *Drosophila* cytonemes were identified that showed signal-specific responses under the influence of either EGF, FGF, Dpp, and Hh [21]. Subsequent studies in the *Drosophila* air sac primordium (ASP), a wing disc-associated tracheal tube that grows in response to Dpp and Branchless (Bnl, an FGF family protein) produced in the wing disc, established that cytonemes are essential conduits of morphogens [21–23].

ASP cells extend signal-specific polarized cytonemes to establish cell–cell membrane contacts with the *bnl* and *dpp* sources in the wing disc (Fig. 1a, b). The ASP cells that express receptors for both Dpp and Bnl sort these receptors into distinct, signal-specific cytonemes (Fig. 1a). Signal-specific cytonemes were found to interact with components of the extracellular matrix (ECM), through which cytonemes navigate [24]. Cytonemes from the ASP that contact the Dpp source in the wing disc localize the Dpp receptor, receive Dpp, and transport receptor-bound ligand from the source to recipient cells. Many proteins, such as activated Diaphanous

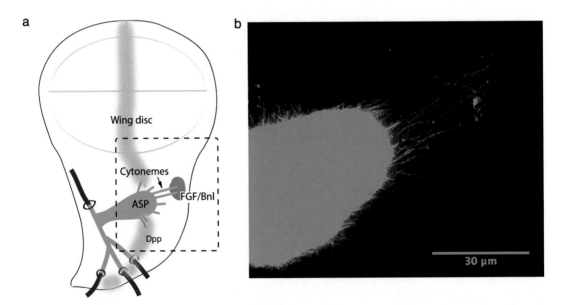

Fig. 1 *Drosophila* ASP cytonemes. (**a**) A drawing depicting the third instar larval wing disc and air sac primordium (ASP). The ASP cells project cytonemes (green) to contact the *bnl* (blue) and *dpp* sources (pink) in the disc for receiving signals. (**b**) Live image of CD8:GFP marked cytonemes emanating from the growing tip region of an ASP; the image in (**b**) corresponds to an area represented by the dashed square in (**a**)

(formin, actin nucleator), Neuroglian (L1-CAM), and Capricious (LRR domain-containing membrane protein involved in neuronal matchmaking) localize at the cytoneme tips and were found to be essential for establishing signaling contacts with the source cells. Removal of their functions from ASP cells affected cytonemes or cytoneme contacts, and consequently led to a reduction in signal transfer and signal transduction in the ASP cells [23, 25]. These results provided first convincing evidence that cytonemes are essential for signal transport and signaling. Importantly, signal-producing cells also send cytonemes to deliver signals to the recipient cells. For example, Hh-producing cells in the *Drosophila* wing imaginal disc and abdominal histoblasts extend cytonemes to deliver Hh and were found to be essential for Hh signaling gradient [26].

Cytonemes and cytoneme-like filopodial projections are abundant in developing tissues of vertebrate and invertebrate models [27, 28]. Almost all families of signaling proteins, including Hedgehog, EGF, WNT/Wingless, Notch/Delta, and FGF are now reported to be localized in cytonemes or cytoneme-like cell projections [29–34]. The *Drosophila* system is an ideal model to characterize this conserved signaling mechanism, given the powerful genetic tools available to engineer and regulate gene expression, standardized genome editing methods, ease of histological studies and imaging, and several signaling systems that are well-characterized. To image the thin and fragile cytonemes in live tissues and visualize their signaling activities at cellular and subcellular resolutions, optimization of a noninvasive imaging method is the most critical requirement. So far, studies on cytonemes are reported from postembryonic developmental stages of *Drosophila* [21, 23, 35]. Here, we describe the methods optimized for imaging cytonemes emanated from embryonic tracheal and *engrailed*-expressing epidermal tissues.

2   Materials

Make all stock buffers in sterile deionized water (dH$_2$O) and avoid contamination of the stock solutions by keeping them in 50 mL aliquots in sealed screw-capped bottles at the recommended temperature.

2.1   Equipment for Fly Pushing

1. Stereomicroscope with dual gooseneck fiber-optic illuminator (e.g., Standard Olympus SZ61 or Nikon SMZ745).
2. *Drosophila* workstation with CO$_2$ pad, as described by [36].
3. Incubator (25 °C, Peltier Refrigerated incubator).
4. Fly stocks: (1) tissue-specific *Gal4* or *LexA* driver lines (2) *UAS* or *LexAop* reporter lines that can tissue-specifically express

membrane localized fluorescent protein (FP) reporter gene. For instance, CD8:GFP. Selection of fluorophore and driver is important (*see* **Note 1**). Stocks can be obtained through public stock centers:

(a) https://bdsc.indiana.edu/index.html.

(b) https://kyotofly.kit.jp/cgi-bin/stocks/index.cgi.
    For example, here we used the following transgenic lines to express FP in tissue pattern: for trachea-specific expression—(1) *breathless-Gal4* (*btl-Gal4*) driver, (2) *UAS-mCD8:GFP* reporter [21, 37]; for epidermal expression—(1) *engrailed-Gal4* driver (*en-Gal4*) driver, (2) *UAS-CD4:IFP2* [38].

2.2 Embryo Collection

1. FlyStuff grape agar premix (*see* **Note 2**).

2. Deionized water (dH$_2$O).

3. Dry baker's yeast. Make a thick paste of required amount of dry yeast with dH$_2$O in a beaker (*see* **Note 3**).

4. 50% sodium hypochlorite: 50% V/V sodium hypochlorite in deionized water. Make fresh.

5. 1× PBS buffer (pH 7.5).

6. PBST (embryo wash): 1% V/V Triton X-100 in 1× PBS (*see* **Note 4**).

7. Petri Dish, 60 × 15 mm^2.

8. Small embryo collection cage (Genesee Scientific, Fig. 2).

9. Metal spatula.

10. Fine-tip paintbrush.

11. Dissection watch glass.

12. Falcon cell strainer (fits 50 mL tube; pore size: 40–100 μm).

2.3 Preparing Live Drosophila Embryos

1. dH$_2$O.

2. Fine-tip paintbrush.

3. Forceps (Dumont no. 5, Fine Science Tools).

4. Microcentrifuge tubes (1.5–2 mL).

5. Glass bottom dishes, 35 mm petri dish, 14 mm glass diameter, No. 1 coverslip (MatTek # P35G-1.5-14-C).

6. Cavity slide (single concavity of 15–18 mm, depth 0.6–0.8 mm; L × W 76 mm × 26 mm, thickness 1.2–1.5 mm).

7. Halocarbon oil 27.

8. Kimwipes.

9. Needle.

10. Stereomicroscope (as described in Subheading 2.1).

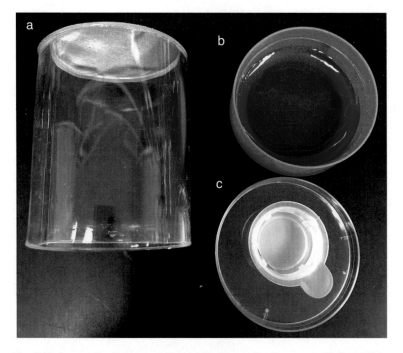

Fig. 2 Embryo collection apparatus. (**a**) Embryo collection cage. (**b**) Grape juice agar plate fits in the plastic cap of the embryo collection cage. (**c**) Cell strainer used for embryo dechorionization

2.4   Preparing Fixed Drosophila Embryos

1. 4% PFA: 4% V/V formaldehyde in PBS. Make fresh. Do not store.
2. Heptane.
3. Methanol (MeOH).
4. 0.1% PBST:0.1% V/V Triton X-100 in 1×PBS.
5. 75% MeOH/25% (0.1% PBST).
6. 50% MeOH/50% (0.1% PBST).
7. 25% MeOH/75% (0.1% PBST).
8. Fine-tip paintbrush.
9. Microcentrifuge tube (1.5–2 mL).
10. Forceps.
11. Orbital shaker.
12. Parafilm M.
13. Needle.
14. Microscope slides.
15. Coverslips (22 mm^2 and 22 × 40 mm^2).

16. Clear nail polish.

17. VECTASHIELD Antifade mounting medium.

18. Timer.

2.5 Image Acquisition and Analysis

1. Laser scanning confocal microscope (e.g., Leica SP5X with a low noise HyD hybrid detector and Z-galvo) either inverted or upright (*see* **Note 5**).

2. Yokogawa CSU-X1 spinning disc confocal microscope (PerkinElmer UltraView Vox) with an EMCCD camera (Hamamatsu) and a piezo-Z.

3. Stage: Universal mounting frame.

4. Long-range apochromat objectives (e.g., Leica 20×/multi-immersion 0.7 NA Plan Apo, 40×/oil 1.25 NA Plan Apo, and 63×/oil 1.4–0.6 NA plan Apo).

5. Multiple laser lines (e.g., 405, 440, 488, 514, 561, 640 nm, for Leica SP5X a white light laser can replace the individual laser lines between 470 and 670 nm).

6. Immersion oil with refractive index 1.518 at 23 °C (Zeiss Immersol™ 518F).

7. Confocal image acquisition software: Leica LASAF software; Volocity Improvision for PerkinElmer.

8. Image analyses software: ImageJ/Fiji.

3   Methods

Visualization of thin dynamic 100–200 nm diameter cytonemes requires a very high signal-to-noise ratio. A high signal can be achieved by expressing fluorescent membrane markers, such as CD8:GFP, or CD8:mCherry, mCherryCAAX, or CD4:IFP2 under a strong tissue-specific *Gal4* or *LexA* driver (*see* **Notes 1** and **6**). For instance, in this method, we marked tracheal cell-membrane by driving CD8:GFP with a trachea-specific *btl-Gal4*, and epidermal engrailed-expressing cells by driving CD4:IFP2 (Infrared fluorescent protein) with *en-Gal4*. Fluorescent-labeled cytonemes are efficiently visualized when they extend over a non-fluorescent and less autofluorescent background. A relatively higher number of cytonemes is observed in live embryos than in the fixed-tissue preparations.

3.1 Genetic Crosses for Tissue-Specific Expression

1. Expand the driver (e.g., *btl-Gal4*) and reporter (e.g., *UAS-CD8:GFP*) lines to collect a large number of flies.

2. Prepare ~50 young female virgin *btl-Gal4* flies, as well as ~30 male *UAS*-CD8:GFP flies (*see* **Note 7**).

3. Anesthetize male and female flies by placing them on CO_2 pad (as recommended in [36] for fly pushing).

4. Put a small scoop of yeast paste with a spatula at the center of the grape juice agar plate.

5. Place females and males together in the small collection cage and close the lid containing the grape juice agar plate. Do not flip the cage when the flies are in anesthetized condition to avoid them sticking to the food (*see* **Note 8**).

6. Flip the cage with the plate side down when the flies wake up.

7. Incubate at 25 °C, which is ideal for egg laying.

8. Replace the old grape juice agar plate with new ones containing a scoop of yeast paste every 12–24 h for at least 3 days. To replace the old plate, flip the cage with its mesh side down facing a CO_2 pad. Once the flies are anesthetized, open the cage lid and replace the old plate.

9. Start embryo collection from 3rd–4th day when most of the female flies start laying eggs (*see* **Note 9**).

3.2 Embryo Collection and Dechorionization

1. Incubate plates for 8–14 h to harvest embryos of all stages. For tracheal cytonemes, mostly the embryos between stage 11–15 are required (*see* **Note 10**). Epidermal cytonemes are visualized from the *engrailed*-expressing cells between embryonic stage 8–16.

2. Follow **step 8** of Subheading 3.1 to collect the plate to be harvested for embryos and replace with a new plate for future collection.

3. Use a small spatula or a brush to remove the yeast paste and dead flies from the agar plate.

4. Add some dH_2O covering the surface of the plate, and use the brush to gently dislodge all the embryos from grape juice agar plate into the water.

5. Transfer them by pouring into the cell strainer and keep them hydrated.

6. Rinse embryos with water for 30 s.

7. Put the cell strainer with embryos in a dissection watch glass dish containing 50% sodium hypochlorite. Dechorionize embryos by slightly rotating the cell strainer till most embryos float up, indicating that they are dechorionized (this takes about 3 min).

8. Rinse embryos thoroughly with flowing dH_2O for at least 1 min to remove residual bleach (*see* **Note 11**).

9. Rinse embryos with embryo wash (1% PBST).

10. Rinse embryos with dH$_2$O to thoroughly wash off all the Triton X-100 (*see* **Note 12**).

11. For **steps 9** and **10**, make sure to wash all the embryos on the side of the strainer down onto the center of the mesh. This facilitates their efficient transfer to the next step.

12. All the procedures are carried out at room temperature.

13. Immediately follow fixation (*see* Subheading 3.3) or live imaging (*see* Subheading 3.4) protocol.

3.3 Embryo Fixation

1. All the following procedures are carried out at room temperature.

2. Prepare fixation solution in a 1.5 mL microcentrifuge tube in the fume hood by mixing 500 μL 4% PFA and 500 μL heptane.

3. Use a paintbrush to transfer the dechorionated embryos from the mesh into a 1.5 mL Eppendorf tube containing the fixation solution.

4. Tightly seal the tube with Parafilm to avoid spillage, and place it on top of the orbital shaker, secure with tape, and shake vigorously at high speed for 20 min.

5. Embryos will be between the two-phase layers. With a pipette, gently remove as much of the lower aqueous phase as possible, without taking the embryos at the interphase.

6. Add 800 μL of 100% MeOH to the tube and immediately shake vigorously for 1 min. This step removes the vitelline membrane.

7. Let the tube stand for a minute when the devitellinized embryos sink to the bottom (*see* **Note 13**).

8. Remove the top layer including the embryos at the liquid interface and with part of MeOH containing embryos that do not sink.

9. Rinse three times with 800 μL MeOH to ensure complete removal of heptane and embryos with partial devitellinization.

10. Embryos in MeOH can be stored at −20 °C. However, we do not recommend this step, because, freshly fixed embryos yield the best fluorescent detection. Directly go to the next step.

11. Rehydrate embryos with three washes of 5 min each: 75% MeOH–25% (0.1% PBST), 50% MeOH–50% (0.1% PBST), 25% MeOH–75% (0.1% PBST).

12. Wash two times in 0.1% PBST, 2 min each.

3.4 Preparing Live Embryo for Imaging

1. Following dechorionization of the embryos (described in **step 11** of Subheading 3.2), transfer them in a small watch glass containing 1×PBS.

2. Monitor under a stereomicroscope and remove the dead embryos with forceps.

3. Place the glass bottom dish on a black background under the stereomicroscope.

4. Use a paintbrush to transfer the embryos to the center of the glass in the glass bottom dish (*see* **Note 14**).

5. Remove the excess liquid from the glass that got transferred along with the embryos.

6. Add a drop of halocarbon oil 27 to cover the embryos completely (*see* **Note 15**).

7. Use needle or forceps to gently spread the embryos in the oil and make a single layer of the embryos. Avoid spreading them close to the edge of the glass. Pipette out extra oil. The indication of a good mounting is when the single-layered embryos firmly sit on the glass, rather than floating in the oil (Fig. 3a).

8. Place a piece of wet kimwipes in the dish surrounding the glass bottom to retain moisture. Put on the lid of the plate (*see* **Notes 16** and **17**).

3.5 Preparing Fixed Embryos for Imaging

1. Put microscope slide on a black background.

2. Remove most of the PBST from the tube, leave about 100 μL of PBST.

3. Use clean scissors or blade to cut the tip off a p200 pipette tip, pipet embryos with PBST and move them to the microscope slide.

4. Remove the extra PBST from the slide, remove bubbles.

5. Add the mounting solution (two drops of VECTASHIELD) on top of the embryos (*see* **Note 18**).

6. Use a needle to spread embryos in the mounting solution, try to center and align the embryos to avoid overlap.

7. Use Kimwipe to remove extra PBST at the edges of the slide.

8. Gently put the coverslip from one side to another to avoid bubble (Fig. 3b).

9. Use nail polish to seal the coverslip. Dry for 2 min.

10. Slides can be stored at 4 °C for future imaging, but freshly prepared samples are the best for detection of cytonemes.

3.6 Image Acquisition and Processing

1. Set up the imaging condition using the confocal microscope software. The laser scanning microscope is suitable for deep tissue low noise imaging; it allows imaging from a selected ROI and zoom-in using zoom factor (most often 2–3×). On the other hand, a spinning-disc microscope enables high-speed, high-resolution, wide-field imaging (*see* **Note 19**).

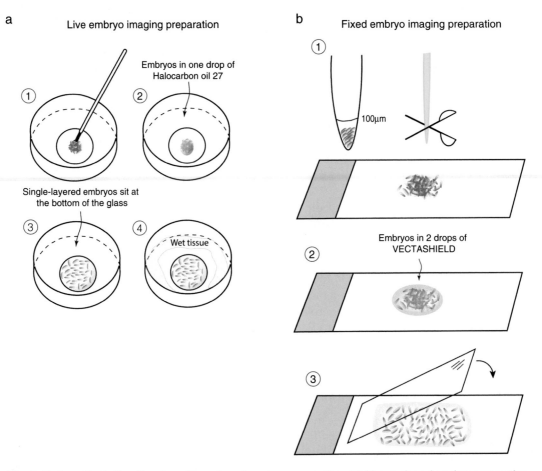

Fig. 3 Cartoon illustrating the steps for embryo imaging preparation. (**a**) Live embryo imaging preparation (description in Subheading 3.4): 1. use a paintbrush to transfer embryos to the center of the cover glass; eliminate excess liquid; 2. add a drop of Halocarbon oil 27 on top of the embryos; 3. gently spread embryos in the oil and make a single layer of embryos; 4. place some wet tissue in the dish surrounding the glass bottom microwell. (**b**) Slide preparation for fixed embryos (description in Subheading 3.5): (1) Leave about 100 μL of PBST in Eppendorf tube with embryos; cut off the tip of a p200 pipette tip, pipette embryos onto a microscope slide; remove the extra PBST liquid from the slide. (2) Add two drops of the VECTASHIELD on top of the embryos. (3) Use a needle to gently spread the embryos in a single layer, gently lower the coverslip (arrow) onto the embryos in VECTASHIELD without including an air bubble

2. Add a drop of immersion oil on 40× or 60× objectives depending on the resolution required. Both objectives are suitable for imaging cytonemes. We recommend starting with using 40× objective with 2–3× zoom factor to avoid FP quenching. 60× imaging is useful for subcellular resolutions.

3. Position a 10–20× air objective in the light path.

4. Place the glass-bottom dish (**step 8** of Subheading 3.4) or the slide (**step 9** of Subheading 3.5) on the universal mounting frame of the microscope stage.

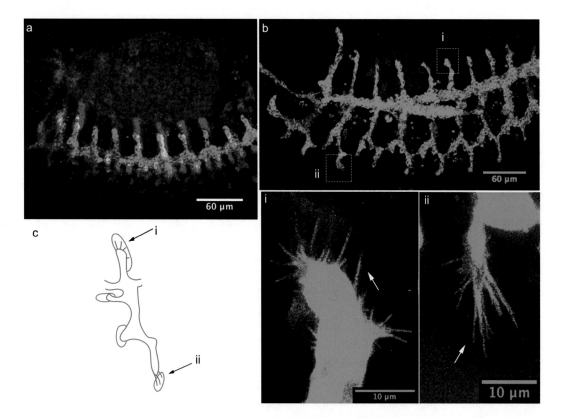

Fig. 4 Examples of cytonemes from fixed and live embryonic trachea. (**a**) Confocal images of a fixed embryo at stage 14 showing dorsal tracheal branches (green), each growing toward a dorsal *bnl* source (red); genotype: *btl-Gal4,UAS*-CD8:GFP/+; *bnl-LexA,lexO*-CherryCAAX/+. (**b**) A stage 14 embryo showing cytonemes from the dorsal (1) and ventral (2) tracheal branches without marking the *bnl*-source; genotype: *btl-Gal4; UAS*-CD8: GFP. Top panel, a z-projection image of 34 optical sections across 51 μm depth of a whole embryo, imaged in fixed condition with a 40× objective in Leica SP5X. Bottom panels, images of cytonemes from a dorsal (1) and a ventral (2) branch of a live stage 14 embryo; these cytonemes project toward the *bnl*-source (*see* panel a); imaging condition: 40× objective with 2× zoom factor, focusing a growing tip of a branch. (**c**) Schematic drawing of a stage 14 trachea metamere (green) and the *bnl*-sources (red), arrows point to the dorsal (1) and the ventral (2) branches as shown in (**b**)

5. Locate an embryo using the 10–20× bright field objective.

6. Switch to epifluorescence and identify the right stage and orientation of the tissue of interest [39] (*see* **Note 20**).

7. Switch to the oil objective already with immersion oil and quickly bring the sample to focus (*see* **Note 21**). Use epifluorescence or bright-field for focusing.

8. Switch to the laser system using the software and quickly select a region of interest (ROI) to image cytonemes in live mode. Usually growing tips of the dorsal and ventral branches project many cytonemes toward the FGF (fibroblast growth factor) source situated ahead of them (Fig. 4) [37]. Cytonemes from

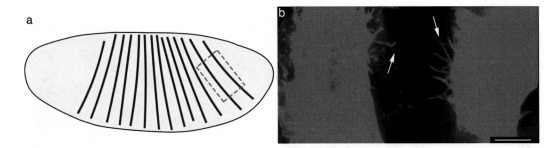

Fig. 5 Example of cytonemes from live embryonic epidermal cells that express *engrailed* (*en-Gal4*). (**a**) Schematic drawing of a stage 12–13 embryo with *engrailed*-expressing stripes. (**b**) CD4:IFP2 expressing cells in the *engrailed* stripes project short cytonemes; 40× magnification with a 2× zoom of ROI function; the region shown in (**b**) corresponds to the dashed square in (**a**); scale bar, 10 μm

cells in the *en* stripes extend toward the neighboring cells (Fig. 5).

9. Start acquiring images. Start with a reasonably low laser power in live mode (**step 8**). For EMCCD camera: adjust exposure time and EM gain according to the level of GFP/RFP expression in the tissues and efficiency of cytoneme detection.

10. For efficient visualization of cytonemes, focus mostly at the edge of the fluorescent tissue (*see* **Note 22**).

11. Open confocal images in Fiji/ImageJ to perform Z-projection of selected sections and adjust brightness and contrast to visualize cytonemes. Convert multicolor images to RGB and save with a scale bar. For a single color, apply the desirable change in intensity and save the image with a scale bar.

4   Notes

1. Membrane marking is critical for efficient detection of the thin 200 nm cytonemes. However, various FP fused proteins also marked cytonemes with variable efficiency [17, 19, 20, 23, 28].

2. Mix one pack of the grape juice agar powder in 450 mL deionized water in a 2-L beaker. Microwave on high for 2 min, stir well, and repeat the microwave step for two more times, 1 min each, followed by stirring to remove all the clumps. If necessary, repeat the heating and stirring steps until a clump-free homogeneous mix is formed. Usually, a homogeneous mix is obtained when the liquid is completely boiled and starting to bubble. Observe to avoid a spillage during microwave.

3. Mix dry baker's yeast with a bit of water in a plastic beaker, mold it with a metal spatula until it forms a homogenous thick paste. Do not make the paste too watery to avoid flies getting stuck in it. Keep the yeast paste at 4 °C, covered with aluminum

foil, for maximum 2 weeks. Fresh yeast paste induces flies to lay more eggs.

4. Triton X-100 is very viscous, so to take up the right amount, cut the narrow end of the micropipette tip and slowly take up the liquid without bubbles. Make sure that the viscous liquid is not sticking to the outer surface of the tip. Mix by a gentle stirring on a magnetic stirrer and avoid bubbles.

5. A standard scanning confocal microscope such as Leica SPE is sufficient to image cytonemes [21]. Motorized x–y stage is not required, but it enables time-lapse imaging from multiple embryos. A stage-top incubator is not required as *Drosophila* embryos develop efficiently at room-temperature between 20 and 25 °C. A fast optical-Z scan is important to capture the cytonemes present deep inside the tissues. Scanning confocal provides better resolution for deep-tissue imaging.

6. Earlier studies used CherryCAAX, CD8:Cherry, CD4:IFP2, CD2:GFP, and CD8:GFP to image cytonemes from larval tissues [21, 23, 24, 38, 40]. We successfully imaged the embryonic cytonemes using all these constructs.

7. The number of parent flies needed depends on the required size of the embryo collection. The recommended size is more than sufficient for live cell imaging. The way of setting up genetic crosses is variable, and depends on the needs and the availability of flies. The cage size can be adjusted accordingly from small to mini or large size (Genesee Scientific). A higher number of parent flies provide better chances to get synchronized embryos within a certain time window. For a repeated collection of the embryos, establishing a genetically combined or recombined fly stock harboring the driver and the reporter genes is convenient and recommended (e.g., recombined *btl-Gal4, UAS*-CD8:GFP homozygous stock on second chromosome [37]).

8. The small size cross can also be set up in tubes for 2 days before transferring them to the cage.

9. Well-fed, healthy, 10-day-old females with big ovary-filled abdomen start to lay many eggs.

10. Tracheal cytonemes were best observed during stages 11–15 when the primary tracheal branches migrate in response to the Branchless signal, and they cover relatively less surface area than the old embryos [41]. Modify the collection window based on the required developmental stage.

11. Sodium hypochlorite is highly toxic and a thorough wash is required for time-lapse imaging.

12. The dechorionated embryos should be processed immediately for imaging or fixation. Do not leave the dechorionated

embryo in the strainer for long. Dehydration in the embryos leads to abnormal cytoneme patterns and tissue morphology leading to death.

13. If less than 50% of the embryos do not sink, for subsequent experiments, try preparing fresh 50% bleach for dechorionation or increase the shaking speed and time in the devitellinization step.

14. Imaging embryos at the junction of the glass-bottom and the plastic disc leads to unwanted diffraction of light.

15. Oil allows oxygen exchange and prevents dehydration.

16. A humid chamber is required, because the high fluidity of halocarbon oil 27 is not effective against dehydration, especially for long duration imaging.

17. An inverted microscope is desirable for long-term imaging using the setup described here. For upright confocal microscopes, adopt an alternative hanging drop method. However, the method is suitable only for a short duration time-lapse imaging. In this method, the live embryos are transferred on a small drop of Halocarbon oil 700 (or 27) that is placed at the center of a coverslip. Align the embryos at the center with a needle. Remove excess oil but keep sufficient amount to cover the embryos. Add four tiny drops of halocarbon oil at the four corners of the coverslip. Place a cavity slide right on top of the coverslip such a way that the embryos on the coverslip are at the center of the concavity and are not squished. Press the slide against the coverslip so that they stick to one another. Flip the slide and temporarily seal the edge of the coverslip with rubber cement (Electron Microscopy Sciences). In this hanging drop preparation, the surface tension of the oil helps the embryo to stick to the coverslip. This alternative method is also suitable for an inverted microscope.

18. It is important to add enough amount of mounting solution to avoid overlapping of embryos. The amount of solution depends on the number of embryos.

19. An optimized condition for imaging cytonemes in scanning confocal microscope: Imaging dimension (XYZCT); X–Y size: 512X512 pixels; Z-volume: 20–40 μm with 0.2–0.35 μm step size (Cytonemes are very thin, about 0.2 μm in thickness); laser power: <20% with white light laser; pinhole size: airy 1; dwell time: set it up during imaging based on the required speed; HyD gain: 100%; scan speed: default (can be bidirectional scan); averaging for 2–3 times. For spinning disc: camera exposure time 20 ms (<100 ms), EM Gain 300. When acquiring time-lapse images, set up a Z volume 3–5 μm up or down from the tissue in focus to accommodate continuous imaging of the tissue in a growing embryo.

20. Embryonic trachea adopts a developmental stage-specific morphology and can reliably be used as a staging marker.

21. Ensure that the oil objectives have sufficient immersion oil to avoid light scatter.

22. The fixed embryos showed less number of cytonemes than the live ones. Few advantages of the fixed embryos are: (1) the embryos are closer to the coverslip, so high-resolution images can be achieved easily, and (2) the fixation process reduces autofluorescence.

Acknowledgments

We thank Dr. T.B. Kornberg and the Bloomington Stock Center for reagents, colleagues and lab members, especially Alex Sohr for reading the manuscript and valuable suggestions, UMD Imaging core facility, and Dr. A.E. Beaven for assistance in the imaging core. Funding from NIH: R00HL114867 and R35GM124878 to S.R.

References

1. Entchev EV, Schwabedissen A, González-Gaitán M (2000) Gradient formation of the TGF-β Homolog Dpp. Cell 103:981–992. https://doi.org/10.1016/S0092-8674(00)00200-2

2. Teleman AA, Cohen SM (2000) Dpp gradient formation in the Drosophila wing imaginal disc. Cell 103:971–980

3. Goentoro LA, Reeves GT, Kowal CP et al (2006) Quantifying the Gurken morphogen gradient in Drosophila oogenesis. Dev Cell 11:263–272. https://doi.org/10.1016/j.devcel.2006.07.004

4. Strigini M, Cohen SM (2000) Wingless gradient formation in the Drosophila wing. Curr Biol 10:293–300

5. Wolpert L (1969) Positional information and the spatial pattern of cellular differentiation. J Theor Biol 25:1–47

6. Rogers KW, Schier AF (2011) Morphogen gradients: from generation to interpretation. Annu Rev Cell Dev Biol 27:377–407. https://doi.org/10.1146/annurev-cellbio-092910-154148

7. Wolpert L (2016) Positional information and pattern formation. Curr Top Dev Biol 117:597–608. https://doi.org/10.1016/bs.ctdb.2015.11.008

8. Müller P, Rogers KW, Yu SR et al (2013) Morphogen transport. Development 140:1621–1638. https://doi.org/10.1242/dev.083519

9. Christian JL (2012) Morphogen gradients in development: from form to function. Wiley Interdiscip Rev Dev Biol 1:3–15. https://doi.org/10.1002/wdev.2

10. Schwank G, Dalessi S, Yang S-F et al (2011) Formation of the long range Dpp morphogen gradient. PLoS Biol 9:e1001111. https://doi.org/10.1371/journal.pbio.1001111

11. Zhou S, Lo W-C, Suhalim JL et al (2012) Free extracellular diffusion creates the Dpp morphogen gradient of the Drosophila wing disc. Curr Biol 22:668–675. https://doi.org/10.1016/j.cub.2012.02.065

12. Yu SR, Burkhardt M, Nowak M et al (2009) Fgf8 morphogen gradient forms by a source-sink mechanism with freely diffusing molecules. Nature 461:533–536. https://doi.org/10.1038/nature08391

13. Dubrulle J, Pourquié O (2004) fgf8 mRNA decay establishes a gradient that couples axial elongation to patterning in the vertebrate embryo. Nature 427:419–422. https://doi.org/10.1038/nature02216

14. Shilo B-Z, Haskel-Ittah M, Ben-Zvi D et al (2013) Creating gradients by morphogen shuttling. Trends Genet 29:339–347. https://doi.org/10.1016/j.tig.2013.01.001

15. Entchev EV, Schwabedissen A, Gonzalez-Gaitan M (2000) Gradient formation of the TGF-beta homolog Dpp. Cell 103:981–991

16. Belenkaya TY, Han C, Yan D et al (2004) Drosophila Dpp morphogen movement is independent of dynamin-mediated endocytosis but regulated by the glypican members of heparan sulfate proteoglycans. Cell 119:231–244. https://doi.org/10.1016/j.cell.2004.09.031

17. Kornberg TB, Roy S (2014) Cytonemes as specialized signaling filopodia. Development 141:729–736. https://doi.org/10.1242/dev.086223

18. Kornberg TB (2017) Distributing signaling proteins in space and time: the province of cytonemes. Curr Opin Genet Dev 45:22–27. https://doi.org/10.1016/j.gde.2017.02.010

19. Ramírez-Weber FA, Kornberg TB (1999) Cytonemes: cellular processes that project to the principal signaling center in Drosophila imaginal discs. Cell 97:599–607

20. Hsiung F, Ramirez-Weber F-A, Iwaki DD, Kornberg TB (2005) Dependence of Drosophila wing imaginal disc cytonemes on decapentaplegic. Nature 437:560–563. https://doi.org/10.1038/nature03951

21. Roy S, Hsiung F, Kornberg TB (2011) Specificity of Drosophila cytonemes for distinct signaling pathways. Science 332:354–358. https://doi.org/10.1126/science.1198949

22. Sato M, Kornberg TB (2002) FGF is an essential mitogen and chemoattractant for the air sacs of the drosophila tracheal system. Dev Cell 3:195–207

23. Roy S, Huang H, Liu S, Kornberg TB (2014) Cytoneme-mediated contact-dependent transport of the Drosophila decapentaplegic signaling protein. Science 343:1244624–1244624. https://doi.org/10.1126/science.1244624

24. Huang H, Kornberg TB (2016) Cells must express components of the planar cell polarity system and extracellular matrix to support cytonemes. elife 5:197. https://doi.org/10.7554/eLife.18979

25. Roy S, Kornberg TB (2011) Direct delivery mechanisms of morphogen dispersion. Sci Signal 4:pt8. https://doi.org/10.1126/scisignal.2002434

26. Rojas-Rios P, Guerrero I, Gonzalez-Reyes A (2012) Cytoneme-mediated delivery of hedgehog regulates the expression of bone morphogenetic proteins to maintain Germline stem cells in Drosophila. PLoS Biol 10:e1001298–e1001213. https://doi.org/10.1371/journal.pbio.1001298

27. Roy S, Kornberg TB (2015) Paracrine signaling mediated at cell-cell contacts. BioEssays 37:25–33. https://doi.org/10.1002/bies.201400122

28. Kornberg TB (2014) Cytonemes and the dispersion of morphogens. Wiley Interdiscip Rev Dev Biol 3:445–463. https://doi.org/10.1002/wdev.151

29. Sanders TA, Llagostera E, Barna M (2013) Specialized filopodia direct long-range transport of SHH during vertebrate tissue patterning. Nature 497:628–632. https://doi.org/10.1038/nature12157

30. Huang H, Kornberg TB (2015) Myoblast cytonemes mediate Wg signaling from the wing imaginal disc and Delta-notch signaling to the air sac primordium. elife 4:e06114. https://doi.org/10.7554/eLife.06114

31. Holzer T, Liffers K, Rahm K et al (2012) Live imaging of active fluorophore labelled Wnt proteins. FEBS Lett 586(11):1638–1644. https://doi.org/10.1016/j.febslet.2012.04.035

32. Stanganello E, Scholpp S (2016) Role of cytonemes in Wnt transport. J Cell Sci 129:665–672. https://doi.org/10.1242/jcs.182469

33. Buszczak M, Inaba M, Yamashita YM (2016) Signaling by cellular protrusions: keeping the conversation private. Trends Cell Biol 26:526–534. https://doi.org/10.1016/j.tcb.2016.03.003

34. Du L, Roy S (2017) Cytonemes mediate formation of a morphogen gradient of FGF during branching morphogenesis of Drosophila trachea. Mol Biol Cell 28:3727 (abstract #M37)

35. Seijo-Barandiarán I, Guerrero I, Bischoff M (2015) In vivo imaging of hedgehog transport in Drosophila epithelia. Methods Mol Biol 1322:9–18. https://doi.org/10.1007/978-1-4939-2772-2_2

36. Stocker H, Gallant P (2008) Getting started: an overview on raising and handling Drosophila. Methods Mol Biol 420:27–44. https://doi.org/10.1007/978-1-59745-583-1_2

37. Du L, Zhou A, Patel A Rao M, Anderson K, Roy S (2017) Unique patterns of organization and migration of FGF-expressing cells during Drosophila morphogenesis. Dev Biol 427:35–48. https://doi.org/10.1016/j.ydbio.2017.05.009

38. Yu D, Gustafson WC, Han C et al (2014) An improved monomeric infrared fluorescent protein for neuronal and tumour brain imaging. Nat Commun 5:3626. https://doi.org/10.1038/ncomms4626

39. Ghabrial A, Luschnig S, Metzstein MM, Krasnow MA (2003) Branching morphogenesis of the Drosophila tracheal system. Annu Rev Cell Dev Biol 19:623–647. https://doi.org/10.1146/annurev.cellbio.19.031403.160043

40. Chen W, Huang H, Hatori R, Kornberg TB (2017) Essential basal cytonemes take up hedgehog in the Drosophila wing imaginal disc. Development 144:3134–3144. https://doi.org/10.1242/dev.149856

41. Sutherland D, Samakovlis C, Krasnow MA (1996) Branchless encodes a Drosophila FGF homolog that controls tracheal cell migration and the pattern of branching. Cell 87:1091–1101

Chapter 4

Measuring Dorsoventral Pattern and Morphogen Signaling Profiles in the Growing Neural Tube

Marcin Zagorski and Anna Kicheva

Abstract

Developmental processes are inherently dynamic and understanding them requires quantitative measurements of gene and protein expression levels in space and time. While live imaging is a powerful approach for obtaining such data, it is still a challenge to apply it over long periods of time to large tissues, such as the embryonic spinal cord in mouse and chick. Nevertheless, dynamics of gene expression and signaling activity patterns in this organ can be studied by collecting tissue sections at different developmental stages. In combination with immunohistochemistry, this allows for measuring the levels of multiple developmental regulators in a quantitative manner with high spatiotemporal resolution. The mean protein expression levels over time, as well as embryo-to-embryo variability can be analyzed. A key aspect of the approach is the ability to compare protein levels across different samples. This requires a number of considerations in sample preparation, imaging and data analysis. Here we present a protocol for obtaining time course data of dorsoventral expression patterns from mouse and chick neural tube in the first 3 days of neural tube development. The described workflow starts from embryo dissection and ends with a processed dataset. Software scripts for data analysis are included. The protocol is adaptable and instructions that allow the user to modify different steps are provided. Thus, the procedure can be altered for analysis of time-lapse images and applied to systems other than the neural tube.

Key words Neural tube, Spinal cord, Tissue development, Morphogen gradient, Pattern formation, Quantitative imaging

1 Introduction

Spinal cord development provides one of the best examples of developmental pattern formation. In this organ, an elaborate and stereotypic pattern of gene expression domains defines the identities of multiple neural progenitor subtypes along the dorsoventral (DV) axis [1]. This pattern is established in response to signaling by antiparallel morphogen gradients [2, 3] in a temporally dynamic manner. During the first 3 days of neural tube development in

Electronic supplementary material: The online version of this chapter (https://doi.org/10.1007/978-1-4939-8772-6_4) contains supplementary material, which is available to authorized users.

Julien Dubrulle (ed.), *Morphogen Gradients: Methods and Protocols*, Methods in Molecular Biology, vol. 1863, https://doi.org/10.1007/978-1-4939-8772-6_4, © Springer Science+Business Media, LLC, part of Springer Nature 2018

mouse and chick, the signaling gradients and the gene expression domains undergo considerable changes. At the same time, the tissue size increases from ~100 μm to ~400 μm. Although many signals and components of the gene network that defines neural tube patterning are known [1, 4–7], quantitative spatiotemporal measurements of their expression are to a large extent still lacking. The size of the tissue and developmental time scale make it difficult to study the temporal dynamics of patterning using live imaging. Hence, one of the best approaches remains the collection of transverse tissue sections through the spinal cord at different developmental time points. Such datasets allow quantifying the mean profiles and variation of signaling activity and gene expression along the DV axis over time [2, 3, 8, 9].

Here we describe how to prepare mouse and chick neural tube sections for immunohistochemistry, imaging, and quantitative temporal analysis of dorsoventral signaling or gene expression profiles. The protocol builds on previous studies [2–4, 6] and is designed for stages between E8 and E11.5 of mouse embryonic development, or Hamburger-Hamilton (HH) stage 9 to HH stage 27 [10] in chick. We discuss key considerations in assigning developmental stage to tissue sections, sample processing, imaging and data analysis that aim to minimize technical error and ensure that protein levels across different samples can be compared.

The first step is staging of the collected embryos and tissue sections. During development, the neural tube extends at the posterior end at the same time as new somites are added every 2 h adjacent to it [11]. Both the neural epithelium and the adjacent somites continue to grow throughout development, but neuroepithelial and somite cells maintain approximate register along the anterior–posterior (AP) axis [12]. Thus, sections taken at the same somite number at different stage can be used to determine the behavior of the tissue over time. In practice, sections are collected from AP positions that encompass 3–4 somites, introducing size variability. However, the fact that anterior positions are developmentally older allows to correct for this variability by reassigning the developmental age of each measured profile based on the DV length of the tissue. The restaged sections can be later grouped into defined time intervals, so that temporal changes in the mean and variance of the expression profiles can be analyzed.

Once the embryos are collected, they are fixed, embedded in gelatin, cryosectioned, and immunostained prior to imaging. To minimize the variability between samples that these steps could introduce, the samples are processed in parallel and the same batch of reagents is used for each time course experiment. In particular, all tissue sections are immunostained together and imaged in the same imaging session using identical settings. The images are then used to quantify the fluorescence intensity (FI) profiles within user-specified regions of interest (ROI), which span the DV length of the neural tube. These profiles represent the average intensity across the ROI width for every pixel along the ROI length. Here we provide two Fiji macro scripts (*see* Subheading 3.3) to facilitate measurement of the FI profiles.

The protocol presented here does not include a step in which immunofluorescence levels are directly calibrated to actual molecule numbers. Such calibration is possible for tagged proteins (*see* [13–15]) and can be easily incorporated into the protocol. Nevertheless, the protein levels can be compared in space and time if the fluorescence intensity levels are proportional to protein levels. This requires linearity of antibody staining and fluorescence detection, which can be achieved with immunohistochemistry and laser scanning confocal microscopy [14–18]. Tests to ensure linear conditions [15, 17, 18] can be applied to the neural tube.

Further processing of the images in Matlab includes subtraction of background fluorescence, defined as the minimum intensity value in each profile, and removal of outlier profiles that deviate significantly from the mean profile for a given time window (*see* also [2, 19]). Outliers most often occur due to damage or distortions in the tissue sections that are caused by the dissection or sectioning procedure. Further corrections that can be optionally implemented include smoothing of the measured profiles and using a maximum projection of a small z-stack of images, rather than individual optical sections. These steps aim to correct for subcellular inhomogeneities of the signal that may arise from using stains that are restricted to the nuclei or cytoplasm. Thus, the signal is effectively averaged at a cellular scale to represent a continuum of DV positions across one cell diameter of the pseudostratified neural epithelium. Here we provide a Matlab script for processing of the measured profiles (*see* Subheading 3.3).

The scripts and analysis steps are easy to modify and customize depending on the purpose of the analysis. The presented method can be directly used to study spatiotemporal changes in gene expression or activity of fluorescent reporters and biosensors. The approach can also be adapted to quantify fluorescence intensity profiles in the consecutive frames of time-lapse recordings or to study correlations between fluctuations of two or more signals [3].

2   Materials

2.1   Equipment and Software

1. Dissection scissors.

2. Dumont #5 and #55 forceps, stainless.

3. Disposable transfer pipettes, 7.5 mL, 15.5 cm.

4. Two tungsten needles, each consisting of approximately 4 cm of sharpened tungsten wire, 0.25 mm diameter, mounted in a pin holder. To sharpen the wires, prepare a beaker with 5 M

KOH. From about 10 cm of wire make several loops, attach to a negative electrode and immerse in the solution. Attach the 4 cm piece of wire to be sharpened to the positive electrode. Set the voltage on a regulated DC power supply unit to a maximum of 4.5 V and connect the electrodes. Dip the tip of the wire into the solution and hold until the desired sharpening or thinning of the wire is achieved (typically about a minute). Bend one of the needles ~0.5 cm from the tip using forceps (to be used for dissection), and leave the other one straight (to be used for embedding).

5. Rocking platform.

6. Disposable base molds (15 × 15 × 5 mm) for embedding.

7. Low-temperature thermometer (-40 °C).

8. Cryostat, Thermo Scientific Microm HM560 or equivalent.

9. ImmunoPen™, Millipore.

10. Plastic 5 microscope slide mailers (Leica) or equivalent.

11. Laser scanning confocal microscope.

12. Fiji software [20], http://fiji.sc/.

13. Matlab software (MathWorks, MA, USA).

2.2 Reagents

1. Phosphate-buffered saline pH = 7.4 (PBS), without calcium and magnesium salts.

2. 4% paraformaldehyde in PBS.

3. 15% sucrose in PBS.

4. 7.5% gelatin from porcine skin, 15% sucrose in PBS. To prepare, warm up PBS to 70 °C in a water bath. Slowly add gelatin from porcine skin (Sigma, G2500) to 7.5% and sucrose to 15% with occasional shaking. Leave on a hotplate with magnetic stirrer until sucrose and gelatin are completely dissolved. A well-mixed solution is essential for good results. Store 5 mL aliquots at -20 °C.

5. Isopentane.

6. Washing buffer: PBS, 0.1% Triton X-100.

7. Blocking buffer: washing buffer, 1% (w/v) bovine serum albumin.

8. Primary antibodies.

9. Fluorescent dye-conjugated secondary antibodies.

10. 5 mg/mL Dapi (4′,6-diamidino-2-phenylindole), in dH_2O.

11. ProLong Gold antifade mountant.

3 Methods

3.1 Mouse/Chick Embryo Preparation

Perform all steps on ice using ice-cold reagents.

1. Dissect embryos in PBS, clearing all extraembryonic tissues (*see* **Note 1**).

2. Count and record the number of somites of each embryo. In wildtype litters, discard embryos that have not developed properly.

3. Remove the embryo heads. Using a plastic Pasteur pipette with a cut tip, transfer the embryos to a 2 mL Eppendorf tube with PBS on ice (for stages corresponding to E10.5 (mouse) / HH17 (chick) or older, use a 5 mL tube). If required, individual embryos can be processed separately (*see* **Note 2**).

4. Wash the collected embryos once with PBS.

5. Remove PBS and fill the tube with 4% paraformaldehyde. Fix on ice with slow rocking for the following amounts of time depending on the embryo stage: mouse E8.5—50 min, E9.5—60 min, E10.5—75 min, E11.5—90 min; chick HH8—11–50 min; HH12—17–60 min; HH18—23–75 min; HH23—27–90 min.

6. Wash two times × 5 min with PBS.

7. Incubate at 4 °C in PBS, 15% sucrose until embryos sink to the bottom of the tube (typically ~1 h for E8.5, and up to overnight for older stages).

8. Defrost gelatin solution in a 42 °C water bath and mix well. Fill a 2 mL Eppendorf tube with 1 mL of gelatin and keep on hot block at 38 °C.

9. Using a tungsten needle with a bent tip, dissect the anterior–posterior region of interest, making a cut perpendicular to the dorsal midline. Perform the dissection in PBS, 15% sucrose (this step can also be performed in PBS prior to the sucrose incubation). Isolate pieces that are approximately 7–10 somites long. For time course analysis, the same somite (e.g., somite #8 corresponding to brachial level in mouse) should be in the middle of the piece (Fig. 1a).

10. Using forceps, carefully transfer the embryo pieces into the gelatin. Wait a few minutes for them to sink to the bottom of the tube. Use separate tubes if embryos are processed separately (*see* **Note 2**).

11. Using a transfer pipette, transfer the embryos to an embedding mold and over-fill with gelatin. Orient the embryo pieces vertically with anterior facing down, ventral facing toward you (Fig. 1b). Arrange the pieces in an asymmetric pattern, so that each embryo can be clearly identified in different sections (Fig. 1b, c, also *see* **Notes 2** and **3**). Keep orienting the

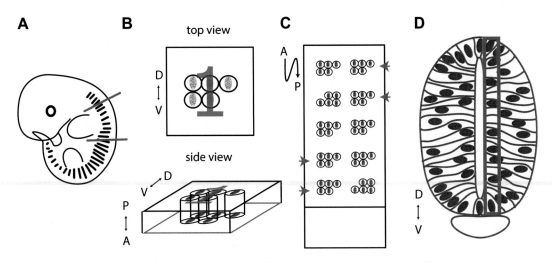

Fig. 1 Sample preparation. (**a**) The brachial region of the embryo is dissected out by cutting along the red lines. (**b**) The dissected pieces are embedded in the indicated orientation in an asymmetric pattern. The number "1" (or any asymmetric symbol) is written on the block to mark the orientation after freezing. (**c**) The block is sectioned in A to P direction and the sections collected on 6–8 microscope slides. Sections corresponding to somite positions in the center of the dissected region are located in corresponding positions on the slides (red arrows). Each embryo can be unambiguously identified across sections. (**d**) Close-up of a transverse section of the neural tube, showing the pseudostratified epithelium. The red rectangle is the ROI used for measuring the fluorescence intensity

pieces with a straight tungsten needle until the gelatin starts to solidify.

12. Once the gelatin has set, put the mold on ice and incubate for 30 min.

13. Remove the gelatin block from the mold and trim the block using a surgical blade. To mark the sample number and orientation, write a number directly on the gelatin block using a permanent marker or pen (test beforehand that the ink does not dissolve during freezing) (Fig. 1b).

14. Pour ~30 mL of isopentane in a small beaker, place a thermometer in the solution and put on dry ice. As soon as the isopentane reaches −40 °C, remove the thermometer and drop the gelatin block in the solution. Wait <1 min until the block is completely frozen, then use tweezers to transfer the frozen block to a tube and store at −80 °C until ready to section. Do not allow the block to stay in isopentane for too long, or it will crack.

15. At the cryostat, mount the block so that sectioning starts from the anterior ventral side of the embryos and the DV axis is aligned orthogonal to the blade (*see* **Note 3**). Cut 12–14 μm slices at −20 °C and collect on 6–8 microscope slides. Alternate slides after every section, so that every anterior–posterior level is equally represented on every slide. Store slides at −80 °C until ready to use.

3.2 Immunohisto-chemistry and Imaging

Carry out the procedures at the specified temperature.

1. Sections from different stages that will be compared have to be stained together and imaged in the same imaging session in order to minimize technical error. Reserve a time slot on the confocal microscope of 4–8 h depending on the number of slides and experience of the user. Imaging should be performed between 2 h and no more than 1 week from staining the slides.

2. Defrost and dry the slides with embryo sections for 10–15 min at room temperature.

3. Draw a hydrophobic boundary at the edge of each slide with ImmunoPen. To remove the gelatin from the sections, immerse the slides in slide containers filled with prewarmed PBS in a 42 °C water bath and incubate for 20 min.

4. Wash once with warm PBS.

5. Wash with cold washing buffer.

6. Transfer the slides in a horizontal position to a chamber lined with wet paper towel for humidification. Immediately cover the slides with blocking buffer and incubate for 1 h at room temperature in a closed chamber. It is critical that solutions are exchanged quickly and the sections are never left to dry during the entire procedure.

7. Prepare a master mix of primary antibody in blocking buffer for all slides. The optimal concentration of antibody should be determined before starting the time course experiment (*see* **Note 4**). Remove the blocking buffer and dispense 200 μL or master mix per slide. Make sure the ImmunoPen border is intact. Incubate at 4 °C overnight. Primary antibody may be reused several times.

8. Wash three times × 5 min in washing buffer.

9. Prepare a master mix of secondary antibody and Dapi in blocking buffer. For secondary antibody concentrations, use manufacturer recommendations. The final concentration of Dapi should be 0.5 μg/ml. Remove washing buffer and dispense 200 μL or master mix per slide. Incubate 2 h at room temperature.

10. Wash 3 times × 5 min in washing buffer.

11. Remove washing buffer. Dispense 2–3 drops (~15μL) of Pro-longGold mounting medium at different positions in the slide. Using forceps, slowly lower a 24 × 60 mm coverslip on the slide, avoiding bubbles. Store at 4°C until ready for imaging (up to 1 week after embedding).

12. At the confocal microscope, adjust the imaging field of interest in a way that fits the largest tissue (latest stage). This approach allows imaging the entire DV length of the neural tube of mouse embryos between E8.5 and E11 of development using a HCX Plan APO CS 40× Oil/1.25 NA objective or equivalent.

13. Adjust the laser power and detection settings so that images from earliest, middle and latest developmental stages are not saturated and no signal above background is lost. Quantify the fluorescence intensity profiles recorded at different settings in order to determine the optimal settings (also *see* **Note 4**).

14. Once the imaging conditions are set, image the sections from all stages, focusing on the sections that were collected at the middle of the block in order to minimize variations in anterior–posterior positions (Fig. 1c). To record the fluorescence across ~1 cell diameter in AP direction, collect z-stacks of three optical slices, 1 μm apart. For subsequent analysis using the provided scripts, the files should be named as "file_name_XXss.tif" (*see* **Note 5**) where XX is a two digit number indicating the somite stage (for stages 1 to 9, use 01 to 09).

3.3 Data Processing   Two Fiji scripts, one Matlab script and a test dataset containing images and the corresponding raw FI profiles are provided for use in conjunction with the following steps (*see* Electronic Supplementary Materials).

1. In Fiji, open the "maximum_projections.ijm" script (*see* Electronic Supplementary Materials) from the File menu and run it. An input folder containing the images saved as .tif files, and an output folder will be requested. The script will save the maximum projections of the files in the user-specified directory without displaying the images.

2. In Fiji, open the "profile_quantification.ijm" script (*see* Electronic Supplementary Materials) from the File menu. Open the maximum projections one by one in Fiji. For each image, run the Fiji script. You will first be prompted to draw an arrow from the floor plate to the roof plate, in order to rotate the image dorsal up, with the central lumen vertically aligned (Figs. 1d and 2a).

3. The second prompt asks the user to specify ROIs for quantification by adding them in the ROI manager. An example ROI of *width* = 12 μm and *height* = "DV length", positioned adjacent to the lumen, is shown in Fig. 2a. Two such ROIs per section can be used for quantifying gene expression patterns and morphogen activity profiles in the neural tube. However, the dimensions, position, and number of ROIs can be chosen by the user depending on the experimental aims. After pressing OK, the FI profiles are recorded as .txt files in a user-specified directory. The mean pixel intensity across the ROI in x-direction is quantified as function of *DVposition* in y-direction. The ventral midline corresponds to *DVposition* = 0. *DVposition* is quantified in units of μm, using the scaling information embedded in the image. Before performing next step,

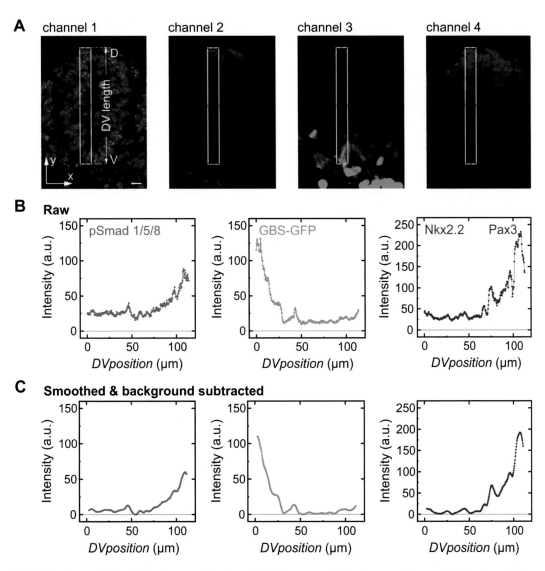

Fig. 2 Fluorescent intensity profile quantification. (**a**) Maximum projection image of the neural tube, dorsal side up, stained for Dapi (channel 1), pSmad (channel 2), GFP (GBS-GFP [6], channel 3), Nkx2.2 and Pax3 (channel 4). The ROI outlined in yellow is 12 μm wide and spans the DV length of the neural tube as indicated. Scale bar, 10 μm. (**b**) FI profiles quantified from the images in A as the mean FI across the ROI width for each *DVposition* along the ROI length (y direction). *DVposition* = 0 corresponds to the ventral boundary of the ROI. (**c**) The FI profiles from B are smoothed with moving average filter of 5 μm and background subtracted

check whether the resulting files contain data from all relevant images (*see* **Note 6**).

4. From this step onward, the data analysis can be implemented by following the steps described below using any appropriate software. Alternatively, the Matlab script "data_analysis.m" provided with this protocol (*see* Electronic Supplementary Materials) can be used and modified as needed. The script

contains an initial section where user-defined values of variables can be specified based on the descriptions provided in the remainder of the protocol. Run script "data_analysis.m". First, the user will be prompted to select the folder with .txt files containing the FI profile quantifications obtained with "profile_quantification.ijm" script. After pressing "Select Folder" all imported data should be in the "profiles_raw" structure in the Matlab workspace (Fig. 2b).

5. The data is smoothed with a moving average filter (Fig. 2c). By default the smoothing window is set to 5 μm, which corresponds roughly to one cell diameter. To modify the size of the smoothing window, change the value of the "smooth_window" variable. The smoothing step can be omitted by commenting out the relevant part of the script (*see* **Note 7**).

6. The background fluorescence intensity is removed for each profile by subtracting the minimum intensity of that profile (*see* **Note 8**).

7. Specify somite stage for each imported file in the "ss_time" array. The script can also automatically retrieve somite stage from the file names ("filenames" array) if the default naming convention is used (*see* Subheading 3.2, **step 14**).

8. Restage the profiles by their DV length. To do this, first the exponential function $L(t) = L_0 \exp(t_{ss}/\tau)$ is fitted to the data, where L is the measured DV length, τ is a fit parameter and the time t_{ss} is determined as the somite stage × 2 h/somite (Fig. 3a). The reassigned time $t(L)$ for each profile is determined via the inverse function, $t(L) = \tau \log(L/L_0)$, and is stored in the "dv_time" array (*see* Fig. 3b, and **Note 9**).

9. To study temporal changes in gene expression, it may be practical to look at the average profiles in defined time intervals. To do this, specify time intervals over which the data should be averaged. The time intervals should be chosen manually (t_1, t_2, t_3, in Fig. 3b) in the "time_intervals" array, or other strategies can be implemented (*see* **Note 10**). Once the profiles are grouped into separate time intervals, the data is independently processed within each interval.

10. To analyze DV positions in units relative to the total DV length, DV positions are divided by the DV length of the profile. Profiles from different stages will have different resolution of positions. To compare profiles between stages, the resolution is unified by linear interpolation (*see* **Notes 7** and **10**).

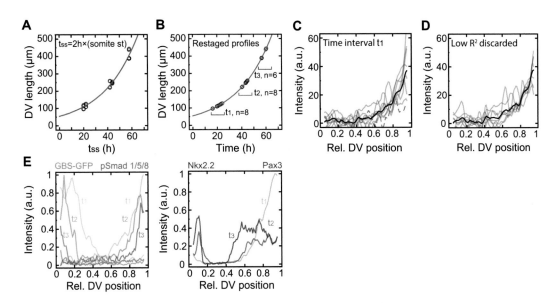

Fig. 3 Mean profile time course for a test dataset. (**a**) DV length as a function of t_{ss}, time derived from somite stage. Fitting $L(t) = L_0 \exp(t_{ss}/\tau)$ to the data (circles) yields parameter values $L_0 = 54.8 \pm 3.8$ μm and $\tau = 29.0 \pm 1.1$ h. (**b**) Restaged profiles (*see* **Note 9**). The profiles were binned into consecutive time intervals denoted by t_1, t_2, t_3. (**c**) Mean profile (black) for all profiles (red, gray) in the t_1 time interval. The profiles were rescaled to relative *DVposition*, the background of the mean profile was subtracted, and the dorsal- and ventral-most 5% of DV positions were excluded. (**d**) R^2 maximization with $R^2 < 0.5$ resulted in discarding the dashed gray profile in C in the first iteration. All profiles had $R^2 > 0.5$ in the second iteration. (**e**) Mean profiles for all time intervals for all analyzed channels. The mean profiles were normalized to the maximal mean intensity in the time course dataset

11. The profiles (in units of relative DV position) are averaged for every time interval. The resulting mean profiles are in the array "profile_mean0".

12. The background of the mean profile for each time interval is subtracted (*see* **Note 11**).

13. The dorsal most and ventral most 5% of DV positions are excluded from subsequent analysis. In some cases these boundary regions can bias the subsequent processing steps (*see* Fig. 3c, **Note 12**).

14. Each profile is linearly rescaled to maximize its similarity to the mean profile, quantified by an R^2 coefficient (see **Note 13**). Profiles with R^2 below a user-defined threshold (variable "R2_threshold" in the script) are discarded (see **Note 13**). A new mean profile is calculated and the procedure iterated until no more profiles are discarded (Fig. 3d).

15. All profiles in the dataset are normalized by a common factor, so that the maximum mean profile intensity in the dataset is equal to 1 (Fig. 3e).

4   Notes

1. Embryo dissection. Mouse: Detailed description available in [21]. Briefly, remove the uterus and transfer to ice-cold PBS. Isolate the decidua by removing the muscle layer of the uterus using forceps. Starting from the mesometrial pole, dissect the deciduum to release the embryo. Remove extraembryonic tissues. Chick: detailed description available in [22]. Briefly, remove 3–5 mL of albumin using a syringe. Cut a window on the upper side of the egg using bent surgical scissors. Cut approximately 1 cm around the embryo, holding the extraembryonic membranes on one side with forceps. Lift and transfer the embryo to a petri dish with ice-cold PBS. Remove all extraembryonic tissues using microscissors and forceps.

2. Litters with mixed genotypes. After dissection, each embryo can be transferred to a separate well of a 48-well plate (for E11.5 – use a 24 well plate) filled with 0.5 mL ice-cold PBS per well. The fixation, washing and sucrose steps are performed in the multiwell plate. Make sure the embryos are fully immersed in buffer at all times. For genotyping, collect the yolk sac for each embryo in an Eppendorf tube with a corresponding number. The yolk sacs can be stored at −20 °C until ready for genotyping. For embedding, each embryo is dissected and transferred to a separate numbered 2 mL Eppendorf tube with ~250 μL gelatin. The embryos are then carefully transferred to the gelatin mold, keeping the order and arranged in a manner that allows unambiguously distinguishing the numbering.

3. Sectioning. Embedding and cryosectioning could introduce deformations of the tissue sections or deviations from orthogonality, i.e., the right angle between the section (DV axis) and the AP axis. To avoid these, a sharp blade without any notches or defects should be used. A blunt blade will tend to squash and fold the tissue sections. The alignment of embryos in the gelatin is also key—straight and precise dissection line helps to keep the embryo pieces upright. At the cryostat, the block should be aligned such that the first sections are cut from the center of the block. Sections cut at an angle that deviates from orthogonality may have DV lengths that are significantly larger and ratios of apicobasal width to DV length that differ from the average for a particular stage. This may lead to incorrect restaging of the sections by DV length (**Note 9**). Significant deviations are likely to become obvious upon inspection of the slide at the microscope, as well as during the automatic discarding of outliers (**Note 13**). However, small deviations can still remain undetected. For many gene expression patterns, analysis of the

FI profile as a function of relative, rather than absolute, position along the DV axis alleviates this problem.

4. Optimizing immunostaining and imaging conditions. Primary antibody concentration and approximate imaging settings should be determined in a preliminary experiment. Slides containing adjacent sections from the same embryos are stained with a dilution series of antibody. The spatial resolution, scanning speed, averaging and bit-depth are selected (in our experience 1024×1024 pixels, pixel dwell time 1.58 μs, 4 line averages and 8-bit image depth produce good results). The slides are then imaged with several settings of laser power and detector gain. The fluorescence intensity profiles are quantified, background subtracted (as described in Subheading 3.3, **steps 1–3, 6**), normalized to the maximum FI of each profile and compared. The optimal conditions are the ones that give the lowest background fluorescence within the tissue without loss of signal. The signal is assessed by inspecting the detectable spatial differences in FI across the profile. Signal that is too low will result in loss of spatial differences in the low range of fluorescence intensities within the profile, whereas too high signal will result in saturation and loss of spatial differences in the high range of FI. Performing this analysis for developmental stages that contain the highest and lowest levels of signal ensures that the analysis spans the full dynamic range of intensity levels within the dataset. Whenever possible, linear relationship between protein levels and fluorescence intensity should be tested with a tagged version of the protein (*see* [16]). Furthermore, confocal detection in the linear range should be ensured using stepwise photobleaching or calibration assays [15].

5. File formats. Microscope-specific file formats can be opened in Fiji as hyperstacks using the Bio-formats plugin. Hyperstacks can then be saved as .tif files. We recommend automating this conversion using a custom Fiji macro (examples can be found through the Fiji help menu; see http://imagej.nih.gov/ij/macros/BatchConvertToJPEG.txt).

6. FI profile files. The intensity profiles are recorded in n output files per active image, where n is the number of specified ROIs. In each file, the first column stores the DV position in units of pixels across the ROI in y-direction. The second column ("DVposition(μm)") stores distance in μm from the ventral midline (*DVposition* $= 0$) across the ROI. The following columns store mean intensity values for the corresponding DV position in the respective channels. Each row corresponds to the mean intensity of pixels across the width of the ROI in x-direction. In addition to the .txt files, maximum projections of the z-stacks and ROIs are saved in corresponding subfolders.

7. Smoothing. Smoothing the FI profiles increases the robustness of subsequent steps of background subtraction and rescaling to relative DV positions. The profiles are rescaled to a predefined number of relative spatial positions (by default 100) by linear interpolation. If the original spatial resolution of the FI profiles is much denser than the resolution in relative units, many points will not be used in the rescaling. Smoothing counteracts this effect by locally averaging the FI values around rescaled positions. The default smoothing window of 5 μm corresponds approximately to 1% DV length for late stages of neural tube development. The smoothing step could affect the magnitude of estimated variance, hence it can be modified or omitted (follow instructions within script to comment it out).

8. Background subtraction. By default the ventral- and dorsal-most 10% of DV positions are excluded for background determination. This improves the subtraction procedure for profiles where the intensity levels are high close to the two poles. This exclusion principle should be modified depending on the expected expression pattern, so that regions containing background fluorescence are not excluded. This can be done in the script by modifying the variables "excludeV_bg = 0.1" and "excludeD_bg = 0.9" accordingly as well as in the inequalities including these variables.

9. Restaging by DV length. The default initial values for the fit parameters are "$L_0 = 100$" and "$tau_0 = 10$". Adjusting these parameters might be necessary for the fitting procedure to converge to the best fit. In the unlikely case that some profiles have smaller DV lengths than L_0 from the fit, the relation $t(L) = \tau \log (L/L_0)$ would result in negative developmental time ($t < 0$) for these profiles. This is corrected by assigning $t = 0$ to these profiles.

10. Time interval binning. For data points that are approximately uniformly distributed within the analyzed time period, binning can be done with fixed time intervals, e.g., every 10 h. However, if some bins contain significantly fewer profiles than the others, this might affect the accuracy of the estimated mean and variance for these bins. In such cases, we recommend collecting more data focusing on the underrepresented intervals (see Note 14). Another strategy [18] is to define the time intervals so that every interval contains the same number of profiles. In this case, care should be taken that some time intervals do not become too large, obscuring relevant temporal changes.

11. Background of mean profiles. This step is optional, as background subtraction at the level of single profiles is often sufficient. However, averaging of profiles within time intervals

reduces fluctuations, hence the background of the mean profile can be more precisely established. By default the same regions are excluded from the analysis as in the background subtraction step at the level of single profiles (*see* **Note 8**).

12. Excluding boundary regions. This step is optional and can be omitted. By default it keeps for analysis only the region between relative DV positions "cutV = 0.05" and "cutD = 0.95". These values correspond approximately to the size of the floor plate and roof plate and can be modified as needed. Note that the relative positions are not rescaled as a result of the exclusion, but the omitted points are simply not plotted.

13. Discarding outliers. This step is optional (follow instructions within the script to leave it out) and should be used when the analysis is targeted to mean profiles. It is used to discard outliers in a systematic way. Both R^2 maximization [2] and Y-alignment [3, 19, 23] can be used. These procedures are similar. In our experience, R^2 maximization results in a smaller percent of discarded profiles (<5% for threshold $R^2 = 0.5$) and fully conserves the mean profile at a given somite stage. By contrast, Y-alignment is more stringent, discarding 10–20% of the profiles for threshold $R^2 = 0.5$. Both procedures are insensitive to small spatial fluctuations, but are not suitable for small sample sizes, large variation between profiles, time intervals that are too large, and nonuniform distribution of the data. In R^2 maximization, the objective function $F(a_1, \ldots, a_k) = \sum_{i=1}^{k} \left[1 - \frac{\sum_x \left(m_x - a_i \, p_x^i \right)^2}{\sum_x (m_x - \bar{m})^2} \right]$ is maximized for each time interval. Here a_i is an unknown scaling factor, k number of profiles in the time interval, p_x^i denotes the fluorescence intensity of the ith profile at relative DV position x, $m_x = \frac{1}{k} \sum_{i=1}^{k} p_x^i$ is the mean intensity at x in that time interval, and $\bar{m} = \frac{1}{DVR} \sum m_x$ is mean of intensities at all positions and all profiles in this time interval, excluding the boundary regions (*see* **Note 12**). The objective function can be interpreted as a sum of R^2 for separate profiles linearly rescaled to best reproduce the mean profile for the corresponding time interval. In practice, the fit parameters $(a_1, \ldots, a_k)$ are first estimated with the least-square method to best reproduce the mean profile. Next, profiles with R^2 below a predefined threshold are discarded one by one. The mean profile without rescaling is recalculated after every step and discarding is repeated until all R^2 are above threshold. The fraction and quality of the discarded profiles should be inspected to determine a threshold R^2 value. The number of samples per time interval should be such that the mean profile does not change significantly if individual profiles are removed after the discarding procedure.

14. Pooling datasets. To increase the sample sizes, it is sometimes necessary to collect time course data in several experiments. The sample preparation and imaging conditions should still be the same for all experiments. To pool together the data from different experiments, first determine the 90th percentile of the fluorescence intensity of each profile χ, which represents a robust estimate of the maximum signal. The separate time-course datasets are then normalized relative to the median χ in each dataset. This normalization typically reduces the difference between the means of two experiments to levels much below the variation coefficients of the individual datasets.

Acknowledgments

We thank J. Briscoe and T. Bollenbach for comments on the manuscript. Funding: IST Austria and European Research Council under European Union's Horizon 2020 research and innovation programme (680037) (MZ, AK).

References

1. Alaynick WA, Jessell TM, Pfaff SL (2011) SnapShot: spinal cord development. Cell 146:178. https://doi.org/10.1016/j.cell.2011.06.038

2. Kicheva A, Bollenbach T, Ribeiro A, Valle HP, Lovell-Badge R, Episkopou V, Briscoe J (2014) Coordination of progenitor specification and growth in mouse and chick spinal cord. Science 329:1466–1468. https://doi.org/10.1126/science.

3. Zagorski M, Tabata Y, Brandenberg N, Lutolf MP, Tkačik G, Bollenbach T, Briscoe J, Kicheva A (2017) Decoding of position in the developing neural tube from antiparallel morphogen gradients. Science 356:1379–1383. https://doi.org/10.1126/science.aam5887

4. Briscoe J, Pierani A, Jessell TM, Ericson J (2000) A homeodomain protein code specifies progenitor cell identity and neuronal fate in the ventral neural tube. Cell 101:435–445. https://doi.org/10.1016/S0092-8674(00)80853-3

5. Oosterveen T, Kurdija S, Alekseenko Z, Uhde CW, Bergsland M, Sandberg M, Andersson E, Dias JM, Muhr J, Ericson J (2012) Mechanistic differences in the transcriptional interpretation of local and long-range Shh Morphogen signaling. Dev Cell 23:1006–1019. https://doi.org/10.1016/j.devcel.2012.09.015

6. Balaskas N, Ribeiro A, Panovska J, Dessaud E, Sasai N, Page KM, Briscoe J, Ribes V (2012) Gene regulatory logic for reading the sonic hedgehog signaling gradient in the vertebrate neural tube. Cell 148:273–284. https://doi.org/10.1016/j.cell.2011.10.047

7. Peterson KA, Nishi Y, Ma W, Vedenko A, Shokri L, Zhang X, McFarlane M, Baizabal JM, Junker JP, van Oudenaarden A, Mikkelsen T, Bernstein BE, Bailey TL, Bulyk ML, Wong WH, McMahon AP (2012) Neural-specific Sox2 input and differential Gli-binding affinity provide context and positional information in Shh-directed neural patterning. Genes Dev 26:2802–2816. https://doi.org/10.1101/gad.207142.112

8. Junker JP, Peterson KA, Nishi Y, Mao J, McMahon AP, van Oudenaarden A (2014) A predictive model of bifunctional transcription factor signaling during embryonic tissue patterning. Dev Cell 31:448–460. https://doi.org/10.1016/j.devcel.2014.10.017

9. Cohen M, Kicheva A, Ribeiro A, Blassberg R, Page KM, Barnes CP, Briscoe J (2015) Ptch1 and Gli regulate Shh signalling dynamics via multiple mechanisms. Nat Commun 6:1–12. https://doi.org/10.1038/ncomms7709

10. Hamburger V, Hamilton H (1951) A series of normal stages in the development of the chick embryo. J Morphol 88:49–92

11. Gomez C, Özbudak EM, Wunderlich J, Baumann D, Lewis J, Pourquié O (2008) Control of segment number in vertebrate embryos.

Nature 454:335–339. https://doi.org/10.1038/nature07020

12. Stern CD, Jaques KF, Lim TM, Fraser SE, Keynes RJ (1991) Segmental lineage restrictions in the chick embryo spinal cord depend on the adjacent somites. Development 113:239–244

13. Lippincott-Schwartz J, Presley JF, Zaal KJ, Hirschberg K, Miller CD, Ellenberg J (1998) Monitoring the dynamics and mobility of membrane proteins tagged with green fluorescent protein. Methods Cell Biol 58:261–281. https://doi.org/10.1016/S0091-679X(08)61960-3

14. Kicheva A, Pantazis P, Bollenbach T, Kalaidzidis Y, Bittig T, Jülicher F, González-Gaitán M (2007) Kinetics of Morphogen gradient formation. Science 315:521–525. https://doi.org/10.1126/science.1135774

15. Kicheva A, Holtzer L, Wartlick O, Schmidt T, González-Gaitán M (2013) Quantitative imaging of morphogen gradients in drosophila imaginal discs. Cold Spring Harb Protoc 8:387–403. https://doi.org/10.1101/pdb.top074237

16. Gregor T, Wieschaus EF, McGregor AP, Bialek W, Tank DW (2007) Stability and nuclear dynamics of the Bicoid Morphogen gradient. Cell 130:141–152. https://doi.org/10.1016/j.cell.2007.05.026

17. Morrison AH, Scheeler M, Dubuis J, Gregor T (2012) Quantifying the Bicoid morphogen gradient in living fly embryos. Cold Spring Harb Protoc 7:398–406. https://doi.org/10.1101/pdb.top068536

18. Dubuis JO, Samanta R, Gregor T (2013) Accurate measurements of dynamics and reproducibility in small genetic networks. Mol Syst Biol 9:639. https://doi.org/10.1038/msb.2012.72

19. Gregor T, Tank DW, Wieschaus EF, Bialek W (2007) Probing the limits to positional information. Cell 130:153–164. https://doi.org/10.1016/j.cell.2007.05.025

20. Schindelin J, Arganda-Carreras I, Frise E, Kaynig V, Longair M, Pietzsch T, Preibisch S, Rueden C, Saalfeld S, Schmid B, Tinevez J-Y, White DJ, Hartenstein V, Eliceiri K, Tomancak P, Cardona A (2012) Fiji: an open-source platform for biological-image analysis. Nat Methods 9:676–682. https://doi.org/10.1038/nmeth.2019

21. Behringer R, Gertsenstein M, Nagy K (2013) Manipulating the mouse embryo: a laboratory manual, fourth edition. Cold Spring Harbor Laboratory Press, New York

22. Bronner-Fraser M (2011) Avian embryology, Methods cell biol, vol 87, 2nd edn, pp 1–409

23. Tkačik G, Dubuis JO, Petkova MD, Gregor T (2015) Positional information, positional error, and readout precision in morphogenesis: a mathematical framework. Genetics 199:39–59. https://doi.org/10.1534/genetics.114.171850

Part II

Analysis of Biophysical Properties and Functions of Morphogens

Chapter 5

Fluorescence Correlation and Cross-Correlation Spectroscopy in Zebrafish

Xue Wen Ng, Karuna Sampath, and Thorsten Wohland

Abstract

There has been increasing interest in biophysical studies on live organisms to gain better insights into physiologically relevant biological events at the molecular level. Zebrafish (*Danio rerio*) is a viable vertebrate model to study such events due to its genetic and evolutionary similarities to humans, amenability to less invasive fluorescence techniques owing to its transparency and well-characterized genetic manipulation techniques. Fluorescence techniques used to probe biomolecular dynamics and interactions of molecules in live zebrafish embryos are therefore highly sought-after to bridge molecular and developmental events. Fluorescence correlation and cross-correlation spectroscopy (FCS and FCCS) are two robust techniques that provide molecular level information on dynamics and interactions respectively. Here, we detail the steps for applying confocal FCS and FCCS, in particular single-wavelength FCCS (SW-FCCS), in live zebrafish embryos, beginning with sample preparation, instrumentation, calibration, and measurements on the FCS/FCCS instrument and ending with data analysis.

Key words FCS, FCCS, Single molecule, Biomolecular interactions, Dissociation constant and affinity

1 Introduction

The understanding of biological mechanisms requires the observation of molecular processes within physiological environments and in particular in multicellular organisms. The observation of these processes against the complex and heterogeneous background in such an environment requires special techniques with high sensitivity. A technique that is used frequently for this purpose is fluorescence correlation spectroscopy (FCS) [1–3]. Any process creates fluctuations in one or more observable parameters. These fluctuations contain information about the frequency and characteristic time scale of the process. However, to detect such fluctuations, one requires an observable parameter that can be measured with high sensitivity, and the fluctuations have to be on the same order as the variance of the signal. This can be achieved when one observes, for

Julien Dubrulle (ed.), *Morphogen Gradients: Methods and Protocols*, Methods in Molecular Biology, vol. 1863, https://doi.org/10.1007/978-1-4939-8772-6_5, © Springer Science+Business Media, LLC, part of Springer Nature 2018

instance, fluorescence signals from small observation volumes. Fluorescence can be detected with very high sensitivity and the detection of single molecules is routine nowadays. Fluorescence has two more important properties. The variance of the fluorescence signal is the same as the average signal and the fluorescence signal is proportional to the fluorophore concentration and thus the volume. Therefore, in small observation volumes the fluorescence fluctuations can be easily measured. FCS uses these fluctuations to obtain information about the underlying processes, which can range from chemical reactions, binding interactions, diffusion to photophysical processes of the fluorophore. The recorded fluorescence fluctuations are then statistically analysed via a temporal autocorrelation function (ACF). Quantitative physical outcomes such as diffusivity, concentrations and the time scales of chemical or photophysical reaction dynamics of the probe molecules can be effectively extracted from the ACF by fitting it with appropriate theoretical models that describe the dimensionality, mode of diffusion, number of components and photophysics of the system. Typically, the diffusion time (τ_D) of the fluorescently labeled molecule, which is the average transit time of the molecules through the observation volume, is defined by the width of the ACF while the concentration of the molecules in the detection volume is inversely proportional to the amplitude of the ACF at zero lag time. These physical parameters serve as potential indicators of the activities and functions of specific biomolecules in regulating biological processes such as ligand–receptor interactions, receptor clustering, protein trafficking, transcription factor–DNA binding, and morphogenesis [4–11]. For example, formation of receptor clusters during the initial phase of signaling at the plasma membrane increases its diffusion time compared to unstimulated conditions while internalization of these molecules that often happen during signaling processing leads to an effective reduction of the concentration at the plasma membrane.

The goal to achieve a more accurate representation of biological processes at the molecular level in physiologically relevant environments motivated biophysicists to quantify intrinsic properties of such processes in live organisms. As opposed to two-dimensional (2D) cell cultures, the three-dimensional (3D), multicellular nature of live organisms allows the retention of proper functionality of biological molecules for the modulation of biological processes. Therefore, the ability to characterize biomolecular dynamics in live organisms along with existing morphological studies bridges the gap between molecular and developmental level events and provides crucial information about the possible actions of specific genes and biomolecules on organism development and onset of diseases. With advancements in both technology and biology, contemporary implementations of FCS in vivo are emerging. Several animal models ranging from invertebrates such as *Caenorhabditis elegans* and *Drosophila melanogaster*, to vertebrates like zebrafish (*Danio rerio*) were studied by FCS to elucidate the mechanisms of processes in the physiological environment of intact organisms from dynamical studies of fluorescently labeled biomolecules [11–16].

Zebrafish is a widely used vertebrate model to understand the fundamental basis underlying biological activities in humans due to it being evolutionarily close to humans and its similar genetic composition, with ~70% of human genes having at least one zebrafish ortholog [17]. The low cost maintenance and small size of zebrafish along with the availability of well-established tools for genetic manipulation of the decoded zebrafish genome have accelerated its use in developmental biology and human disease modeling [18, 19]. Furthermore, due to its optical transparency in early embryonic stages, zebrafish can be effectively implemented into fluorescence imaging-based studies, making it ideal for FCS measurements [20]. However, despite the optical clarity of zebrafish embryos, there are still some limitations associated with FCS measurements in live zebrafish embryos. These include the distortion of the observation volume from the scattering of light by thick tissues of the zebrafish as a result of refractive index mismatch between the tissues and water, which limits the penetration depth usable for effective FCS measurements. Additionally, the contribution of autofluorescence mainly from the skin of the zebrafish embryo as background noise lowers the signal-to-noise ratio (SNR). We have characterized these limitations by defining the accessible penetration depth for confocal FCS in live zebrafish embryos and by determining the contribution of autofluorescence at different parts of the zebrafish embryo [21]. Common practices to increase penetration depth and reduce autofluorescence include the use of zebrafish larvae with genetically or chemically (1-phenyl-2-thiourea (PTU)) suppressed pigmentation [22–25] and employing two-photon excitation (TPE) with appropriate fluorophores (high molecular brightness with TPE) [21, 26]. Typically, the penetration depth accessible for one-photon excitation and two-photon excitation FCS is 50–80 μm and around 200 μm respectively. FCS parameters can be more accurately quantified at deeper penetration depths by combining adaptive optics, which corrects for optical aberrations by means of a deformable mirror, with FCS measurements [27, 28].

One of the first applications of FCS in zebrafish is the determination of flow velocities and direction of blood flow in the blood vessels of live zebrafish embryos via a modified version of FCS, known as line-scan FCS, suitable for flow measurements [29, 30]. These FCS results were combined with developmental studies by fluorescence imaging of control and mutant zebrafish transgenic lines to map the stages of liver growth with respect to the blood circulation and vasculogenesis in zebrafish [30]. In a two-focus line scan FCS study, the formation and maintenance of the Fgf8 morphogen gradient was determined in live zebrafish embryos by measuring diffusion and concentrations of enhanced green fluorescent protein (EGFP) tagged Fgf8 fusion protein (Fgf8-EGFP) at various distances from the source of Fgf8-EGFP proteins in the extracellular space [11]. Yu et al. proposed a source–sink mechanism for the regulation of the gradient where Fgf8 proteins undergo free diffusion away from the source, spread through the extracellular space and are eventually taken up by target cells which act as the sink mediated by receptor–ligand interactions and endocytosis.

A recent confocal FCS study from our group linked cerebellar development of live Wnt3-EGFP transgenic zebrafish embryos with the molecular dynamics of secreted signaling protein, Wnt3, and highlighted the importance of membrane localization and secretion of palmitoylated Wnt3 for proper development of the brain [15]. By using a Porcupine inhibitor (Wnt-C59) to inhibit the post-translational palmitoylation of Wnt3 at the endoplasmic reticulum, our results revealed a reduced population of membrane-localized and secreted Wnt3 that led to abnormal brain development in the embryos. Furthermore, we extended this study to elucidate the subresolution membrane organization of Wnt3 in the same transgenic line using an imaging mode of FCS known as single plane illumination microscopy-FCS (SPIM-FCS) and the related FCS diffusion law [31–33]. Our results indicate that Wnt3 associates with membrane domains (cholesterol, sphingomyelin, and GM1-dependent) in the cerebellar cell membrane and such domain confinement is reduced upon inhibition of Wnt3 palmitoylation by Wnt-C59 in an inhibitor dose dependent manner [31, 34]. This demonstrates that the inhibition of palmitoylation not only reduces membrane expression of Wnt3 but also alters its subresolution organization, which possibly has functional implications. Our latest results also highlight the correlation of Wnt protein association with lipid component-specific plasma membrane domains with receptor binding and downstream Wnt signaling activation [34].

Besides diffusion measurements, biomolecular interactions in live samples can be quantified by fluorescence cross-correlation spectroscopy (FCCS), the dual-color extension of FCS. FCCS analysis is conducted by temporally cross-correlating the fluorescence signals of spectrally distinct fluorescently labeled species detected in separate wavelength-selective detection channels. In the event of biomolecular interactions and binding, the distinctly fluorescently labeled particle species will move together, resulting in an elevated cross-correlation function (CCF) between the fluorescence fluctuations of both species. There are several variations of FCCS such as dual-color FCCS (DC-FCCS), single-wavelength FCCS (SW-FCCS), two-photon excitation FCCS (TP-FCCS), pulsed interleaved excitation FCCS (PIE-FCCS), quasi-PIE

FCCS, and, more recently, Imaging FCCS. In DC-FCCS, two lasers are used to excite different fluorescent species individually and their individual fluorescence signals are detected in separate channels [35, 36]. A limitation of DC-FCCS is the difficulty to precisely align and overlap the excitation volumes of two lasers with different wavelengths due to chromatic aberration and different sizes of their excitation volumes. In this case, SW-FCCS removes the need to align the two excitation volumes together by employing a single laser to excite the different fluorescently labeled species simultaneously, thus minimizing the issue of imperfect overlap between observation volumes [37–39]. However, the choice of fluorophore pairs for SW-FCCS is restricted to those that have overlapping excitation spectra but well-separated emission spectra. This is contrary to TP-FCCS which also excites with a single laser source in a two-photon absorption process but has a wider availability of fluorophore pairs with overlapping two-photon excitation spectra as compared to fluorophore pairs with overlapping single photon excitation spectra in SW-FCCS, albeit its disadvantage of lower molecular brightness attributed to high photobleaching and saturation [40–44].

A major issue in FCCS is the spectral cross talk between two channels mainly due to overlap between the emission spectra of both fluorophore species which leads to an elevated cross-correlation if the data sets are not properly cross talk and background corrected. In light of this, PIE-FCCS was introduced to remove cross talk between spectrally distinct fluorescent species by alternating the excitation of the fluorophores in the nanosecond range with different pulsed lasers, as opposed to the continuous wavelength lasers used in the aforementioned FCCS modalities, to ensure that the excitation and emission cycles of each fluorophore do not overlap with one another [45]. Hence, the CCFs generated are not contaminated with cross talk. A cross talk elimination method similar to PIE-FCCS is the quasi-PIE FCCS, which requires only a single-pulsed laser and a continuous-wave laser where statistical filtering is conducted on the fluorescence lifetime of the fluorophore excited by the pulsed laser, making it more economical than PIE-FCCS [46, 47]. Lastly, spatial multiplexing detection of biomolecular interactions was achieved by Imaging FCCS either in the total internal reflection fluorescence microscopy (TIRF) mode or the SPIM mode, generating spatial maps of biomolecular interaction [47–51]. The methodology of FCCS measurements in live system is extensively reviewed here [36, 52].

The influence of the physiological environment on the biomolecular interaction of Cdc42, a small Rho-GTPase, and IQGAP1, a scaffold protein involved in the organization of the actin cytoskeleton, was determined by SW-FCCS [53]. In this study, Cdc42 was found to bind with higher affinity to IQGAP1 in muscle fiber cells of live zebrafish embryos (apparent dissociation constant (K_d) of ~100 nM) than in cultured CHO-K1 cells (apparent K_d of

~1000 nM), illustrating the significance of probing biological interactions in live organisms to accurately quantify the processes occurring in vivo.

In a recent application of SW-FCCS to elucidate the mechanism of the regulation of morphogen gradients in Wang et al., we determined the diffusivity of the zebrafish Nodal proteins Squint (Sqt/Ndr1) and Cyclops (Cyc/Ndr2) and reported their respective binding affinities to the major Nodal receptor Acvr2b and the Nodal inhibitor Lefty2. We then compared these to the signaling range of the factors, and estimated the degradation rates of the Nodal proteins [16]. Finally, we conducted theoretical and computer simulations of the Nodal morphogen gradients with simulation parameters derived from experimental results of the same work and several other works. From these experimental and simulation outcomes, the Nodal morphogen gradient was shown to be influenced by multiple factors which include diffusivity, extracellular interactions with Nodal receptors and inhibitors and possibly additional unidentified molecules, and ligand degradation of the Nodal ligands.

Here, we describe conventional confocal FCS and SW-FCCS approaches in studying biomolecular dynamics and interactions, respectively, in live zebrafish embryos at high spatiotemporal resolution. We will start off with sample preparation, then the instrumentation required to conduct FCS and SW-FCCS measurements, followed by the steps needed for calibration and measurements on the system, and lastly, possible issues that could arise and troubleshooting protocols.

2   Materials

In this section, we list out the materials and apparatus needed for proper maintenance of zebrafish embryos (*see* **Note 1**), generation of fluorescent protein (FP) fusion recombinant plasmid DNA and capped mRNA constructs (*see* **Note 2**), microinjection of these constructs into zebrafish embryos, mounting of zebrafish embryos for confocal FCS or SW-FCCS (herewith denoted as FC(C)S) measurements (*see* **Note 3**) and instrumentation of confocal FC(C)S. All solutions are prepared in double deionized water with resistivity of 18.2 MΩ cm at 25 °C.

2.1   Zebrafish Maintenance and Mounting

2.1.1   Buffer and Temperature Conditions for Zebrafish Embryos

1. 30% Danieau's solution: 17.4 mM NaCl, 0.21 mM KCl, 0.12 mM $MgSO_4 \cdot 7H_2O$, 0.18 mM $Ca(NO_3)_2$, 1.5 mM HEPES, pH 7.6.

2. Egg water: 60 μg/mL "Instant Ocean" sea salts stock solution.

3. E3 medium: 5 mM NaCl, 0.17 mM KCl, 0.33 mM $CaCl_2$, 0.33 mM $MgSO_4$, pH adjusted to 6.8–6.9.

4. Incubator set at 28.5 °C.

2.1.2 Generation of Fluorescent Protein (FP) Fusion Recombinant Plasmid DNA Constructs and Capped mRNA	1. cDNA synthesis kit. 2. Polymerase Chain Reaction (PCR) cloning kit. 3. Designed primers. 4. DNA ligase. 5. Restriction enzymes. 6. Bacterial transformation kit. 7. Gel electrophoresis apparatus. 8. Gene sequencing equipment. 9. In vitro transcription kit for capped mRNA synthesis.
2.1.3 Microinjection of Recombinant DNA or mRNA into Zebrafish Embryos	1. 1–1.5% low-melting-point agarose: Dissolve and melt either 1 g (1%) or 1.5 g (1.5%) of low-melting-point agarose powder in 100 mL of 30% Danieau's solution, Egg water, or E3 medium. 2. Glass Pasteur pipette or plastic dropper. 3. Petri dish. 4. Stereomicroscope. 5. Micromanipulator. 6. Micropipette. 7. Automated pressure-regulated injection system. 8. Chamber mount: Wedge-shaped troughs or wells molded onto 1–1.5% low-melting-point agarose. 9. Tabletop fluorescence microscope.
2.1.4 Zebrafish Embryo Mounting for Confocal Imaging and FC(C)S Measurements	1. Embryo medium: 30% Danieau's solution, Egg water, or E3 medium. 2. 0.7–1% low-melting-point agarose: Dissolve and melt either 0.7 g (0.7%) or 1 g (1%) of low-melting-point agarose powder in 100 mL of 30% Danieau's solution, Egg water, or E3 medium. 3. Melanin inhibitor 0.003% (w/v) 1-phenyl-2-thiourea (PTU) in embryo medium. 4. Anesthetic agent 0.05% (w/v) ethyl 3-aminobenzoate methanesulfonate (tricaine) in embryo medium. 5. 2 mg/mL pronase in embryo medium. 6. Stereomicroscope. 7. A pair of Number 5 Dumont forceps or two 27G syringe needles or a Number 5 Dumont forceps and a 27G syringe needle. 8. Glass Pasteur pipette or plastic dropper. 9. Heat block or microwave oven: Set temperature at 42 °C.

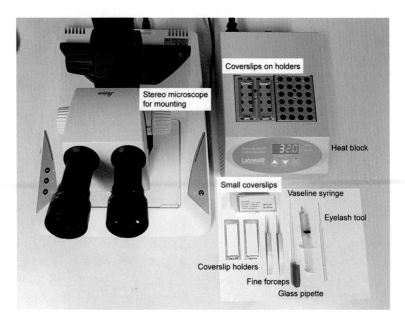

Fig. 1 Tools and equipment for mounting zebrafish embryos

10. No. 1 or No. 1.5 35 mm glass-bottom dishes.

11. Microloader pipette tip or fishing wire loop or "eyelash tool" (a small piece of fishing wire or an eyelash fixed to the end of a Pasteur pipette with paraffin wax). Figure 1 illustrates several tools that are commonly used for zebrafish embryo mounting.

2.2 Confocal FC(C)S Measurements

2.2.1 Instrumentation for Confocal FC(C)S Experiments

The instrumentation for confocal FC(C)S experiments is essentially an extension of the confocal microscope with an additional FC(C)S detection module coupled to the existing system (Fig. 2). This extension is commercially available (LSM upgrade kit SymPhoTime 64 (Picoquant), ConfoCor 2 and 3 (Zeiss), AlbaFCS (ISS)) or can be custom-built by the user based on their specific needs [35, 36, 54, 55]. The individual units required to perform FC(C)S measurements in the confocal system were outlined by our group previously [52] (*see* **Note 4**).

1. Laser excitation sources: 488 nm, 515 nm, and 543 nm lasers.

2. Objective lens: 60×, NA 1.2, water immersion objective.

3. Excitation dichroic mirror.

4. Pinhole.

5. Dichroic beam splitter mirror.

6. Emission filter(s).

7. Hardware or software correlator.

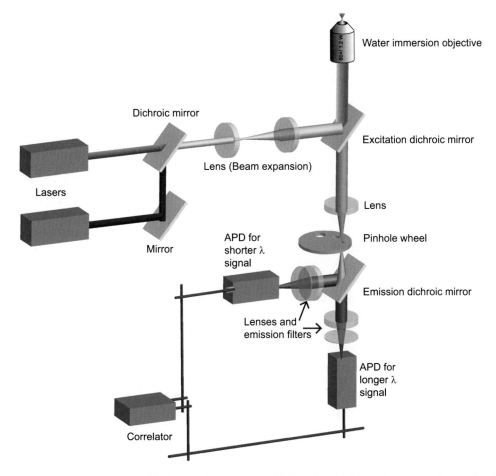

Fig. 2 Schematics for confocal FCCS setup. Components added to the regular confocal setup for confocal FC(C)S measurements include an emission dichroic mirror to split the emitted fluorescence according to the emission wavelength of the two fluorophores, and a second detection path including a lens, emission filter, and avalanche photodiode (APD). In FCCS, the correlator will cross-correlate the fluorescence signals of the two separate wavelength channels

2.2.2 Confocal FC(C)S Calibration

1. Calibration dye.
2. Dimethyl sulfoxide (DMSO).
3. 1× phosphate buffered saline (PBS).
4. UV–visible spectrophotometer.

3 Methods

One of the most important factors for successful confocal FC(C)S measurements in live zebrafish embryos is the proper preparation of the zebrafish samples. This includes zebrafish maintenance, generation of appropriate RNA and DNA constructs and their microinjection into zebrafish embryos at different stages, generation of transgenic lines and sample mounting strategies for FC(C)S measurements. Furthermore, the choice of fluorophores for FC(C)S is crucial for quantification of dynamic properties. Fluorophores that have high photostability, high molecular brightness, and ease of incorporation into live zebrafish without compromising the original function of the biomolecules are ideal candidates for FC(C)S measurements. In addition to these properties, fluorophore pairs for SW-FCCS measurements require optimally overlapping excitation spectra and well-separated emission spectra to ensure sufficient signal from both fluorophores while being excited by a single laser line. The available literature that addresses the type of fluorophores suitable for FC(C)S measurements in live zebrafish can be found here [20, 52, 56]. For zebrafish measurements, the common choice of fluorophores is fluorescent proteins (FPs). As proteins can be genetically tagged with FPs, this allows for the generation of fluorescently labeled transgenic lines or transient transfection by microinjection of DNA or RNA into early stage embryos. In this section, we focus on the generation of FP fusion recombinant plasmid DNA or capped mRNA constructs and their microinjection into early stage zebrafish embryos, FC(C)S calibration procedure, zebrafish mounting strategy for FC(C)S measurements and data analysis for FC(C)S measurements in live zebrafish embryos. The preparation of transgenic lines is explained elsewhere [57, 58].

3.1 Zebrafish Sample Preparation

For the sample preparation of zebrafish embryos, we will address the following topics: generation of DNA and mRNA constructs, microinjection of DNA or mRNA constructs into early stage embryos, and lastly, mounting of embryos for confocal FC(C)S measurements.

3.1.1 Generation of Fluorescent Protein (FP) Fusion Recombinant Plasmid DNA Constructs

FP fusion DNA constructs are created to specifically label the protein of interest by genetically encoding the FP's gene sequence directly together with the gene sequence of the protein of interest. The constructs are then introduced into the zebrafish embryos for protein translation by microinjection (see Subheading 3.1.3). In the following, we provide step-by-step guidelines on the general method for constructing FP fusion recombinant plasmid DNA and capped mRNA constructs via molecular cloning.

1. Identify the gene sequences corresponding to the protein of interest in the zebrafish with the aid of zebrafish genomic databases available online (http://www.ensembl.org/Danio_rerio/Info/Index, https://zfin.org/).

2. Isolate total RNA from zebrafish embryos using well-established RNA isolation methods available in literature and commercial kits [59–61].

3. Synthesize cDNA with carefully designed primers by reverse transcription of isolated RNA using reverse transcriptase in commercially available cDNA synthesis kits with their respective protocols. cDNA is then amplified via PCR (*see* **Note 5**).

4. Select suitable vectors containing the required FP sequence at the desired region (N- or C-terminal), restriction sites at the multiple cloning site (MCS) for insertion of the gene sequence of interest and antibiotic resistance gene (*see* **Note 6**).

5. Amplify the cDNA by PCR with the designed primers that include the restriction site(s) sequence and the Kozak sequence in the forward primer (*see* **Note 6**) to create an insert that comprises the restriction sites which matches with the vector. Assembly of the selected vector, containing the FP gene sequence, with the insert by restriction enzyme digestion and ligation or other available methods generates the FP fusion recombinant DNA plasmid construct. Upon confirmation of the presence of insert in the recombinant plasmid by colony PCR and the gene sequence of the insert DNA by gene sequencing, the recombinant plasmid can be amplified by bacterial transformation and purified by maxiprep.

3.1.2 Generation of Fluorescent Protein (FP) Fusion Capped mRNA

As an example, we highlight the use of the mMessage mMachine kit for the synthesis of capped mRNA. Capping mRNA at the 5′ end of the mRNA with a guanosine nucleotide cap stabilizes the mRNA and makes it less susceptible to degradation by phosphatases.

1. Generate and linearize the plasmid DNA construct that contains the insert (*see* Subheading 3.1.1) and a RNA polymerase promoter site, upstream of the insert, specific to the type of mMessage mMachine kit used, for example SP6 RNA polymerase promoter site specific to mMessage mMachine SP6 kit, by cutting at one of the endonuclease restriction sites at the MCS downstream of the insert with the corresponding restriction enzyme.

2. Transcribe a capped mRNA from the linearized DNA with the specific RNA polymerase in the mMessage mMachine kit. The steps for capped mRNA transcription from linearized DNA is as detailed in the manufacturer's user guide for the mMessage mMachine kit.

3.1.3 Microinjection of FP Fusion Recombinant Plasmid DNA or Capped mRNA Constructs into Zebrafish Embryos

1. Load an appropriate dose of mRNA/DNA optimized for FC (C)S measurements, usually in the range of pg–ng, into a micropipette for injection into a single cell or the yolk of the zebrafish embryo at around 1–64 cell stage [11, 15, 16, 21] (*see* **Note 7**). A stereomicroscope is used to visualize the entire microinjection process.

2. Align zebrafish embryos along the chamber mount on a Petri dish. Use a micromanipulator to facilitate the movement of the micropipette tip for ease of mRNA/DNA injection into the yolk or the cytoplasm of a single cell of the fertilized zebrafish embryos. An automated pressure-regulated injection system which passes the pressure through a micropipette holder is utilized to control injection rates.

3. After microinjection, transfer the zebrafish embryos into a clean dish containing fresh embryo medium (30% Danieau's solution, egg water or E3 medium) and incubate at 28.5 °C for optimal growth.

4. Screen for fluorescence expression in later stages of the embryos (>4 h post fertilization (hpf)) a few hours after microinjection under a tabletop fluorescence microscope. Select healthy embryos with suitable expression levels for confocal FC(C)S measurements.

3.1.4 Zebrafish Embryo Mounting for Confocal FC (C)S Measurements

Good sample mounting strategy not only ensures the accuracy of confocal FC(C)S measurements but also their reproducibility. Here, we describe the zebrafish embryo mounting procedure for confocal FC(C)S measurements.

1. Dechorionate zebrafish embryos under a stereomicroscope with a pair of Number 5 Dumont forceps or two 27G syringe needles or one 27G syringe needle and one Number 5 Dumont forceps, using one as a support to hold the embryo in place and the other to tear the chorion apart and free the embryo. Alternatively, zebrafish embryos can be treated with pronase in embryo medium until their chorions turn brittle and rinsed 3–4 times with embryo medium to remove their chorions and residual pronase as described in [60]. Dechorionation is not required for later stage embryos, around 3 dpf onward, by which time most embryos would have hatched out of their chorions.

2. Add 0.003% (w/v) PTU into embryo medium between 20 and 30 hpf of the zebrafish embryos to inhibit melanin formation and thus pigmentation at later developmental stages.

3. Early stage embryos (until ~20 hpf) can be mounted in agarose directly after dechorionating. At later stages of development (~22 hpf onward), the zebrafish embryo's heartbeat initiates along with muscle contraction and involuntary muscle twitching, necessitating the use of anesthetics [62, 63]. For this, dechorionated zebrafish embryos are incubated in 0.05% (w/v) tricaine anesthetic agent in embryo medium. Depending on the growth stage of the embryo, the incubation time in tricaine differs (*see* **Note 8**). In general, young zebrafish embryos require a longer duration of incubation as compared

to older zebrafish embryos due to the inverse relationship of drug penetration efficiency of tricaine with zebrafish age. Typically, 2–3 dpf embryos are incubated in 0.05% (w/v) tricaine for 20–30 min prior to confocal FC(C)S measurements.

4. Mount zebrafish embryos in 0.7–1% low-melting-point agarose on a glass-bottom dish under a stereomicroscope. First, add a small drop of liquefied low-melting-point agarose (~37 °C) to the center of the glass-bottom dish. Place the zebrafish embryo into the small drop of low-melting-point agarose located at the center of the glass-bottom dish. Subsequently, add more liquefied low-melting-point agarose on the embryo. A microloader pipette tip or fishing wire loop or "eyelash tool" is used to orient the embryo within the agarose. Place the low-melting-point agarose in a heat block set to around 42 °C to maintain its liquid state (*see* **Note 9**). Ideally, the location of interest on the embryo for measurement should be placed closest to the coverslip to minimize refractive index mismatch between the water immersion objective, coverslip, low-melting-point agarose, and the embryo, and for minimal scattering of light by the low-melting-point agarose. For example, if we are interested in conducting measurements on the cerebellum of the zebrafish embryo, the dorsal part of the embryo can be placed flat against the coverslip or the embryo can be tilted with its head area against the coverslip for more direct exposure to the laser focal spot (Fig. 3a, b). Young zebrafish embryos at blastula and gastrula stages (2.25–5.25 hpf) can be mounted directly into the 0.7–1% low-melting-point agarose on the glass-bottom dish without the need for anesthetization. The embryo's position can be oriented with the animal pole facing downward on the coverslip, toward the microscope objective as shown in Fig. 3c.

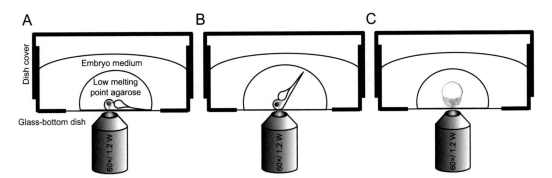

Fig. 3 Various zebrafish mounting strategies for confocal FC(C)S measurements. (**a**) Zebrafish embryo at ~2 dpf mounted dorsally on the glass-bottom dish with laser spot focused on the brain. (**b**) Zebrafish embryo mounted in a tilted position with laser spot focused on the brain. (**c**) Zebrafish embryo at the blastula or gastrula stage (2.25–5.25 hpf) mounted with the animal pole facing the coverslip of the glass-bottom dish

3.2 Confocal FC(C)S Calibration Procedure

Before conducting FC(C)S measurements, it is important to characterize the effective volumes of the detection channel(s) and the position of the pinhole to ensure maximum SNR achieved by the confocal system. In addition, for SW-FCCS measurements, calibration is conducted to determine the effective volumes of both detection channels and the cross-correlation volume. Here, we describe the calibration procedures for confocal FC(C)S to determine the effective volumes of the detection channel(s).

3.2.1 Selection of Calibration Dye for Confocal FC(C)S Calibration

1. For confocal FCS calibration, select a calibration dye with known diffusion coefficient (D) which matches the excitation and emission spectra of the fluorophore used in the actual experiments [64]. For example, Atto488 calibration dye with known D of 400 $\mu m^2/s$ in water at 25 °C can be used for the calibration of the confocal system when green fluorescent protein (GFP) is the fluorophore used in the experiments [65].

2. For confocal FCCS calibration, select a calibration dye with known diffusion coefficient (D) which possesses an emission spectrum that encompasses the emission spectra of the fluorophore pair for simultaneous alignment of the effective volumes of both detection channels and the cross-correlation volume. For example, Rhodamine 6G ($D \sim 410$ $\mu m^2/s$ [65]) is used to calibrate for the confocal SW-FCCS system with excitation by a single 514 nm laser line which is used to excite the fluorophore pair(s) GFP/EGFP and monomeric red fluorescent protein (mRFP)/mCherry in the actual experiment, and its emission spectrum enables the splitting of its fluorescence into the two detection channels for the alignment and determination of their effective volumes from the individual ACFs of each channel and the cross-correlation volume. Individual calibration dyes with emission wavelengths corresponding to the detection wavelengths of the respective detection channels can also be used to calibrate for the effective volumes for SW-FCCS.

3.2.2 Preparation of Calibration Dye Solution

1. Prepare a stock solution of the calibration dye by dissolving the dye powder in dimethyl sulfoxide (DMSO) and store at −20 °C without light exposure.

2. Determine the concentration of the stock solution by UV–visible spectrophotometry.

3. Since the observation volume in FCS is typically <1 fL, prepare an optimal working concentration of the dye solution of around 1–10 nM, corresponding to 0.6–6 molecules per fL, to generate optimal fluorescence fluctuations in the focal volume for good SNR.

4. Prepare the working solution of the dye by diluting the dye stock solution with 1× phosphate buffered saline (PBS). Note that the final working dye solution should have <0.1% of DMSO to achieve an aqueous environment for FCS/FCCS calibration with negligible influence from the organic solvent DMSO.

3.2.3 Optimization of Instrumental Conditions

1. Place ~30 μL of working dye solution on a No. 1 or 1.5 glass cover slide (0.13–0.16 mm and 0.16–0.19 mm in thickness respectively, refer to Subheading 3.2.6 for the objective correction collar adjustment procedure to correct for the thickness of the cover slide) located above the objective.

2. Focus the laser beam into the solution at a point that is around 20 μm above the cover slide.

3. Select appropriate excitation dichroic beam splitters with respect to the laser line(s) used along the illumination pathway.

4. For the detection pathway, in the case of FCS, select an emission filter, preferably a band-pass filter with steep cut-ons and cut-offs (*see* **Note 4**), to encompass the emission maxima of the fluorophore to collect the maximum fluorescence signal.

5. For FCCS, select a dichroic beam splitter to split the fluorescence signal according to the fluorophore pair used into the respective detection channels and choose a band-pass emission filter for each of the detection channels to spectrally filter for the emission wavelength range of the respective fluorophores in the fluorophore pair.

3.2.4 Laser Power for Confocal FC(C)S Measurements

One of the most crucial factors to consider in FC(C)S measurements is the laser power used. It is important to work with laser powers below the saturation levels of the fluorophores to prevent enlargement and distortion of the observation volume that will lead to erroneous FCS outcomes from the ACF and CCF fits derived from FCS fitting models typically defined with a three-dimensional (3D) Gaussian intensity profile of the focal volume [66–70]. This can be determined by characterizing the laser power dependence of the fluorophore brightness (η) in counts per particle per second (cps).

1. Generate a plot of η against laser power by calculating the molecular brightness of a standard calibration dye, which is the ratio of the average fluorescence intensity and number of particles derived from the amplitude of the ACF (*see* Subheading 3.4.3), at varying laser powers. At low laser powers, η increases linearly with laser power until it reaches a saturation during which η does not change with laser power and eventually η decreases at very high laser powers due to photobleaching.

2. Conduct FC(C)S measurements at laser powers falling within the linear regime of the η against laser power plot. Since the SNR of a FCS measurement depends on the particle brightness [71], it is not recommended to use a laser power at the lower side of the linear regime of the η against laser power plot. However, high laser powers lead to photobleaching and will modify the shape of the ACF and CCF, causing inaccurate determination of dynamics and concentrations. Therefore, there is a trade-off between maximizing SNR and minimizing photobleaching. The laser power should be optimized to achieve an appropriate compromise between both parameters.

3. In our laboratory, we commonly use laser powers of 5–20 µW, measured before the objective, for FC(C)S measurements in live cells and organisms [11, 15, 39, 53]. Molecular brightness of 800–1500 cps for fluorescent fusion proteins in zebrafish embryos was reported to be sufficient for the quantification of biomolecular dynamics and interactions by SW-FCCS [53]. In another work, a criterion of a least 500 cps was recommended for proper FCS measurements [72]. In comparison, the molecular brightness of the calibration dye is significantly higher (~20 kcps) than FPs, generating higher quality ACFs due to higher SNR.

3.2.5 Adjustment of Pinhole Position for Optimal Alignment of Detection Volume

To generate femtoliter-sized observation volumes suitable for FC (C)S measurements, multiple pinholes of physical sizes ranging from 60 to 300 µm available in the confocal system were utilized [55]. Upon optimization to obtain the highest SNR, the 150 µm pinhole size, which corresponds to 1.6 airy unit for an excitation wavelength of 514 nm, a stage magnification of 3×, and a 60×, NA 1.2, water immersion objective, was selected to yield the highest η for a given dye as compared to the other pinhole sizes with the same objective. Pinhole sizes ranging from 0.9 to 1.8 airy unit in the confocal setup are shown to generate femtoliter-sized observation volumes and thus recommended for FC(C)S measurements [73]. A pinhole of 1 airy unit just captures the central peak of the image of a point in a microscope system. In our setup, the combination of a pinhole size of 150 µm, a 60×, NA 1.2, water immersion objective and a 3× stage magnification generates an effective volume of ~0.5 fL in the green channel on excitation by a 514 nm laser line [39, 53].

1. Adjust the pinhole size and position in commercial confocal microscopes by following the instruction manual provided by the manufacturer for the respective make and model of the microscope used. Generally, commercial confocal microscopes offer straightforward adjustment of the pinhole size and position via software control for each laser line used. Manual adjustment of the pinhole position is more common in custom-built confocal microscopes.

2. Perform adjustments of the pinhole position in the x-, y- and z-directions to maximize fluorescence intensity captured in the detection channel emitted by the particular calibration dye used upon excitation by the respective laser line.

3. On attaining maximum fluorescence intensity in the detection channel, conduct a FCS measurement to generate an ACF in the hardware or software correlator which is fitted by a home-written FCS fitting program in Igor Pro (WaveMetrics, Lake Oswego, OR; available at http://www.dbs.nus.edu.sg/lab/BFL/confocal_FCS.html), or freely (PyCorrFit) [74] or commercially available fitting softwares to obtain the structure factor K (ratio of the axial (z_0) and radial (ω_0) distances of the laser focal spot at $1/e^2$ value of the maximum intensity at the focus of the observation volume which defines the shape of the observation volume) (*see* Subheading 3.2.7) and η in cps (*see* Subheading 3.2.4).

4. Further adjust the pinhole position in minute steps and repeat FCS measurements until a K value of 3–8 and the maximum η is achieved. In addition to the criteria highlighted above, SW-FCCS requires an aptly selected calibration dye with an emission spectrum that covers the emission spectra of the fluorophore pair when excited by the single laser line. For example, Rhodamine 6G is a suitable dye in this context since it provides adequate signal to both detection channels when excited at 514 nm. A proper alignment should split the fluorescence signal of Rhodamine 6G in both detection channels where FCCS measurements will generate two ACFs and a CCF which ideally should overlap with each other, indicating perfect overlap in the detection volumes of both channels.

3.2.6 Objective Collar Adjustment to Correct for Cover Slide Thickness

Most high NA detection objectives used for confocal FC(C)S measurements possess a coverslip correction collar which corrects for the thickness of different cover slides.

1. After optimizing the pinhole and detection pathway position, adjust the correction collar to obtain the maximum η of the calibration dye upon conducting FCS measurements and data fitting.

2. Adjust the objective correction collar every time before experiments with respect to the cover slide/glass-bottom dish thickness (No. 1: 0.13–0.16 mm, No. 1.5: 0.16–0.19 mm) used for the actual experiments. No. 1 and 1.5 glass cover slides or glass-bottom dishes are regularly used since their thickness fall within the correction range of the objective's correction collar (0.13–0.21 mm).

| 3.2.7 Determination of Effective Observation Volume for the Detection Channel(s) | The effective observation volumes of each detection channel can be calculated from the diffusion time ($\tau_{D,\text{calibration dye}}$) and K values of calibration dye(s) with known D extracted from the FCS fits of the respective channels with the following equations, |

$$D = \frac{\omega_0^2}{4\tau_D} \tag{1}$$

$$K = \frac{z_0}{\omega_0} \tag{2}$$

$$V_{\text{eff}} = \pi^{3/2}\omega_0^2 z_0 = K(4\pi\tau_D D)^{3/2} \tag{3}$$

For confocal FCCS measurements, the cross-correlation effective volume ($V_{\text{eff, GR}}$) is given by:

$$V_{\text{eff,GR}} = \left(\frac{\pi}{2}\right)^{3/2}\left(\omega_{0,G}^2 + \omega_{0,R}^2\right)\left(z_{0,G}^2 + z_{0,R}^2\right)^{1/2} \tag{4}$$

where $\omega_{0,G}$, $\omega_{0,R}$, $z_{0,G}$, and $z_{0,R}$ are the radial and axial distances of the laser focal spot at $1/e^2$ value of the maximum intensity at the focus of the observation volumes of the respective green and red detection channels that can be experimentally determined from calibration dye(s) of known D.

3.3 Confocal FC(C)S Zebrafish Measurement Procedure

After calibrating the system for FC(C)S measurements with the calibration dye (*see* Subheading 3.2), zebrafish measurements are next conducted on the calibrated system. The following steps describe the procedure of a confocal FCS or SW-FCCS measurement in live zebrafish embryos.

3.3.1 Placement of Mounted Zebrafish Embryos on the Confocal FC(C)S System

1. After the low-melting-point agarose embedding the zebrafish embryo solidifies, place the glass-bottom dish on top of the sample stage of the confocal microscope.

2. Locate and focus on the zebrafish embryo immobilized within the solidified low-melting-point agarose by transmitted light illumination through the microscope's eye piece.

3. Position and focus the structure of interest within the zebrafish embryo at the center of the microscope's field of view (*see* **Note 10**).

4. Next, illuminate the fluorescently labeled region of the structure with the excitation laser line, image and focus at the imaging module of the setup with the corresponding dichroic mirrors and filters. For confocal SW-FCCS, ensure that fluorescence of the fluorophore pair can be individually detected in two spectrally distinct photomultiplier tube (PMT) channels at the imaging module.

1. Select an appropriate set of dichroic mirrors and/or filters with reference to the excitation wavelength of the laser beam and the fluorescence emission of the fluorophore/fluorophore pair in the FC(C)S module of the confocal system. It is important to use the same set of dichroic mirrors and/or filters for both dye calibration (Subheading 3.2) and experiments.

2. In the scanning and imaging mode of the confocal system, focus the region of interest (ROI) and take an image. Switch to the nonscanning (point) mode of the system and select a suitable point within the ROI to perform FCS measurements. Take note that it is imperative to measure within the penetration depth of around 80 μm, accessible by one-photon excitation, in the live zebrafish sample to minimize distortion of the observation volume and reduce scattering of light by thick tissues [21] (see **Note 11**). Direct the fluorescence emission to the FC(C)S module, monitor the correlation functions computed in real time by the hardware or software correlator (s) and conduct FC(C)S measurements.

3. For measurements on the membrane, place the point at the center of the cell in the embryo in the imaging mode. Next, switch to FC(C)S mode, move the focus of the detection objective upward or downward and monitor the fluorescence intensity and fluctuations until the largest intensity and fluctuations are detected. This indicates that the laser spot is focused on the upper or lower membrane respectively as the largest fluorescence fluctuations originate from fluorophores on the membrane (see **Note 12**). Another way to locate the membrane is to use the imaging mode of the confocal system to focus on the upper or lower membrane.

4. The usual count rates for precise generation of ACF(s) and CCF curves is 5–300 kHz and the recommended concentration range of fluorophores for FC(C)S measurements is >1 pM and <100 nM [75] (see **Note 13**). Modulate count rates by controlling the expression levels of fluorescently labeled proteins in the zebrafish embryos. Fusion proteins exogenously introduced into the early-stage zebrafish embryo by microinjection have tunable protein expression levels through the amount of mRNA/DNA injected into the embryos, which is generally positively correlated. This should be optimized to provide count rates of highest SNR at nonsaturating laser powers adapted for FC(C)S measurements. This is not the case for transgenic lines which stably express FP fusion proteins at a single expression level suitable for imaging. However, the fluorescence expression levels optimized for imaging often lead to count rates too high for FCS. This can be resolved by generating multiple transgenic lines with different expression levels suitable for both imaging and FCS [15].

5. The measurement time required for FC(C)S experiments in live zebrafish embryos is a compromise between maximizing SNR and reducing photobleaching and sample movement. This translates to measurement time long enough to gather sufficient statistics to plot proper ACF and CCF curves while limiting the duration of the experiment to reduce photobleaching, photodamage and sample movement effects of the live embryo (*see* **Notes 14** and **15**). Measurement times of 15–30 s are commonly used in live zebrafish embryos which yield good SNR for FC(C)S experiments with minute effects of photobleaching and sample movement [15, 16]. The tradeoff between acquiring sufficient statistics per embryo to photodamaging the embryo governs the number of measurements to be taken for each embryo. To be conservative, we typically measure around 5–10 embryos with 3–5 measurement points per embryo, each point having three repeats, to ensure sufficient statistics of data over different fish and different positions.

6. The background count rate of the system should be determined and recorded for accurate determination of concentrations and binding/dissociation constants during FC(C)S data analysis. Measure the background count rate from wild-type zebrafish embryos not injected with mRNA/DNA FP fusion constructs for a single detection channel (FCS) or two detection channels (SW-FCCS). Reduce background contribution by switching off lights before every experiment.

3.4 Data Analysis for FC(C)S Measurements in Live Zebrafish Embryos

A detailed description of the data fitting and analysis of FC(C)S measurements are provided in several articles [36, 52, 76]. The raw data of the FC(C)S experiments can either be computed and fitted directly in commercial software packages provided by the manufacturers (Picoquant SymPhoTime 64, Zeiss ConfoCor 2/3, ISS AlbaFCS) or exported to and fitted with self-written FCS/FCCS data analysis programs in Origin, Igor Pro, Mathematica, and Matlab that contain a Levenberg–Marquardt nonlinear least-squares fitting algorithm and the relevant FC(C)S fit models [77]. Here, we provide an overview of the data analysis process for FC(C)S measurements.

3.4.1 Determination of Appropriate Range of Lag Time for FC(C)S Fitting and Correction for Detector Afterpulsing

Firstly, determine a suitable range of lag time (τ) to fit the correlation functions, bearing in mind to incorporate the diffusion dynamics, interactions, and decay of the correlation function while filtering out the contribution from detector afterpulsing, a detector artifact, which contaminates the photophysical properties (e.g., triplet state kinetics) contribution of the fluorophore/fluorophore pair in the microsecond time scale to the ACF [36, 52]. Detector afterpulsing is an artifact where a detector generates false photon events after the detection of an actual photon. It occurs in the microsecond time scale and therefore influences the fluorophore's photodynamics contribution to the ACF. The photodynamics of the fluorophores can only be accounted for in the ACF fit only if the effect of detector afterpulsing is discriminated. There are two ways to remove the influence of detector afterpulsing. One way is to split the emission of a single fluorophore species equally with a 50:50 beam splitter into two detection channels consisting of the same set of filters and conduct a cross-correlation between both channels to obtain the CCF [36]. As detector afterpulsing is a random process, the fluctuations arising from the afterpulsing of both detectors are not correlated, thus eliminating afterpulsing effects upon cross-correlation analysis. Note that this removes the afterpulsing contribution to the ACFs but the counts from afterpulsing still contribute as uncorrelated background which leads to uncertainty in the amplitude of the CCF. The cross-correlation method is mainly suited for FCS measurements since it requires two detectors to generate a single detector afterpulsing-corrected CCF for a single emission wavelength. Another method, fluorescence lifetime correlation spectroscopy (FLCS), to remove detector afterpulsing effects is to apply statistical filtering using the fluorophore's fluorescence lifetime, accessible by a pulsed laser and a time correlated single photon counting (TCSPC) module which is commercially available as an FCS upgrade kit to confocal microscopes [78–80], to separate the actual fluorescence signal from spurious photon events generated by afterpulsing [81, 82]. FLCS removes both the afterpulsing contribution to the ACFs and background counts from afterpulsing events which result in more accurate ACF and CCF amplitudes. FLCS is applicable to both FCS and FCCS measurements.

3.4.2 FC(C)S Fitting Process and Selection of Fitting Models

Next, appropriate fitting models should be selected based on the principle of Occam's razor. Prior knowledge of the system under study such as its dimensionality, i.e., 2D fitting models for membrane measurements and 3D models for measurements in the cytosol, nuclei and solution, and possible number of components should be accounted for when choosing a fitting model. Parameters with known values such as the K factor derived from FC(C)S calibration (*see* Subheading 3.2) should be fixed during FC(C)S fitting to reduce the number of free parameters and thus uncertainty in the fits. The appropriate model can thereafter be chosen by comparing the residuals of the fits derived from two competing models through a statistical F-test [83] or by Bayesian model selection [70, 84, 85] (*see* **Note 16**). To determine the apparent D of the labeled biomolecules in the zebrafish embryos, we can use the relationship of $D \propto 1/\tau_D$ to calculate for D_{sample} from the known $D_{calibration\ dye}$ and the $\tau_{D,calibration\ dye}$ given by FCS calibration, and the $\tau_{D,sample}$ derived from the FCS fits.

3.4.3 Quantification of Biomolecular Interactions by Confocal FCCS

For studies on the interaction of biomolecules in live zebrafish, the ratio of the amplitudes of the CCF and ACF (i.e., $G_{CCF, X}(0)/G_{ACF, G\ or\ R}(0)$), termed the q factor, serves as an indication of the degree of binding when compared with the ratios for positive (tandem proteins) and negative (noninteracting proteins) controls (*see* Notes 17 and 18) and is written as [47, 86]:

$$G_{ACF,G}(0) = \frac{1}{N_G} = \frac{1}{N_g + N_{gr}} \tag{5}$$

$$G_{ACF,R}(0) = \frac{1}{N_R} = \frac{1}{N_r + N_{gr}} \tag{6}$$

$$G_{CCF,X}(0) = \frac{1}{N_X} = \frac{N_{gr}}{(N_g + N_{gr})(N_r + N_{gr})} \tag{7}$$

$$
\begin{aligned}
q &= \frac{N_{gr}}{\min\left[(N_g + N_{gr}); (N_r + N_{gr})\right]} \\
&= \max\left[\frac{N_{gr}}{N_g + N_{gr}}; \frac{N_{gr}}{N_r + N_{gr}}\right] \\
&= \max\left[\frac{G_{CCF,X}(0)}{G_{ACF,R}(0)}; \frac{G_{CCF,X}(0)}{G_{ACF,G}(0)}\right] \\
&= \frac{\max[N_G; N_R]}{N_X}
\end{aligned}
\tag{8}
$$

where $G_{ACF, G}(0)$, $G_{ACF, R}(0)$, and $G_{CCF, X}(0)$ are the experimentally determined amplitudes of the ACFs in the green and red detection channels and the CCF respectively, N_G, N_R, and N_X are the apparent particle numbers derived from the amplitudes of the ACFs in the green and red detection channels and the CCF respectively, and N_g, N_r, and N_{gr} are the particle numbers of the unbound green molecules, unbound red molecules and bound red–green complex molecules respectively. The q factor reports the ratio of the double-labeled particles to that of the less abundant single-labeled particles as a measure of the upper limit of the amount of particles in complexes. It provides a qualitative measure of the degree of interaction between two molecules and is easily derived by normalizing the CCF amplitude ($G_{CCF, X}(0)$) by the lower of the two ACF amplitudes ($G_{ACF, G}(0)$ or $G_{ACF, R}(0)$) (Fig. 4). In order to obtain a more accurate measure of the biomolecular interactions between two proteins, the amplitudes of the ACFs and CCF should be corrected for background, difference in molecular brightness between the red and green fluorophores, cross talk, and Förster resonance energy transfer (FRET). Also note that it is important to select suitable red FPs as there are tendencies for mRFP and mCherry to aggregate in live cells and zebrafish which will lead to erroneous quantification of biomolecular interactions by FCCS (*see* Note 19).

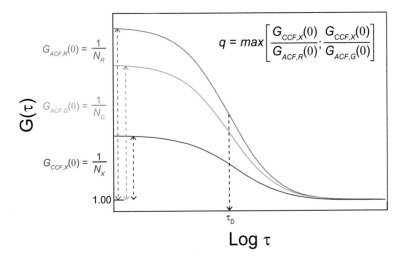

Fig. 4 Illustration of ACFs and CCF from a SW-FCCS measurement and the derivation of the q factor from the ratio of the amplitudes of the CCF and ACF. τ_D is the diffusion time, $G_{\mathrm{ACF},\,G}(0)$, $G_{\mathrm{ACF},\,R}(0)$, and $G_{\mathrm{CCF},\,X}(0)$ are the experimentally determined amplitudes of the ACFs in the green and red detection channels and the CCF respectively, and N_G, N_R, and N_X are the apparent particle numbers derived from the amplitudes of the ACFs in the green and red detection channels and the CCF respectively

3.4.4 Background Correction for ACF and CCF Amplitudes

The significant background contribution from the autofluorescence of cells and tissues in in vivo systems coupled with the low molecular brightness of FPs necessitate the need to conduct a background correction to the amplitudes of the ACFs and CCF in confocal FC(C)S measurements which would otherwise be underestimated. First, the average background intensities of both channels, B_G and B_R for the green and red detection channels respectively, are measured in wild-type embryos that are not microinjected with any DNA/mRNA constructs [16]. The amplitudes of the ACFs and CCF are then background corrected by accounting for the background intensities for the respective detection channels [16, 39]. In order to incorporate the background terms into the amplitudes of the ACFs and CCF, the amplitudes are expressed in terms of the molecular brightness and number of particles for each detection channel to yield:

$$G_{\mathrm{ACF},G}(0) = \frac{\eta_{g,G}^2 \left(N_g + N_{gr}\right)}{\left[\eta_{g,G}\left(N_g + N_{gr}\right) + B_G\right]^2} \tag{9}$$

$$G_{\mathrm{ACF},R}(0) = \frac{\eta_{r,R}^2 \left(N_r + N_{gr}\right)}{\left[\eta_{r,R}\left(N_r + N_{gr}\right) + B_R\right]^2} \tag{10}$$

$$G_{\text{CCF,X}}(0) = \frac{\eta_{g,G}\eta_{r,R}N_{gr}}{\left[\eta_{g,G}(N_g + N_{gr}) + B_G\right]\left[\eta_{r,R}(N_r + N_{gr}) + B_R\right]}$$

(11)

$\eta_{g,\,G}$ and $\eta_{r,\,R}$ are the molecular brightness, in cps, of the green and red fluorophores detected in the green and red channels, respectively. Assuming that the molecular brightness of the fluorophores remain unchanged upon complexation, background corrected $\eta_{g,\,G}$ and $\eta_{r,\,R}$ can be determined individually in control experiments by expressing single-labeled fluorophores (either green or red FPs) in zebrafish embryos to yield the following expressions:

$$\eta_{g,G} = \frac{\langle I_{g,G}\rangle}{N_{g,G}}\left[\frac{\langle I_{g,G}\rangle}{(\langle I_{g,G}\rangle - B_G)}\right]$$

(12)

$$\eta_{r,R} = \frac{\langle I_{r,R}\rangle}{N_{r,R}}\left[\frac{\langle I_{r,R}\rangle}{(\langle I_{r,R}\rangle - B_R)}\right]$$

(13)

$\langle I_{g,\,G}\rangle$ and $\langle I_{r,\,R}\rangle$ are the average measured intensities of the single-labeled green and red FPs in the green and red channels, respectively. $N_{g,\,G}$ and $N_{r,\,R}$ are the particle numbers of single-labeled green and red FPs computed in the ACFs of the green and red channels, respectively. The molecular brightness of fluorophores can also be extracted from FCS data by another method known as the photon counting histogram (PCH) analysis [87]. In PCH analysis, the photon counts from the intensity time trace utilized for FCS data analysis are plotted as a histogram and fitted with PCH theoretical models to extract the molecular brightness of the fluorophores. Thereafter, the background corrected values of N_g, N_r, and N_{gr} can be solved from Eqs. 9–11 upon obtaining the respective molecular brightness and background intensities.

3.4.5 Cross Talk Correction for ACF and CCF Amplitudes

Spectral cross talk occurs when there is a signal leakage from the green fluorophore into the red detection channel or from the red fluorophore into the green detection channel due to overlap between the emission spectra of both the green and red fluorophores. The major contributor of cross talk is the signal leakage from the green fluorophore into the red detection channel. In contrast, signal leakage from the red fluorophore into the green detection channel is generally less significant. Cross talk leads to an elevated CCF amplitude due to the signal from the green fluorophore detected in the green channel cross-correlating with its own signal that bled through into the red channel and vice versa, thus resulting in false positive cross-correlation. This can be corrected by accounting for the cross talk contributions from the green fluorophore into the red channel and from the red fluorophore into the green channel as other fluorescent species in the mathematical expressions of the ACF and CCF amplitudes [39].

$$G_{ACF,G}(0) = \frac{\eta_{g,G}^2 N_g + \eta_{r,G}^2 N_r + \left(\eta_{g,G} + \eta_{r,G}\right)^2 N_{gr}}{\left[\eta_{g,G} N_g + \eta_{r,G} N_r + \left(\eta_{g,G} + \eta_{r,G}\right) N_{gr} + B_G\right]^2} \tag{14}$$

$$G_{ACF,R}(0) = \frac{\eta_{r,R}^2 N_r + \eta_{g,R}^2 N_g + \left(\eta_{r,R} + \eta_{g,R}\right)^2 N_{gr}}{\left[\eta_{r,R} N_r + \eta_{g,R} N_g + \left(\eta_{r,R} + \eta_{g,R}\right) N_{gr} + B_R\right]^2} \tag{15}$$

$$G_{CCF,X}(0) = \frac{\left(\eta_{g,G} + \eta_{r,G}\right)\left(\eta_{r,R} + \eta_{g,R}\right) N_{gr} + \eta_{g,G}\eta_{g,R} N_g + \eta_{r,R}\eta_{r,G} N_r}{\left[\eta_{g,G} N_g + \eta_{r,G} N_r + \left(\eta_{g,G} + \eta_{r,G}\right) N_{gr} + B_G\right]}$$
$$\times \left[\eta_{r,R} N_r + \eta_{g,R} N_g + \left(\eta_{r,R} + \eta_{g,R}\right) N_{gr} + B_R\right]^{-1} \tag{16}$$

$\eta_{g,R}$ and $\eta_{r,G}$ are the molecular brightness from the cross talk contribution of the green fluorophore into the red channel and the red fluorophore into the green channel respectively. Both can be individually obtained by conducting control experiments on zebrafish embryos expressing single-labeled green/red FPs and detecting the bleed through of the green/red FP signal into the red/green detection channel.

$$\eta_{g,R} = \frac{\langle I_{g,R} \rangle}{N_{g,R}} \left[\frac{\langle I_{g,R} \rangle}{\left(\langle I_{g,R} \rangle - B_R\right)} \right] \tag{17}$$

$$\eta_{r,G} = \frac{\langle I_{r,G} \rangle}{N_{r,G}} \left[\frac{\langle I_{r,G} \rangle}{\left(\langle I_{r,G} \rangle - B_G\right)} \right] \tag{18}$$

where $\langle I_{g,R} \rangle$ and $\langle I_{r,G} \rangle$ are the average measured intensities of the single-labeled green and red FPs that cross talked into the red and green channels respectively, and $N_{g,R}$ and $N_{r,G}$ are the particle numbers of single-labeled green and red FPs detected in the red and green channels respectively. On obtaining the cross talk contributions in terms of $\eta_{g,R}$ and $\eta_{r,G}$, accurate values of background and cross talk corrected N_g, N_r, and N_{gr} can be solved from Eqs. 14–16. As mentioned earlier in Subheading 1, one way to remove cross talk is to employ PIE-FCCS [45].

3.4.6 Correction for Förster Resonance Energy Transfer (FRET) Between FPs

The occurrence of FRET between FPs that are in close proximity with each other, especially for positive FCCS controls using tandem proteins that concurrently express two FPs in a fusion protein, will lead to a decrease in the CCF amplitude. This can be corrected by introducing correction factors (q_g and q_r) to account for the effects of FRET between FPs [39].

$$G_{ACF,G}(0) = \frac{\eta_{g,G}^2 N_g + \eta_{r,G}^2 N_r + \left(q_g \eta_{g,G} + q_r \eta_{r,G}\right)^2 N_{gr}}{\left[\eta_{g,G} N_g + \eta_{r,G} N_r + \left(q_g \eta_{g,G} + q_r \eta_{r,G}\right) N_{gr} + B_G\right]^2}$$

(19)

$$G_{ACF,R}(0) = \frac{\eta_{r,R}^2 N_r + \eta_{g,R}^2 N_g + \left(q_r \eta_{r,R} + q_g \eta_{g,R}\right)^2 N_{gr}}{\left[\eta_{r,R} N_r + \eta_{g,R} N_g + \left(q_r \eta_{r,R} + q_g \eta_{g,R}\right) N_{gr} + B_R\right]^2}$$

(20)

$$G_{CCF,X}(0) = \frac{\left(q_g \eta_{g,G} + q_r \eta_{r,G}\right)\left(q_r \eta_{r,R} + q_g \eta_{g,R}\right) N_{gr} + \eta_{g,G} \eta_{g,R} N_g + \eta_{r,R} \eta_{r,G} N_r}{\left[\eta_{g,G} N_g + \eta_{r,G} N_r + \left(q_g \eta_{g,G} + q_r \eta_{r,G}\right) N_{gr} + B_G\right]}$$
$$\times \left[\eta_{r,R} N_r + \eta_{g,R} N_g + \left(q_r \eta_{r,R} + q_g \eta_{g,R}\right) N_{gr} + B_R\right]^{-1}$$

(21)

where q_g is the correction factor which accounts for the reduction of molecular brightness of the donor (green FP) as a result of FRET and is derived by taking the ratio of $\eta_{g,G}$ obtained from zebrafish embryos expressing tandem FPs or coexpressing green and red FPs in fusion proteins with $\eta_{g,G}$ obtained from embryos expressing single-labeled green FP in the green channel. On the other hand, q_r is the correction factor which accounts for the increase in molecular brightness of the acceptor due to sensitized emission resulting from FRET and is quantified by taking the ratio of $\eta_{r,R}$ obtained from embryos expressing tandem FPs or coexpressing green and red FPs in fusion proteins with $\eta_{r,R}$ obtained from embryos expressing single-labeled red FP in the red channel. An alternative method to correct for FRET is to perform PIE-FCCS measurements to reassign the amount of photons gained by the acceptor (red FP) due to sensitized emission back to the donor channel which donor photons (green FP) were lost due to FRET [39, 45].

3.4.7 Determination of Dissociation Constant K_d

From the background, cross talk and FRET corrected values of N_g, N_r and N_{gr}, the dissociation constant K_d can be calculated from the concentrations of the unbound green molecules (C_g), unbound red molecules (C_r), and bound red–green complex molecules (C_{gr}).

$$c_g = \frac{N_g}{N_A V_{eff,G}}$$

(22)

$$C_r = \frac{N_r}{N_A V_{\text{eff,R}}} \qquad (23)$$

$$C_{gr} = \frac{N_{gr}}{N_A V_{\text{eff,GR}}} \qquad (24)$$

$$K_d = \frac{C_g C_r}{C_{gr}} \qquad (25)$$

N_A is Avogadro's constant and $V_{\text{eff, G}}$ and $V_{\text{eff, R}}$ are the effective observation volumes of the green and red channels respectively.

The presence of nonfluorescent FPs due to FP maturation issues, FPs in dark states and photobleaching of FPs was observed to influence the quantification of K_d and should be corrected for more accurate determination of K_d [39]. This is done so by determining the probabilities of the green (p_g) and red (p_r) FPs being fluorescent. p_g and p_r can be deduced by calculating the ratio of the experimentally determined molecular brightness of dimeric green and red FPs with twice the experimentally determined molecular brightness of monomeric green and red FPs respectively. The steps to utilize the molecular brightness of dimeric and monomeric FPs to quantify for the presence of nonfluorescent FPs is extensively detailed in [88]. Consequently, the K_d, corrected for the presence of nonfluorescent FPs, can be deduced by using the apparent concentrations of the unbound green molecules ($C_{g,\ app}$), unbound red molecules ($C_{r,\ app}$) and bound red–green complex molecules ($C_{gr,\ app}$) expressed in terms of p_g and p_r.

$$C_{g,\text{app}} = p_g C_g + p_g(1 - p_r)C_{gr} \qquad (26)$$

$$C_{r,\text{app}} = p_r C_r + p_r\left(1 - p_g\right)C_{gr} \qquad (27)$$

$$C_{gr,\text{app}} = p_g p_r C_{gr} \qquad (28)$$

Essentially, to ensure accurate quantification of biomolecular interactions in vivo by the determination of K_d, all of the above-mentioned factors should be systematically accounted for. Documentation of the quantification process for binding studies in live systems using SW-FCCS can be found here [39, 53]. An important criterion for biomolecular interaction studies by SW-FCCS is that the τ_D values for all ACFs and CCF of interacting partners should match for at least one diffusive component which corresponds to the complex that is common and detected in all correlation functions. As an example, we highlight the overall strategy for conducting interaction studies by SW-FCCS in live blastula stage zebrafish embryos for a nodal receptor (Acvr2b-mCherry) and ligand (Cyclops-EGFP/Squint-EGFP) pair in Fig. 5. Briefly, Acvr2b-mCherry nodal receptor capped RNA was first microinjected into the zebrafish embryo at one cell stage. After embryonic growth and expression of Acvr2b-mCherry, capped RNA of Cyclops-EGFP or

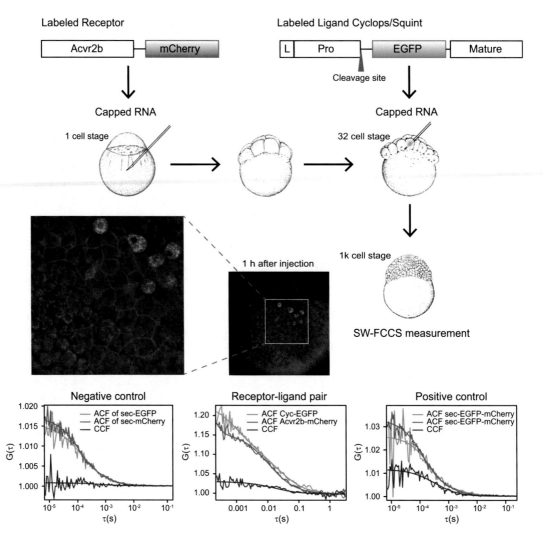

Fig. 5 Overall strategy for conducting interaction studies by SW-FCCS in live blastula stage zebrafish embryos for a nodal receptor (Acvr2b-mCherry) and ligand (Cyclops-EGFP/Squint-EGFP) pair. The ACFs and CCFs for negative control secreted-EGFP (sec-EGFP) and secreted-mCherry (sec-mCherry), receptor–ligand pair Acvr2b-mCherry and Cyclops-EGFP (Cyc-EGFP), and positive control tandem secreted-EGFP-mCherry (sec-EGFP-mCherry) are illustrated. Note that the time scale of diffusion of the receptor–ligand pair is different as the receptor in the membrane diffuses much slower than the ligand in the extracellular space

Squint-EGFP nodal ligand was next microinjected into a single cell at 32 cell stage. SW-FCCS measurements were subsequently conducted at various distances from source cells expressing Cyclops-EGFP or Squint-EGFP at the animal pole of the zebrafish embryo at 1000 cell stage and the interactions between Acvr2b-mCherry receptor and Cyclops-EGFP or Squint-EGFP ligand were quantified. From measurements conducted in live blastula stage zebrafish embryos, a flat and elevated CCF is observed for the negative and positive controls of secreted-EGFP and secreted-mCherry and tandem secreted-EGFP-mCherry, respectively, corresponding to the minimum and maximum $G_{CCF, X}(0)/ G_{ACF, G \text{ or } R}(0)$ ratios achievable. An intermediate CCF is observed for the experimental data of Acvr2b-mCherry receptor and Cyclops-EGFP ligand, indicating a certain degree of binding above the negative control.

4 Notes

As with all other techniques, there are several important points to take note of and caveats associated with conducting FC(C)S measurements in live zebrafish embryos which are addressed in this section.

1. The proper maintenance and preparation of zebrafish samples is crucial to accurately probe their biomolecular properties by FC(C)S at physiological conditions. In addition, measurement artefacts from samples are minimized with proper sample preparation.

2. The most common method of introducing fluorescent labels to target proteins in the zebrafish embryos is to microinject FP fusion plasmid DNA constructs or mRNA into early stage embryos to utilize the zebrafish's intrinsic transcription and translation mechanisms to transcribe the injected DNA or RNA and translate them into specific fusion proteins [89, 90]. By selecting a suitable zebrafish promoter, expression of the fusion proteins can be localized to specific regions of interest (cerebellar cells, muscle cells, etc.), making FC(C)S measurements highly specific. Moreover, encoding the gene sequence of FP into that of the protein of interest ensures specific fluorescent labeling of the protein. Therefore, it is important to have access to commercially available molecular cloning tool kits in order to generate the FP fusion DNA or capped mRNA constructs required for FC(C)S measurements.

3. Both confocal imaging and FC(C)S measurements require the same sample mounting procedure. Upon screening for positive fluorescence expression with the appropriate expression levels needed for FC(C)S experiments under a tabletop fluorescence microscope, the selected embryos are mounted for measurements.

4. Briefly, laser excitation sources are beam expanded and introduced to the back focal aperture of a high numerical aperture (NA) water immersion objective (60×, NA 1.2, Olympus, Singapore) to generate a focused laser spot on the sample. The use of a water immersion objective is advantageous since its refractive index matches with the aqueous environment in live zebrafish embryos. Fluorescence emitted from the excited

spot on the sample is then collected back to the same objective, passed through an excitation dichroic mirror and spatially filtered via a pinhole. Notably, the pinhole spatially removes the out of focus light and thereby provides optical sectioning to shape the observation volume. Thereafter, the emitted fluorescence travels across an emission filter to spectrally filter out background and scattered light from the signal. A band-pass emission filter with steep cutons/cutoffs is preferable as it can remove background from Raman scattering of water as opposed to longpass emission filters and can efficiently transmit and reject fluorescence and background respectively. For SW-FCCS, the fluorescence is split into two detection channels by a dichroic beam splitter mirror (typically 560 dichroic mirror for GFP/red fluorescent protein (RFP) or mCherry fluorophore pair) selected to minimize cross talk. Each of the detection channels contains their respective emission filters (e.g., 545AF35 (Omega Optical, VT, USA) for GFP channel and 615DF45 (Omega Optical, VT, USA) for RFP or mCherry channel) to further select for the required emission wavelength range and reduce cross talk between both channels. After spectral filtering by emission filters, the fluorescence is captured by highly sensitive photon counting detectors such as avalanche photodiodes (APDs). APDs possess low dark count rate and excellent time resolution, limited only by the detector's dead time of ~50 ns. In the case of FCS, only a single detection channel and thus detector is utilized to capture fluorescence emitted from a single fluorophore source. In contrast, two detectors are employed for FCCS to detect spectrally separated fluorescence emission from the corresponding detection channels, with their respective emission filters, of each fluorophore of the fluorophore pair. Upon detection of the fluorescence signals by the APDs, a hardware or software correlator is used to compute the ACF by autocorrelating the fluorescence signal of a single detection channel for FCS or the CCF by cross-correlating synchronized fluorescence signals of two different detection channels for FCCS.

5. cDNA synthesis from RNA is recommended as cDNA is more stable as compared to RNA which is vulnerable to RNAse cleavage, resulting in its eventual degradation.

6. Generally, a Kozak consensus sequence is introduced immediately upstream of the start codon and restriction sites of the forward primer to enable initiation of translation of the protein of interest [91].

7. Depending on the nature of the study, either capped mRNA or DNA can be microinjected into zebrafish embryos. Microinjection of mRNA is suitable for early embryonic stage studies since the onset of ectopic protein expression occurs at early

developmental stages of the zebrafish embryos (few hpf) as a result of direct translation of proteins from the injected mRNA as compared to plasmid DNA. Moreover, injected mRNA has a more widespread distribution than that of DNA, thus ensuring a larger fraction of cells within the embryo to express the exogenous protein. However, the expression derived from injected mRNA does not last throughout development due to degradation of mRNA. Therefore, DNA microinjection is preferred for studies conducted on embryos at later developmental stages (few days post fertilization (dpf)) because of its stability, which provides longer lasting protein expression in comparison to mRNA. The disadvantage of DNA microinjection, however, is that the integration of the DNA is in a smaller fraction of cells compared to mRNA due to the additional transcriptional process from DNA to mRNA which reduces the DNA's incorporation efficiency into cells. Moreover, there may be mosaic expression as a result of rapid cell division of early stage embryos and nonuniform distribution of injected DNA. Hence, it is important to identify the optimal conditions suitable for the type of study conducted on zebrafish embryos [89]. Detailed description of the microinjection procedure is given here [89, 90, 92].

8. The incubation time of the zebrafish embryos in 0.05% (w/v) tricaine anesthetic agent should be monitored and optimized for effective anesthetization during the course of the experiment (*see* Subheading 3.1.4). Prolonged incubation of the embryos in tricaine will kill them while insufficient incubation will lead to premature recovery from the anesthetic effects of tricaine and undesirable movements of the embryos during measurement. In addition, the pH of tricaine should be adjusted to ~7 as tricaine is acidic and thus harmful to embryos.

9. The temperature of the low-melting-point agarose used to set the zebrafish embryo on the glass-bottom dish should be maintained above its gelling temperature for effective manipulation of the embryo's orientation but not too high that it will injure and kill the embryo (*see* Subheading 3.1.4). A good practice is to use a heat block and set it at around 42 °C to maintain the low-melting-point agarose in the liquid phase. Note that the gelling temperature of the low-melting-point agarose increases with increasing percentage of agarose. After adding the liquefied low-melting-point agarose on the embryo, orient the embryo quickly and precisely with the microloader pipette tip, fishing wire loop or "eyelash tool" before the agarose hardens.

10. The height at which the anesthetized zebrafish embryo is mounted in low-melting-point agarose on the glass-bottom dish should not exceed the working distance of the objective

on the confocal system to ensure that the structure of interest is accessible for confocal FC(C)S measurements. In addition, the structure of interest to be measured in the embryo should be mounted as close to the surface of the glass-bottom dish as possible to reduce scattering of the excitation light by the low-melting-point agarose.

11. It is crucial to determine the accessible penetration depth of the confocal setup to perform accurate FC(C)S measurements in live zebrafish embryos. Measuring beyond the accessible penetration depth will lead to inaccurate determination of diffusion dynamics, concentrations and biomolecular interactions due to the distortion of the observation volume caused by the scattering of the excitation light by thick tissues of the zebrafish. The accessible penetration depth can be assessed by performing FC(C)S measurements in live zebrafish embryos and comparing the diffusion times (τ_D) and maximum cross-correlation values of the positive control at different sample depths. The usable penetration depth is generally defined within the range of sample depth where the τ_D and maximum cross-correlation of the positive control remain consistent within the margins of error [21]. Any changes in τ_D and maximum cross-correlation values with respect to the penetration depth indicate a distortion in the observation volume caused by spherical and chromatic aberrations originating from the refractive index mismatch between tissues and water which results in a change in the size and shape of the observation volume and a reduction in the overlap of the observation volumes of the green and red channels. In this case, FC(C)S fitting models no longer hold true for the data sets that are measured in a distorted observation volume at deeper penetration depths. This leads to erroneous τ_D, concentration and cross-correlation values to be determined from the ACFs and CCF and obscures quantitative FC(C)S measurements. Note that the penetration depth is dependent on the opacity of the tissues in the embryo which changes at different locations and embryonic stages of the zebrafish. Thus, the accessible penetration depth should be defined at the location of interest and at the appropriate embryonic stage. To increase accessible penetration depth, optical clarity of the embryo can be improved by phenyl-2-thiourea (PTU) treatment which inhibits melanin formation and suppresses pigmentation. Alternatively, genetically modified transparent transgenic lines can be used [22–25].

12. It should be noted that the maximum fluorescence intensity does not always coincide with the focused z-position of the membrane which corresponds to the minimum diffusion time [93–96]. This could be due to several factors such as optical aberrations arising from cover slide thickness discrepancies

and/or refractive index mismatch between the imaging medium and water, and the saturation of the observation volume with high laser intensities which lead to an increase and distortion of the effective observation volume [93, 95, 96]. A method to resolve this issue is to perform z-scan FCS measurements at nonsaturating laser intensities at various z-positions along the membrane to determine the focused z-position on the membrane that corresponds to the minimum diffusion time upon fitting the plot of diffusion time against z-position with a quadratic function [93, 96].

13. High count rates due to high fluorophore concentrations may reduce the strength of the fluorescence fluctuations within the detection volume while low count rates refers to signal close to background, both resulting in the generation of noisy correlation functions [71]. The nonlinear recordings of the APDs (>700 kHz) at high count rates will also lead to erroneous FCS results [36], especially in terms of the particle concentration. In addition, overexposing the APDs to high intensities of light for a substantial period of time may lead to detector damage. Therefore, high count rates should be avoided at all times.

14. If photobleaching occurs during confocal FC(C)S measurements, it could manifest as an additional decay in the ACF/CCF curves which will lead to artefactual values of the fitted parameters. Moreover, photobleaching would reduce the count rates of the zebrafish sample and result in inaccurate determination of the concentration of fluorescently labeled proteins. This is especially detrimental for confocal FCCS measurements since the photobleaching behavior and rates of different fluorescent proteins may differ which will lead to unreliable interaction studies. Thus, it is important to minimize photobleaching by utilizing low laser powers while collecting sufficient count rates for accurate FC(C)S data analysis. In addition, fluorescent protein variants with higher photostability can be employed to reduce the effect of photobleaching [97]. Note that cryptic photobleaching, which does not lead to an observable decrease in the intensity trace, may still occur at low laser powers and cause a reduction in the τ_D [98]. In this case, one should check for the consistency of τ_D at low laser powers.

15. Sample movement during confocal FC(C)S measurements is highly undesirable and may lead to artificial components in the ACF and CCF. The movement of zebrafish embryos usually occurs due to recovery of the embryos from the anesthetic effects of tricaine. This can be alleviated either by limiting the experimental duration per embryo to within the effective period of tricaine (depending on the age of the embryo) or

by adding 0.05% (w/v) tricaine in embryo medium into the glass-bottom dish containing the agarose-mounted zebrafish embryo for longer experiment durations. Another movement within the live zebrafish embryo is the beating of its heart. Avoid performing FC(C)S measurements at regions close to the heart as the heartbeat of the embryo will cause a periodic time trace within the ACF/CCF [21].

16. One of the most prominent issues in FC(C)S data fitting is the ambiguity in model selection statistical methods such as F-test and maximum entropy model that generate similar residuals of fits for multiple models. Furthermore, overfitting of the data with excessively complex models does not provide a true interpretation of experimental outcomes, albeit giving lower residuals of fits. These issues can be circumvented by applying Bayesian model selection to FC(C)S fitting as a robust and unbiased approach [70, 84, 85]. Bayesian model selection essentially selects the least complex model that most appropriately describes the given data set by employing the Bayesian inference procedure to calculate for the model probabilities and parameter estimates of each given model.

17. Although the expected ratio of the CCF and ACF amplitudes ($G_{CCF, x}(0)/ G_{ACF, G\ or\ R}(0)$) for a positive control of tandem EGFP-mCherry fusion proteins is theoretically 1, this is not achievable in practice due to several factors including the non-perfect overlap of the observation volumes for the red and green detection channels, the presence of nonfluorescent FPs, and FRET between the FPs. These factors are extensively discussed in Subheading 3.4 and refs. 39, 52. The largest factor is typically the problem of nonfluorescent FPs and this is, based on experience, more of a problem for red FPs than for the green ones. Therefore, despite the smaller observation volume of the green channel, one observes nevertheless a larger amplitude for the red channel.

18. In FCCS measurements, the ratio of the amplitudes of the CCF and ACF ($G_{CCF, x}(0)/ G_{ACF, G\ or\ R}(0)$), indicative of the degree of binding between red and green FP-labeled proteins, is influenced by the concentration ratios of the red and green FP-labeled proteins. Unequal concentration ratios of the red and green FP-labeled proteins result in an overestimation of the degree of binding. For example, if the concentration of the green FP-labeled proteins is much higher than the concentration of the red FP-labeled proteins, the $G_{CCF, x}(0)/ G_{ACF, G}(0)$ will be higher than when there are similar concentrations of green and red FP-labeled proteins. Therefore, it is recommended to conduct FCCS measurements with similar concentrations of red and green FP-labeled proteins, ideally within a factor of 2, to avoid artifacts such as cross talk and

overestimation of the amount of proteins in complexes [53]. For measurements in live zebrafish, the concentrations of the red and green FP-labeled proteins can be controlled by adjusting the expression levels of both proteins via the amount of DNA or mRNA microinjected into early stage embryos.

19. The process of labeling proteins of interest with fluorescent proteins may alter their original behavior due to factors such as aggregation of fluorescent proteins and artificial protein clustering induced by fluorescent proteins [99, 100]. Red fluorescent proteins such as mRFP and mCherry expressed in some fusion proteins tend to aggregate in live cells [101]. Our group has also observed a similar tendency of mCherry aggregation in the motor neurons of hb9:mCherry expressing live zebrafish embryos (unpublished data). Such aggregation of fluorescent proteins leads to a misrepresentation of the biomolecular dynamics, interactions and functions of the proteins of interest by FC(C)S measurements. Therefore, functional studies should be performed on fusion proteins that are genetically labeled with fluorescent proteins to ensure the retention of the protein's original function. Furthermore, to overcome the aggregation issue of mRFP and mCherry, brighter and monomeric red fluorescent protein variants can be used to replace mRFP/mCherry [97]. In our group, we tested mApple as a counterpart to EGFP. Despite the somewhat larger spectral overlap of the two proteins, the higher photostability of mApple resulted in stronger cross-correlations (unpublished data).

Acknowledgment

X.W.N. is supported by the NUS graduate research scholarship. T.W. acknowledges funding by the Ministry of Education of Singapore (grant number MOE2016-T3-1-005). Work in the laboratory of K.S. is supported by Warwick Medical School and the BBSRC. K.S. thanks Andreas Zaucker and Scott Clarke for the image of the experimental setup for mounting zebrafish embryos.

References

1. Magde D, Elson EL, Webb WW (1972) Thermodynamic fluctuations in a reacting system-measurement by fluorescence correlation spectroscopy. Phys Rev Lett 29:705–708

2. Elson EL, Magde D (1974) Fluorescence correlation spectroscopy. I. Conceptual basis and theory. Biopolymers 13:1–27

3. Magde D, Elson EL, Webb WW (1974) Fluorescence correlation spectroscopy. II. An experimental realization. Biopolymers 13:29–61

4. Pramanik A, Olsson M, Langel Ü et al (2001) Fluorescence correlation spectroscopy detects galanin receptor diversity on insulinoma cells. Biochemistry 40:10839–10845

5. Pramanik A, Rigler R (2001) Ligand-receptor interactions in the membrane of cultured cells monitored by fluorescence correlation spectroscopy. Biol Chem 382:371–378

6. Meissner O, Häberlein H (2003) Lateral mobility and specific binding to GABA_A receptors on hippocampal neurons monitored by fluorescence correlation spectroscopy. Biochemistry 42:1667–1672

7. Pick H, Preuss AK, Mayer M et al (2003) Monitoring expression and clustering of the ionotropic $5HT_3$ receptor in plasma membranes of live biological cells. Biochemistry 42:877–884

8. Herrick-Davis K, Grinde E, Cowan A, Mazurkiewicz JE (2013) Fluorescence correlation spectroscopy analysis of serotonin, adrenergic, muscarinic, and dopamine receptor dimerization: the oligomer number puzzle. Mol Pharmacol 84:630–642

9. Saito K, Ito E, Takakuwa Y et al (2003) In situ observation of mobility and anchoring of PKCβI in plasma membrane. FEBS Lett 541:126–131

10. White MD, Angiolini JF, Alvarez YD et al (2016) Long-lived binding of Sox2 to DNA predicts cell fate in the four-cell mouse embryo. Cell 165:75–87

11. Yu SR, Burkhardt M, Nowak M et al (2009) Fgf8 morphogen gradient forms by a source-sink mechanism with freely diffusing molecules. Nature 461:533–536

12. Petrášek Z, Hoege C, Hyman AA, Schwille P (2008) Two-photon fluorescence imaging and correlation analysis applied to protein dynamics in C. elegans embryo. Proc SPIE 6860:68601L

13. Petrášek Z, Hoege C, Mashaghi A et al (2008) Characterization of protein dynamics in asymmetric cell division by scanning fluorescence correlation spectroscopy. Biophys J 95:5476–5486

14. Abu-Arish A, Porcher A, Czerwonka A et al (2010) High mobility of bicoid captured by fluorescence correlation spectroscopy: implication for the rapid establishment of its gradient. Biophys J 99:L33–L35

15. Teh C, Sun G, Shen H et al (2015) Modulating the expression level of secreted Wnt3 influences cerebellum development in zebrafish transgenics. Development 142:3721–3733

16. Wang Y, Wang X, Wohland T, Sampath K (2016) Extracellular interactions and ligand degradation shape the nodal morphogen gradient. eLife 5:e13879

17. Howe K, Clark MD, Torroja CF et al (2013) The zebrafish reference genome sequence and its relationship to the human genome. Nature 496:498–503

18. Rasooly RS, Henken D, Freeman N et al (2003) Genetic and genomic tools for zebrafish research: the NIH zebrafish initiative. Dev Dyn 228:490–496

19. Veldman MB, Lin S (2008) Zebrafish as a developmental model organism for pediatric research. Pediatr Res 64:470–476

20. Weber T, Köster R (2013) Genetic tools for multicolor imaging in zebrafish larvae. Methods 62:279–291

21. Shi X, Teo LS, Pan X et al (2009) Probing events with single molecule sensitivity in zebrafish and Drosophila embryos by fluorescence correlation spectroscopy. Dev Dyn 238:3156–3167

22. Henion PD, Raible DW, Beattie CE et al (1996) Screen for mutations affecting development of Zebrafish neural crest. Dev Genet 18:11–17

23. Lister J, Robertson C, Lepage T et al (1999) nacre encodes a zebrafish microphthalmia-related protein that regulates neural-crest-derived pigment cell fate. Development 126:3757–3767

24. Antinucci P, Hindges R (2016) A crystal-clear zebrafish for in vivo imaging. Sci Rep 6:29490

25. Karlsson J, von Hofsten J, Olsson P-E (2001) Generating transparent zebrafish: a refined method to improve detection of gene expression during embryonic development. Mar Biotechnol 3:522–527

26. Benninger RKP, Piston DW (2013) Two-photon excitation microscopy for the study of living cells and tissues. Curr Protoc Cell Biol. Chapter 4 Unit 4:11.1–24

27. Leroux C-E, Wang I, Derouard J, Delon A (2011) Adaptive optics for fluorescence correlation spectroscopy. Opt Express 19:26839–26849

28. Leroux C-E, Monnier S, Wang I et al (2014) Fluorescent correlation spectroscopy measurements with adaptive optics in the intercellular space of spheroids. Biomed Opt Express 5:3730–3738

29. Pan X, Yu H, Shi X et al (2007) Characterization of flow direction in microchannels and zebrafish blood vessels by scanning fluorescence correlation spectroscopy. J Biomed Opt 12:14034

30. Korzh S, Pan X, Garcia-Lecea M et al (2008) Requirement of vasculogenesis and blood circulation in late stages of liver growth in zebrafish. BMC Dev Biol 8:84

31. Ng XW, Teh C, Korzh V, Wohland T (2016) The secreted signaling protein Wnt3 is associated with membrane domains in vivo: a SPIM-FCS study. Biophys J 111:418–429

32. Wawrezinieck L, Rigneault H, Marguet D, Lenne P-F (2005) Fluorescence correlation spectroscopy diffusion laws to probe the submicron cell membrane organization. Biophys J 89:4029–4042

33. Ng XW, Bag N, Wohland T (2015) Characterization of lipid and cell membrane organization by the fluorescence correlation spectroscopy diffusion law. Chimia 69:112–119

34. Sezgin E, Azbazdar Y, Ng XW et al (2017) Binding of canonical Wnt ligands to their receptor complexes occurs in ordered plasma membrane environments. FEBS J 284:2513–2526

35. Schwille P, Meyer-Almes F-J, Rigler R (1997) Dual-color fluorescence cross-correlation spectroscopy for multicomponent diffusional analysis in solution. Biophys J 72:1878–1886

36. Bacia K, Schwille P (2007) Practical guidelines for dual-color fluorescence cross-correlation spectroscopy. Nat Protoc 2:2842–2856

37. Hwang LC, Wohland T (2004) Dual-color fluorescence cross-correlation spectroscopy using single laser wavelength excitation. ChemPhysChem 5:549–551

38. Liu P, Sudhaharan T, Koh RML et al (2007) Investigation of the dimerization of proteins from the epidermal growth factor receptor family by single wavelength fluorescence cross-correlation spectroscopy. Biophys J 93:684–698

39. Foo YH, Naredi-Rainer N, Lamb DC et al (2012) Factors affecting the quantification of biomolecular interactions by fluorescence cross-correlation spectroscopy. Biophys J 102:1174–1183

40. Schwille P, Heinze KG (2001) Two-photon fluorescence cross-correlation spectroscopy. ChemPhysChem 2:269–272

41. Kim SA, Heinze KG, Waxham MN, Schwille P (2004) Intracellular calmodulin availability accessed with two-photon cross-correlation. Proc Natl Acad Sci U S A 101:105–110

42. Kim SA, Heinze KG, Bacia K et al (2005) Two-photon cross-correlation analysis of intracellular reactions with variable stoichiometry. Biophys J 88:4319–4336

43. Swift JL, Heuff R, Cramb DT (2006) A two-photon excitation fluorescence cross-correlation assay for a model ligand-receptor binding system using quantum dots. Biophys J 90:1396–1410

44. Hwang LC, Wohland T (2007) Recent advances in fluorescence cross-correlation spectroscopy. Cell Biochem Biophys 49:1–13

45. Müller BK, Zaychikov E, Bräuchle C, Lamb DC (2005) Pulsed interleaved excitation. Biophys J 89:3508–3522

46. Macháň R, Kapusta P, Hof M (2014) Statistical filtering in fluorescence microscopy and fluorescence correlation spectroscopy. Anal Bioanal Chem 406:4797–4813

47. Yavas S, Macháň R, Wohland T (2016) The epidermal growth factor receptor forms location-dependent complexes in resting cells. Biophys J 111:2241–2254

48. Krieger JW, Singh AP, Garbe CS et al (2014) Dual-color fluorescence cross-correlation spectroscopy on a single plane illumination microscope (SPIM-FCCS). Opt Express 22:2358–2375

49. Krieger JW, Singh AP, Bag N et al (2015) Imaging fluorescence (cross-) correlation spectroscopy in live cells and organisms. Nat Protoc 10:1948–1974

50. Szalóki N, Krieger JW, Komáromi I et al (2015) Evidence for homodimerization of the c-Fos transcription factor in live cells revealed by FRET, SPIM-FCCS and MD-modeling. Mol Cell Biol 35:3785–3798

51. Pernuš A, Langowski J (2015) Imaging Fos-Jun transcription factor mobility and interaction in live cells by single plane illumination-fluorescence cross correlation spectroscopy. PLoS One 10:e0123070

52. Ma X, Foo YH, Wohland T (2014) Fluorescence cross-correlation spectroscopy (FCCS) in living cells. In: Engelborghs Y, Visser AJWG (eds) Fluorescence spectroscopy and microscopy. Methods and protocols, Methods in molecular biology. Humana Press, Totowa, NJ, pp 557–573

53. Shi X, Yong HF, Sudhaharan T et al (2009) Determination of dissociation constants in living zebrafish embryos with single wavelength fluorescence cross-correlation spectroscopy. Biophys J 97:678–686

54. Sengupta P, Balaji J, Maiti S (2002) Measuring diffusion in cell membranes by fluorescence correlation spectroscopy. Methods 27:374–387

55. Pan X, Foo W, Lim W et al (2007) Multifunctional fluorescence correlation microscope for intracellular and microfluidic measurements. Rev Sci Instrum 78:53711

56. Shi X, Foo YH, Korzh V et al (2010) Applications of fluorescence correlation spectroscopy in living zebrafish embryos. In: Karuna S, Sudipto R (eds) Live imaging zebrafish – insights into development and disease. World Scientific Publishing, Singapore, pp 69–103

57. Higashijima S, Okamoto H, Ueno N et al (1997) High-frequency generation of

transgenic zebrafish which reliably express GFP in whole muscles or the whole body by using promoters of zebrafish origin. Dev Biol 192:289–299

58. Burket CT, Montgomery JE, Thummel R et al (2008) Generation and characterization of transgenic zebrafish lines using different ubiquitous promoters. Transgenic Res 17:265–279

59. Peterson SM, Freeman JL (2009) RNA isolation from embryonic zebrafish and cDNA synthesis for gene expression analysis. J Vis Exp (30):1–5

60. Westerfield M (2000) The zebrafish book. A guide for the laboratory use of zebrafish (Danio rerio), 4th edn. University of Oregon Press, Eugene

61. Linney E, Dobbs-McAuliffe B, Sajadi H, Malek RL (2004) Microarray gene expression profiling during the segmentation phase of zebrafish development. Comp Biochem Physiol C 138:351–362

62. Stainier DY, Lee RK, Fishman MC (1993) Cardiovascular development in the zebrafish. I. Myocardial fate map and heart tube formation. Development 119:31–40

63. Malone MH, Sciaky N, Stalheim L et al (2007) Laser-scanning velocimetry: a confocal microscopy method for quantitative measurement of cardiovascular performance in zebrafish embryos and larvae. BMC Biotechnol 7:40

64. Rüttinger S, Buschmann V, Krämer B et al (2008) Comparison and accuracy of methods to determine the confocal volume for quantitative fluorescence correlation spectroscopy. J Microsc 232:343–352

65. Kapusta P (2010) Absolute diffusion coefficients: compilation of reference data for FCS calibration. PicoQuant Appl Note 0–1

66. Hess ST, Webb WW (2002) Focal volume optics and experimental artifacts in confocal fluorescence correlation spectroscopy. Biophys J 83:2300–2317

67. Gregor I, Patra D, Enderlein J (2005) Optical saturation in fluorescence correlation spectroscopy under continuous-wave and pulsed excitation. ChemPhysChem 6:164–170

68. Nagy A, Wu J, Berland KM (2005) Characterizing observation volumes and the role of excitation saturation in one-photon fluorescence fluctuation spectroscopy. J Biomed Opt 10:44015

69. Buschmann V, Krämer B, Koberling F, et al (2009) Quantitative FCS: determination of the confocal volume by FCS and bead

scanning with the MicroTime 200. PicoQuant Appl Note 1–8

70. Sun G, Guo S-M, Teh C et al (2015) Bayesian model selection applied to the analysis of fluorescence correlation spectroscopy data of fluorescent proteins in vitro and in vivo. Anal Chem 87:4326–4333

71. Koppel DE (1974) Statistical accuracy in fluorescence correlation spectroscopy. Phys Rev A 10:1938–1945

72. Mütze J, Ohrt T, Schwille P (2011) Fluorescence correlation spectroscopy in vivo. Laser Photon Rev 5:52–67

73. Rigler R, Mets Ü, Widengren J, Kask P (1993) Fluorescence correlation spectroscopy with high count rate and low background: analysis of translational diffusion. Eur Biophys J 22:169–175

74. Müller P, Schwille P, Weidemann T (2014) PyCorrFit-generic data evaluation for fluorescence correlation spectroscopy. Bioinformatics 30:2532–2533

75. Sezgin E, Schwille P (2011) Fluorescence techniques to study lipid dynamics. Cold Spring Harb Perspect Biol 3:a009803

76. Kim SA, Heinze KG, Schwille P (2007) Fluorescence correlation spectroscopy in living cells. Nat Methods 4:963–973

77. Marquardt DW (1963) An algorithm for least-squares estimation of nonlinear parameters. J Soc Ind Appl Math 11:431–441

78. Kapusta P, Wahl M, Benda A, et al (2006) Fluorescence lifetime correlation spectroscopy. PicoQuant Appl Note 1–4

79. Wahl M (2014) Time-correlated single photon counting. PicoQuant Tech Note 1–14

80. Becker W (2017) The bh TCSPC handbook, 7th edn. Becker & Hickl GmbH, Berlin

81. Enderlein J, Gregor I (2005) Using fluorescence lifetime for discriminating detector afterpulsing in fluorescence-correlation spectroscopy. Rev Sci Instrum 76:33102

82. Kapusta P, Macháň R, Benda A, Hof M (2012) Fluorescence lifetime correlation spectroscopy (FLCS): concepts, applications and outlook. Int J Mol Sci 13:12890–12910

83. Meseth U, Wohland T, Rigler R, Vogel H (1999) Resolution of fluorescence correlation measurements. Biophys J 76:1619–1631

84. He J, Guo S-M, Bathe M (2012) Bayesian approach to the analysis of fluorescence correlation spectroscopy data I: Theory. Anal Chem 84:3871–3879

85. Guo S-M, He J, Monnier N et al (2012) Bayesian approach to the analysis of fluorescence correlation spectroscopy data II:

Application to simulated and in vitro data. Anal Chem 84:3880–3888

86. Kohl T, Haustein E, Schwille P (2005) Determining protease activity in vivo by fluorescence cross-correlation analysis. Biophys J 89:2770–2782

87. Chen Y, Müller JD, So PTC, Gratton E (1999) The photon counting histogram in fluorescence fluctuation spectroscopy. Biophys J 77:553–567

88. Macdonald PJ, Johnson J, Chen Y, Mueller JD (2014) Brightness experiments. In: Engelborghs Y, Visser AJWG (eds) Fluorescence spectroscopy and microscopy. Methods and protocols, Methods in molecular biology. Humana Press, Totowa, NJ, pp 699–718

89. Xu Q (1999) Microinjection into zebrafish embryos. In: Guille M (ed) Molecular methods in developmental biology. Xenopus and zebrafish, Methods in molecular biology. Humana Press, Totowa, NJ, pp 125–132

90. Holder N, Xu Q (1999) Microinjection of DNA, RNA, and protein into the fertilized zebrafish egg for analysis of gene function. In: Sharpe Ivor Mason PT (ed) Molecular embryology. Methods and protocols, Methods in molecular biology, pp 487–490

91. Kozak M (1986) Point mutations define a sequence flanking the AUG initiator codon that modulates translation by eukaryotic ribosomes. Cell 44:283–292

92. Rosen JN, Sweeney MF, Mably JD (2009) Microinjection of zebrafish embryos to analyze gene function. J Vis Exp 25:e1115

93. Benda A, Beneš M, Mareček V et al (2003) How to determine diffusion coefficients in planar phospholipid systems by confocal fluorescence correlation spectroscopy. Langmuir 19:4120–4126

94. Humpolíčková J, Gielen E, Benda A et al (2006) Probing diffusion laws within cellular membranes by Z-scan fluorescence correlation spectroscopy. Biophys J 91:L23–L25

95. Weiß K, Enderlein J (2012) Lipid diffusion within black lipid membranes measured with dual-focus fluorescence correlation spectroscopy. ChemPhysChem 13:990–1000

96. Heinemann F, Betaneli V, Thomas FA, Schwille P (2012) Quantifying lipid diffusion by fluorescence correlation spectroscopy: a critical treatise. Langmuir 28:13395–13404

97. Cranfill PJ, Sell BR, Baird MA et al (2016) Quantitative assessment of fluorescent proteins. Nat Methods 13:557–562

98. Macháň R, Foo YH, Wohland T (2016) On the equivalence of FCS and FRAP: simultaneous lipid membrane measurements. Biophys J 111:152–161

99. Landgraf D, Okumus B, Chien P et al (2012) Segregation of molecules at cell division reveals native protein localization. Nat Methods 9:480–482

100. Gahlmann A, Moerner WE (2014) Exploring bacterial cell biology with single-molecule tracking and super-resolution imaging. Nat Rev Microbiol 12:9–22

101. Katayama H, Yamamoto A, Mizushima N et al (2008) GFP-like proteins stably accumulate in lysosomes. Cell Struct Funct 33:1–12

Chapter 6

FRAP Analysis of Extracellular Diffusion in Zebrafish Embryos

Gary H. Soh and Patrick Müller

Abstract

Morphogens are signaling molecules that provide positional information to cells during development. They must move through embryonic tissues in order to coordinate patterning. The rate of a morphogen's movement through a tissue—its effective diffusivity—affects the morphogen's distribution and therefore influences patterning. Fluorescence recovery after photobleaching (FRAP) is a powerful method to measure the effective diffusion of molecules through cells and tissues, and has been successfully employed to examine morphogen mobility and gain important insights into embryogenesis. Here, we provide detailed protocols for FRAP assays in vitro and in living zebrafish embryos, and we explain how to analyze FRAP data using the open-source software PyFRAP to determine effective diffusion coefficients.

Key words Fluorescence Recovery After Photobleaching, FRAP, Zebrafish, Morphogens, Extracellular signaling molecules, Developmental biology

1 Introduction

Gradients of signaling molecules known as morphogens have long been proposed to direct the formation of tissues during embryogenesis by providing positional information [1, 2]. In the classical model of morphogen-mediated patterning, morphogens diffuse from localized morphogen-producing source cells into the surrounding tissue to form concentration gradients [3, 4]. Morphogen mobility has been demonstrated to be crucial in several patterning contexts. For example, hindering the mobility of the secreted morphogen Dpp abolishes its ability to pattern developing *Drosophila* wings [5]. Differences in the mobility of extracellular signaling molecules are also thought to be important in patterning processes. During zebrafish germ layer patterning, poorly diffusive Nodal signals form short-range gradients that induce and pattern endoderm and mesoderm, whereas Nodal signaling is antagonized by highly diffusive Leftys to allow ectoderm formation [6, 7]. Similarly, the diffusion of BMP and its antagonist Chordin is critical to

Julien Dubrulle (ed.), *Morphogen Gradients: Methods and Protocols*, Methods in Molecular Biology, vol. 1863,
https://doi.org/10.1007/978-1-4939-8772-6_6, © Springer Science+Business Media, LLC, part of Springer Nature 2018

pattern the dorsal–ventral axis during zebrafish development, and recent measurements of BMP and Chordin diffusion coefficients clarified the mechanism by which gradients of these proteins are established [7–9]. Therefore, measuring the mobility of signaling molecules is crucial to understand the dynamics of the biological processes they control.

Fluorescence Recovery After Photobleaching (FRAP) assays were developed more than 40 years ago [10, 11] and have been used extensively to assess the diffusion of molecules in living cells and tissues [12]. FRAP experiments measure effective diffusion, which takes into account diffusion hindrance by geometrical obstacles in the tissue as well as interactions with binding partners. Such interactions can significantly slow down the movement of diffusing molecules, and effective diffusion coefficients are typically smaller than the molecules' free unhindered diffusivities [2]. In FRAP assays, fluorescent molecules are bleached in a selected region by exposure to a strong laser pulse, and the movement of unbleached molecules from surrounding areas into the bleached region is recorded by quantitative time lapse microscopy (Fig. 1). The faster the fluorescent molecules diffuse, the faster the bleached region regains fluorescence (Fig. 1a, b). The average intensity within the bleached region over time is calculated from the images, and mathematical models of diffusive processes are fitted to the data to determine the diffusivity of the fluorescent molecules. Since the bleached region can be tailored to different samples, FRAP can be used to measure effective diffusion in complex tissues of many shapes and sizes, including those undergoing morphogen-mediated patterning [2, 6, 13–15].

Here, we provide detailed protocols and data analysis methods for FRAP experiments to measure the effective diffusivity of extracellular molecules in living zebrafish embryos at blastula stages. To ensure accurate measurements, two controls are crucial. First, a linear relationship between fluorophore concentration and intensity must be established because analysis methods assume that a change in fluorescence intensity is due to a proportional change in fluorophore concentration. Second, the ability of the experimental system to accurately determine diffusivities should be validated using an in vitro system with a defined geometry and fluorescent molecules of known diffusion coefficients.

Once the accuracy of the experimental setup has been confirmed, in vivo measurements of extracellular diffusion can be executed. In addition to explaining how to carry out FRAP experiments, we describe how to use the versatile, open-source software PyFRAP [16] to process raw images and compute diffusion coefficients. PyFRAP interpolates the first postbleach image onto a three-dimensional mesh approximating the shape of the sample for numerical simulations of fluorescence recovery. This allows the software to properly account for sample geometry and potential

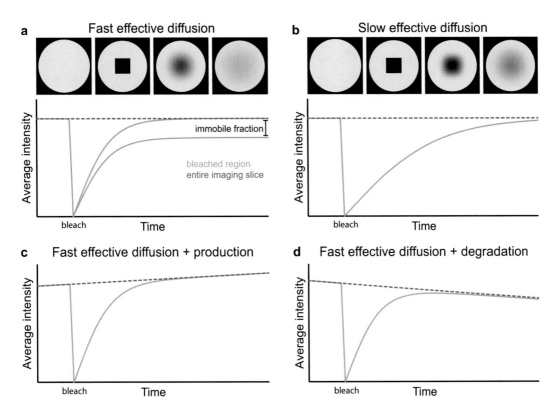

Fig. 1 Overview of Fluorescence Recovery After Photobleaching (FRAP) experiments to measure extracellular diffusion in zebrafish embryos. (**a**) Schematic of a FRAP experiment with a highly diffusive secreted molecule. A cross-section ("imaging slice") through the zebrafish blastoderm is shown. Fluorescence in embryos uniformly expressing a secreted signaling molecule (light green) is bleached (black square), and average fluorescence intensities in the bleached region (green line) and in the entire imaging slice (red dashed line) are monitored. Failure of the curves describing intensity changes in the bleached region and in the entire imaging slice to converge at long time scales is indicative of an immobile fraction that cannot recover by diffusion [6]. (**b**) Schematic of a FRAP experiment with a poorly diffusive molecule. The recovery is slower than in panel (**a**). (**c** and **d**) Schematic of FRAP experiments with a highly diffusive molecule that continues to be produced (**c**) or that is significantly degraded (**d**) throughout the experiment

experimental artifacts discussed in detail below. The solution from the numerical simulation is then fitted to the experimental data. Additional reaction kinetics such as production or degradation (Fig. 1c, d) can also be taken into account to compute accurate effective diffusion coefficients.

2 Materials

2.1 Molecules for Fluorescent Samples

1. Recombinant green fluorescent protein (GFP, diffusion coefficient $D = 96 \pm 2 \ \mu m^2/s$ [16]).

2. Solid bovine serum albumin (BSA).

3. Fluorescein-coupled dextrans with known diffusivities and with excitation–emission spectra similar to the fluorescent protein to be analyzed (*see* **Note 1**).

4. mRNA encoding a fluorescently tagged secreted protein (*see* **Note 2**).

2.2 Embryo Manipulation

1. Glass pipettes with flame-polished tips (*see* **Note 3**).

2. Pipette pump (*see* **Note 3**).

3. Dissection needle or eyelash glued to a glass pipette (*see* **Note 4**).

4. E3 medium: 5.03 mM NaCl, 0.17 mM KCl, 0.33 mM $CaCl_2$, 0.33 mM $MgSO_4$, 0.1% (w/v) methylene blue [17].

5. 6-well tissue culture plates coated with 2% agarose in E3 medium.

6. Microinjection apparatus with a micrometer calibration slide (*see* **Note 5**).

7. 5 mg/mL pronase dissolved in E3 medium.

8. Microinjection dish to hold embryos during microinjection (described in The Zebrafish Book [18], https://zfin.org/zf_info/zfbook/chapt5/5.1.html).

9. Binocular dissection stereomicroscope for manipulating and orienting embryos.

10. Small glass petri dish for pronase-mediated embryo dechorionation. The glass petri dish has to be small enough to fit into a 200 mL glass beaker.

11. 200 mL glass beaker for washing dechorionated embryos.

12. Incubator set to a temperature of 28 °C to incubate zebrafish embryos.

2.3 Imaging

1. 35 mm uncoated glass bottom microscopy dishes, No. 1.5 Coverslip, 10 mm glass diameter.

2. Rectangular coverslips, 60 mm × 24 mm, No. 1.5, 0.17 mm thickness.

3. Square coverslips, 24 mm × 24 mm.

4. Transparent plastic block (*see* **Note 6**), approximately 2.7 cm × 1.5 cm × 3 mm (Fig. 2). Create a small well in the center ~700 μm in diameter and ~100 μm in depth with a dental drill such as the Gates Glidden drill #2.

5. Heavy mineral oil.

6. E3 medium without methylene blue: 5.03 mM NaCl, 0.17 mM KCl, 0.33 mM $CaCl_2$, 0.33 mM $MgSO_4$ [17].

7. Molten 1% low melting point agarose in E3 medium in a 1.5 mL microcentrifuge tube.

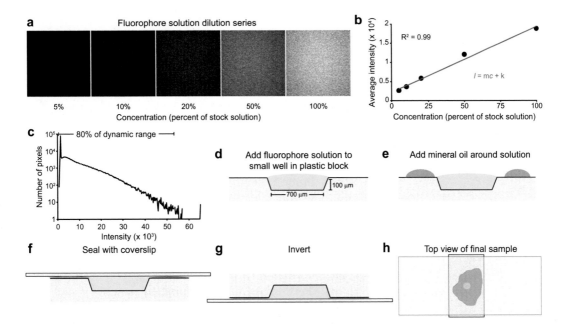

Fig. 2 FRAP control experiments. (**a–c**) Dilution series to determine whether the relationship between fluorescence intensity and fluorophore concentration is linear. (**a**) A dilution series (5%, 10%, 20%, 50%, 100%) is made from a fluorescent dextran solution that produces similar intensity as in vivo samples with the selected confocal microscope settings for in vivo FRAP. The images show fluorescence intensities of a 70 kDa fluorescein-labeled dextran dilution series made from a 25 µg/mL stock solution (= 100%, corresponding to a concentration of 360 nM). (**b**) The average intensities of the images are then measured (e.g., using Fiji [19]) and plotted against the fluorophore concentration. This should yield a linear fit (red) with an R^2 value of at least 0.9. (**c**) The optimal signal of the stock solution should fill about 75–80% of the 16-bit dynamic range of the confocal microscope settings with similar intensity as the in vivo sample. (**d–h**) In vitro FRAP experiments using molecules with previously measured diffusion coefficients. (**d**) Well geometry for in vitro experiments and FRAP sample preparation. First, the fluorophore solution is pipetted into the plastic block well. (**e**) Mineral oil is pipetted around the well, so that it completely surrounds the well. (**f**) A rectangular coverslip is then placed over the well. The fluorescent solution must remain completely surrounded by the mineral oil after the coverslip has been placed. (**g**) Invert the sample for imaging on an inverted confocal microscope. (**h**) Schematic of the final sample preparation

8. Heat block at 42 °C for molten agarose (place next to dissection stereomicroscope).

9. Inverted confocal microscope system with lasers and filters appropriate for the fluorophore to be imaged (*see* **Note 7**).

2.4 Analysis Software

1. PyFRAP [16]. Download the free Python-based software from https://mueller-lab.github.io/PyFRAP.

2. Fiji [19]. Download the commonly used free software for biological image processing from https://fiji.sc/#download.

3   Methods

All procedures are carried out at room temperature.

3.1   Control Experiments

3.1.1   Test of Confocal Microscope Settings

Before executing a FRAP experiment, it is crucial to confirm a linear relationship between the detected intensity and concentration of a fluorescent solution over a wide range of input concentrations (Fig. 2a–c). This ensures that the intensity measured by the microscope corresponds to the concentration of the fluorophore.

1. Prepare a dilution series of a fluorescent dextran with excitation-emission spectra similar to the fluorescent protein to be analyzed at concentrations of 90, 180, 360, and 720 nM.

2. Pipette 2 µL of the dextran solution onto the center of a rectangular coverslip, and then place a square coverslip over it.

3. Image the solutions with the confocal settings to be used for the in vivo FRAP experiments. Important parameters are the laser power, the detector gain, and offset. These settings will depend on the fluorescence intensity of the sample as well as the microscope (*see* **step 4** in Subheading 3.2.2).

4. Determine the concentration that yields a signal filling about 75–80% of the 16-bit dynamic range of the selected confocal microscope settings (Fig. 2c) with similar intensity as the in vivo sample (*see* Subheading 3.2).

5. Using that concentration, prepare a 2×, 5×, 10×, and 20× dilution series of the solution (Fig. 2a). For example, if the concentration determined in **steps 1–4** is 360 nM, prepare 180, 72, 36, and 18 nM dilutions.

6. Image the solutions generated in **step 5** (Fig. 2a) and measure their average fluorescence intensities. Fluorescence intensity can be measured with Fiji [19]. Open the image and go to Edit → Selection → Select All, then Analyze → Measure.

7. Plot average intensity against concentration and fit a linear trend line to the data using the equation $I = mc + k$, where I is the fluorescence intensity, m is the slope, c is the concentration, and k the background fluorescence (Fig. 2b). This can be done for example in Excel: Organize the data in columns, then select Insert → Scatter. Go to Layout → Trendline → Format trendline, and select "*Linear*" and "*Display R-squared value on chart.*" Generally, an R^2 value larger than 0.9 indicates a good fit.

8. If the linear equation fits the data well (R^2 value >0.9), then the imaging settings are appropriate. If the intensity values plateau at high or low concentrations, the signal is too strong or too weak, respectively. To address this problem, adjust the microscope settings or change the amount of injected mRNA for in vivo FRAP experiments (*see* Subheading 3.2).

*3.1.2 In Vitro FRAP
Sample Preparation*

After establishing a linear relationship between fluorophore concentration and intensity, in vitro control experiments using molecules with known diffusion coefficients (*see* **Note 1**) in a defined geometry can be used to validate the experimental setup (Fig. 2d–h). Incorrect diffusion coefficients may indicate problems with the experimental setup.

1. Prepare aqueous solutions containing fluorescent molecules with known diffusion coefficients. When using a fluorescent protein (such as 4 μM recombinant GFP, *see* Subheading 2.1), add 5% bovine serum albumin (BSA, *see* **Note 8**). The BSA blocks potential protein binding sites in the plastic well. 100–500 nM solutions of fluorescein-labeled dextrans of different sizes can also be used to carry out the positive controls (*see* **Note 1**).

2. Place the plastic block with a small well on a dissection stereomicroscope.

3. Transfer 2 μL of fluorescent solution into the small well until it slightly flows over (Fig. 2d).

4. Pipette a few microliters of mineral oil in a donut shape around the small well, such that the oil completely surrounds the fluorophore solution (Fig. 2e). Leave a small gap between the mineral oil and the well (*see* **Note 9**).

5. Place a rectangular coverslip over the solution (Fig. 2f), such that the sides of the plastic block protrude (Fig. 2h) and it can be manually transported without moving the coverslip.

6. Carefully flip the plastic block over, so that the sample can be imaged on an inverted confocal microscope (Fig. 2g). The mineral oil should cause the coverslip and the plastic block to stick together via capillary action. If they fall apart, more mineral oil is needed.

7. Place the plastic block on the confocal microscope with the coverslip facing the objective (Fig. 2h).

8. Add immersion solution onto the objective. We use an LD LCI Plan-Apochromat 25×/0.8 NA Imm Korr DIC objective (Zeiss) and immersion oil (Immersol™ W, $n = 1.334$ at 23 °C, Zeiss).

9. *See* Subheading 3.2.2 for instructions on performing the FRAP experiment.

3.2 FRAP Measurements of Secreted Molecules In Vivo

3.2.1 Zebrafish Embryo Sample Preparation

1. Inject mRNA encoding the fluorescently tagged secreted molecule directly into the cell of zebrafish embryos at the one-cell stage, or inject purified recombinant fluorescent protein into the extracellular space of zebrafish embryos around high or dome stage [6, 20, 21]. The amount to be injected has to be empirically determined, but a good starting point is 30 pg of mRNA or 500 pg of fluorescent protein.

2. Incubate the injected zebrafish embryos at 28 °C until they reach oblong stage.

3. Proteolytically dechorionate the embryos in bulk by incubation in 10 mL of 1 mg/mL pronase in a glass petri dish for 7–10 min at room temperature (*see* **Note 10**). A movie of this procedure has been published recently [20]. Embryos can also be proteolytically dechorionated immediately before or after injection. Alternatively, embryos can be dechorionated manually using fine forceps.

4. Carefully pour the embryos into a 200 mL glass beaker filled with E3 medium.

5. Wait for the embryos to settle at the bottom of the glass beaker, and decant approximately 80% of the E3 medium without pouring out the embryos.

6. Rinse the embryos by pouring E3 medium into the side of the glass beaker until the beaker is nearly filled.

7. Repeat **steps 5** and **6**, then decant 80% of the E3 medium. This will cause the chorions to fall apart and release the embryos.

8. Using a flame-polished glass pipette connected to a pipette pump, transfer the embryos into a glass petri dish or an agarose-coated 6-well plate filled with E3 medium.

9. Using a glass pipette with a flame-polished tip, transfer a single zebrafish embryo into molten 1% low melting point agarose (LMA) incubated at 42 °C in a heat block (Fig. 3a) without carrying over a large amount of E3 medium into the LMA.

10. Quickly remove excess medium from the glass pipette.

11. With the same glass pipette, carefully pick up the zebrafish embryo along with some molten LMA (Fig. 3a) and place it on a glass bottom microscopy dish (Fig. 3a′).

12. Using a dissection needle (or an eyelash glued to a glass pipette) and a dissection stereomicroscope, carefully rotate the embryo so that the animal pole is pressed onto the glass bottom (Fig. 3a″). Ensure that the embryo maintains this position until the LMA solidifies (Fig. 3a‴).

13. Once the LMA is fully solidified, pour E3 medium without methylene blue onto the glass bottom dish to keep the agarose hydrated.

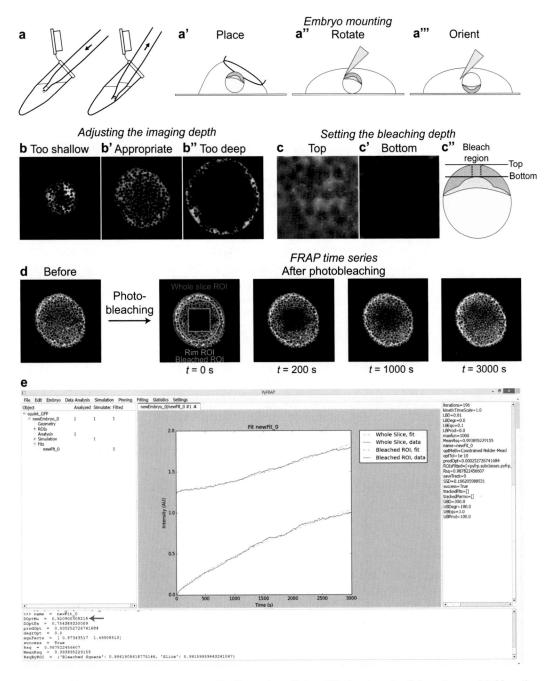

Fig. 3 FRAP experiments to measure effective extracellular diffusion in zebrafish embryos. (**a**) Mounting zebrafish embryos for FRAP experiments using an inverted confocal microscope. Use a glass pipette to submerge an embryo in molten 42 °C-warm low melting point agarose (LMA), then withdraw it along with some LMA. (**a′**) Place the embryo with molten LMA on a glass bottom microscopy dish. (**a″**) Position the embryo with a dissection needle, such that the animal pole (gray) is pressed against the glass. (**a‴**) Maintain the embryo in this position until the agarose solidifies. (**b** and **c**) Identifying optimal in vivo FRAP imaging settings. (**b**) Adjusting the imaging depth in a zebrafish embryo around dome stage expressing the fluorescently tagged signaling molecule Squint-GFP [6]. The imaging depth of the sample must not be too shallow

Some microscope software has built-in FRAP settings to automatically execute the bleaching and subsequent image acquisition. For maximum flexibility independent of the microscope system, here we describe how to perform FRAP experiments without automated FRAP settings using a Zeiss LSM 780 confocal microscope.

1. Use the lowest magnification objective to find the sample and bring it into focus.

2. Switch to the objective with the desired magnification. We use a $25\times$ objective because an entire zebrafish embryo fits in the field of view, but lower or higher magnifications can also be used depending on the size of the sample.

3. Adjust the confocal microscope settings by changing the laser power and detector gain. Start with a detector gain of 800 using the shortest pixel dwell time and laser power of 0.1%, and adjust the laser power (to a maximum of 5% to avoid photobleaching) until a clear image with minimal speckle is seen. The signal should fill about 75–80% of the 16-bit dynamic range (Fig. 2c). If a laser power of 5% is insufficient, start increasing the pixel dwell time. Note that a high pixel dwell time will increase the time needed to acquire each image, which will be problematic if fluorescence recovery is fast. If the image is still too dim, increase the amount of mRNA injected for the next round of experiments. Before the first FRAP experiment, perform the control experiments described in Subheading 3.1 to ensure that the fluorescence output with the selected imaging settings is in the middle of the linear range (Fig. 2b).

4. Determine the z-plane in which fluorescence recovery will be imaged. For the in vitro samples, a good position is approximately halfway down the well ($\sim50\ \mu m$). The optimal imaging

Fig. 3 (continued) (**b**) or too deep (**b″**). An optimal compromise (**b′**) is an imaging plane in which the center remains clearly visible and the sample fills most of the field of view. (**c–c″**) Adjusting the bleaching conditions. When performing the bleaching z-scan, start from the top of the sample (**c**) and move the focus deep into the tissue until almost no fluorescence can be detected (**c′**). The schematic in (**c″**) indicates the top and bottom bleaching positions in a blastula stage zebrafish embryo. (**d** and **e**) Example of expected results. FRAP analysis of Squint-GFP diffusion in a zebrafish embryo around dome stage previously injected at the one-cell stage with 30 pg Squint-GFP-encoding mRNA [6]. (**d**) Raw images from a FRAP time series before and after photobleaching. The regions of interest (ROI) for selection in PyFRAP are highlighted in the first postbleach image. PyFRAP uses these ROIs to identify the entire sample ("Whole slice ROI," red) and the bleached region ("Bleached ROI, green), and to define an annulus ("Rim ROI," orange) for the estimation of fluorophore concentration outside of the field of view and in regions below the imaging plane. (**e**) PyFRAP GUI output from analysis of the FRAP experiment in (**d**). The curves describing the recovery in the bleached region (green) and the fluorescence intensity changes in the entire imaging slice (red) do not converge due to a significant immobile fraction (compare to Fig. 1a). The diffusion coefficient of Squint-GFP is displayed in the GUI console at the bottom (red arrow)

plane for zebrafish embryos must be deep enough to cover a large portion of the embryo but shallow enough to ensure homogenous illumination of the imaging plane (Fig. 3b–b″, *see* **Note 11**).

5. Set this *z*-position as the reference point. The microscope software must be able to rapidly return to this position after bleaching.

6. Determine the depth of the *z*-position relative to the surface of the sample. Record this information for subsequent data analysis (*see* Subheading 3.2.3).

7. Acquire two images to provide a record of the sample before bleaching; use the same time interval between images as in **step 13**.

8. Zoom in so that the region to be bleached completely fills the field of view (*see* **Note 12**). Assuming the sample fills the image at 1× zoom, bleach by zooming in using a factor of around 2.5×.

9. Move to the surface of the sample. Set this as the first *z*-position (Fig. 3c).

10. Move deep into the tissue, ideally the lower end of the sample. Set this as the last *z*-position (Fig. 3c′–c″). The z-stack should cover about 80–100 μm.

11. Bleach the volume by acquiring a z-stack at highest laser power from the first to the last *z*-position. A longer pixel dwell time and a larger number of scanned slices can be used to compensate for low laser power, but bleaching should not take longer than a few minutes. Ideally, the bleached area should be at least 30% dimmer after bleaching. If bleaching is insufficient, use a stronger laser (e.g., an argon laser with a total power output of 250 mW across its emission wavelengths) or a more bleachable fluorophore (e.g., fluorescein is more bleachable than Alexa 488, and Dendra2 is more bleachable than GFP).

12. Immediately after bleaching, return to the reference *z*-plane set in **step 5**, e.g., using the "*return home*" option in the Zeiss Zen software (*see* **Note 13**). Set the laser power to the percentage used for the prebleaching image and use the same imaging conditions and zoom factor as in **step 7**.

13. Acquire a time series of 300 images (see snapshots before and after bleaching in Fig. 3d). The appropriate imaging time interval depends on the mobility of the fluorescent molecule—faster molecules require shorter intervals between images—and must be determined empirically. One frame per second is generally a good timescale to try initially (*see* **Notes 14–16**).

14. Save the images. To facilitate data analysis using PyFRAP (see below), add the identifiers "*pre*," "*bleach*," and "*post*" to the prebleach, bleach, and postbleach image series, respectively.

3.2.3 FRAP Data Analysis   PyFRAP [16] is an open-source software for analyzing FRAP data sets on Windows, MacOS X, and Linux operating systems (Fig. 3e). Below, we provide basic instructions to use the software with default parameters. For a full description of PyFRAP's versatile data analysis options, please refer to the GitHub wiki (https:// github.com/mueller-lab/PyFRAP/wiki).

1. Download PyFRAP from https://mueller-lab.github.io/ PyFRAP.

2. Follow the instructions for installing PyFRAP from https:// github.com/mueller-lab/PyFRAP/wiki/Installation#short.

3. Open the GUI by double-clicking on runPyFRAP.bat (Windows), runPyFRAP.command (MacOS X), or runPyFRAP.sh (Linux).

4. Go to → File → New Molecule and enter a name for the molecule.

5. Go to → Edit → PyFRAP Wizard. This will open a wizard, which systematically guides the user through the analysis steps.

6. Select "*Create embryo from microscope data*" if the microscope saves the data acquired in Subheading 3.2.2 in .czi or .lsm file formats.

   OPTIONAL: It is also possible to select "*Create embryo from already prepared data*" to load a TIFF stack containing the FRAP recovery series (see https://github.com/mueller-lab/ PyFRAP/wiki/FirstSteps for details).

7. Select "*Change*" and go to the folders containing the prebleach and postbleach files.

8. Fill in the required parameters. Essential parameters are the resolution (μm/pixel), imaging depth (μm), and frame interval (s). Once finished, select "*Done.*" When using .czi or .lsm formats, resolution and frame intervals are recognized automatically.

9. The "Select Geometry" popup will appear. Select the option that best suits the sample. PyFRAP provides several preloaded sample geometries. The zebraFishDomeStage geometry is appropriate for zebrafish embryo samples, while the cylinder or cone geometries suit the in vitro samples. Select "*Done.*"

10. In the next dialog box, several parameters for the geometry will appear. Values for these parameters will be automatically populated by the software during subsequent image analysis steps (*see* below). Select "*Done.*"

11. A new dialog will pop up with three different options for ROI (region of interest) creation. These ROIs will be used to identify the entire sample as well as the bleached region. Select "*Use ROI Wizard.*"

12. First, create an ROI containing the entire sample ("Whole Slice ROI," Fig. 3d). Use the drop-down menu to choose the appropriate shape for the ROIs. The default is a circle (radial-Slice). Select "*Create ROI.*" Draw the ROI by left-clicking on the picture or adjusting the parameters in the dialog box. The ROI can be moved using the arrow keys, and the diameter of the ROI can be increased/decreased by pushing Ctrl and up/down arrow keys at the same time. Right-click on the image to cancel the selection and redraw the ROI if necessary. Select "*Done.*"

13. Next, create an ROI containing the bleached area ("Bleached ROI," Fig. 3d). Use the drop-down menu to choose the appropriate shape for the ROIs. The default is a square, but it can also be changed to other polygons or a circle. Select "*Create ROI*" and draw the ROI.

14. Create an ROI for the rim of the sample ("Rim ROI," Fig. 3d). An annulus covering the rim of the sample is used to estimate the fluorophore concentration outside of the field of view and in regions below the imaging plane, and the selected rim radius should be approximately 80% of the slice ROI. Select "*Create ROI*" and draw the ROI.

15. The "Edit Geometry" popup will appear again. Select the first option "*Grab from ROI.*" Select the ROI that contains the entire sample (This should be named "*Slice.*"), then select "*Done.*" Do the same for the second "*Grab from ROI,*" click "*Slice,*" then select "*Done.*"

16. The "Mesh Settings" popup will appear. Change Element Size (px) to 17. Select "*Done.*" This will generate a mesh with the shape of the sample to discretize the data for numerical simulation and will take several minutes to finish. Increasing the mesh element size value can drastically increase the computation time, but a finer mesh may be necessary depending on the imaging data.

17. The "Analysis Settings" popup will appear. Select "*Done*" to use the default settings (*see* **Note 17**).

18. The "Simulation Settings" popup will appear. Enter a value of 1000 for the simulation timesteps. Select "*Done.*" This will start a numerical simulation of the fluorescence recovery based on the fluorescence distribution of the first time point. Depending on the settings, the simulation can take around 15–30 min to finish.

19. The "Fit Settings" popup will appear. Select "*Add*" to add the "Slice" and "Bleached Square" ROIs. If you are working with in vivo samples, tick "*Fit Production*" and "*Fit Degradation*" (*see* **Note 18**). If production and degradation parameters are known [6, 9, 20, 22], they can also be directly entered here. Select "*Done.*"

20. The "Ideal Pinning" popup will appear. Select "*Done.*" This will fit the simulated recovery curve to the measured recovery data.

21. Go to → Fitting → Print fit results to display the fitted curve and other results of the analysis.

22. The diffusion coefficient is given by DOptMu in $\mu m^2/s$ in the text output of the console at the bottom (Fig. 3e) and in the panel on the right of the PyFRAP GUI.

4 Notes

1. The following fluorescein-coupled dextrans from Thermo Fisher have well characterized diffusion coefficients [16]: 70 kDa ($D = 27 \pm 5$ $\mu m^2/s$)—Catalog No. D1823, 40 kDa ($D = 45 \pm 11$ $\mu m^2/s$)—Catalog No. D1844, 10 kDa ($D = 83 \pm 8$ $\mu m^2/s$)—Catalog No. D1821, 3 kDa ($D = 170 \pm 22$ $\mu m^2/s$)—Catalog No. D3305.

2. mRNA can be produced by in vitro transcription from suitable purified and linearized plasmid DNA using the mMESSAGE mMACHINE kit from Thermo Fisher.

3. We recommend using Fisherbrand Disposable Borosilicate Glass Pasteur Pipets (Catalog No. 13-678-20A) and 10 mL Bel-Art SP Scienceware Pipette Pump Pipetters (Catalog No. 13-683C) to handle zebrafish embryos. The glass pipettes can be flame-polished to prevent the tip from scratching the embryos.

4. Sometimes new dissection needles are too sharp and will puncture the embryo. They can be blunted by grinding the needle tip repeatedly on a hard stone until the tip becomes flat. Take care not to bend the tip while blunting it.

5. There are various types of microinjection setups and approaches (e.g., the micromanipulation system by Narishige, as well as micropipette holders and pressure injection regulators from World Precision Instruments). Detailed protocols for microinjection have been published previously [23, 24]. The injection volume has to be adjusted by measuring the injection drop in mineral oil on a micrometer calibration slide.

6. A plastic block such as the PLEXIGLAS XT (allround) Clear 0A000 GT with 3 mm thickness from Evonik Industries is appropriate.

7. We use a Zeiss LSM780 confocal microscope with a GaAsP array. The filter sets and lasers depend on the desired fluorophore used. A range of 494–542 nm is useful to image GFP and fluorescein. A pinhole diameter yielding a slice thickness of 5 μm works well for in vitro and in vivo FRAP experiments.

8. Aqueous solutions of fluorescent proteins such as recombinant GFP must contain 5% bovine serum albumin to prevent GFP from sticking to the plastic block walls (similar to blocking a membrane during Western blotting).

9. It is crucial that the mineral oil completely surrounds the aqueous sample when the coverslip is placed over the plastic block. If not, the very small volume of aqueous sample will quickly evaporate.

10. Dechorionated embryos are easily damaged by contact with plastic or air. Keep dechorionated embryos in glass dishes or plastic dishes coated with 2% agarose.

11. Zebrafish embryos are spherical, and signal intensity will diminish more rapidly in the center than at the sides since the tissue is thicker in the center. Therefore, image at an optimal depth where the signal intensities in the center and on the sides are similar (Fig. 3b′).

12. The size of the bleached area must not be too small or too big. If it is too small, the signal will recover too quickly to be properly measured. If it is too big, bleaching will take too long, and recovery might start before imaging. While the software PyFRAP accounts for the finite sample geometries in FRAP experiments, an overly large bleaching area will greatly affect analysis with other software packages, which assume that the fluorescent pool is infinitely large.

13. Image acquisition must start within seconds after bleaching to ensure a significant perturbation in the fluorescence profile and an extensive difference between the unbleached image and the first postbleach image.

14. Molecules with large diffusion coefficients should be imaged at a faster frame rate (e.g., GFP with a diffusion coefficient of ~100 $\mu m^2/s$ in vitro should be imaged at a rate of one frame per second for 5 min), while images of molecules with a smaller diffusion coefficient can be recorded with a slower frame rate (e.g., Squint-GFP with a diffusion coefficient of ~3 $\mu m^2/s$ in vivo can be imaged around one frame every 10 s for 50 min [6]). However, production and degradation of the fluorescent protein can significantly contribute to changes in

fluorescence intensity over longer imaging times (Fig. 1c, d) and have to be taken into account for data analysis (also *see* **Note 18**). In case of high pixel dwell times, it is important to keep in mind that the time needed to record each frame must be less than the time step between each frame to ensure that the frame rate reported by the imaging software is accurate.

15. The sample could experience inadvertent photobleaching during imaging of the recovery. PyFRAP can correct for bleaching during imaging (also *see* **Note 18**), but this should be minimized if possible. To assess potential photobleaching, image the sample for 300 frames with the shortest possible interval and compare the first and last images. If the average intensity of the last image is lower than the first, bleaching is an issue and the laser power should be reduced. Alternatively, acquire 300 images rapidly and determine whether the data series shows differences compared to an experiment with only 30 images taken over the same time period.

16. Potential artifacts in FRAP experiments might arise from inhomogeneous bleaching and overexpression. PyFRAP can correct for inhomogeneous bleaching since the software simulates fluorescence recovery using the first image of the potentially inhomogeneously bleached sample as the initial condition. In some instances, overexpression artifacts can be uncovered by repeating FRAP experiments with different amounts of injected mRNA.

17. PyFRAP offers several methods to correct imaging artifacts for FRAP data. If images are noisy, a Gaussian or median filter can be used for smoothing and denoising. For some microscope settings, the center of the sample is more strongly illuminated than the peripheral regions. This inhomogeneous illumination can be accounted for in PyFRAP using a prebleach image or a "flattening" data set [16]. A prebleach image can be used for in vitro samples and requires the input of a sample image before bleaching. A flattening data set can be used for any sample type. This is a homogenously fluorescent sample as described in Subheading 3.1.1 (**steps 1–4**), which can be used to detect and correct for inhomogeneous illumination.

18. During the recovery process, production and degradation/clearance of the fluorescent protein in living samples can additionally influence fluorescence recovery (Fig. 1c, d). This effect becomes significant when imaging over a long time (e.g., around 50 min). PyFRAP can account for these processes and output the corrected diffusion coefficient if the options "*Fit Production*" and "*Fit Degradation*" are selected. Select "*Fit Degradation*" to take bleaching into account if this is an issue during recovery and if this cannot be prevented by changing the imaging conditions.

Acknowledgments

We thank Katherine W. Rogers, Alexander Bläßle, David Mörsdorf, and Hannes Preiß for useful discussions. This work was supported by the Max Planck Society and ERC Starting Grant 637840.

References

1. Wolpert L (1969) Positional information and the spatial pattern of cellular differentiation. J Theor Biol 25(1):1–47. https://doi.org/10.1016/S0022-5193(69)80016-0

2. Müller P et al (2013) Morphogen transport. Development 140(8):1621–1638. https://doi.org/10.1242/dev.083519

3. Crick F (1970) Diffusion in embryogenesis. Nature 225(5231):420–422. https://doi.org/10.1038/225671b0

4. Rogers KW, Schier AF (2011) Morphogen gradients: from generation to interpretation. Annu Rev Cell Dev Biol 27:377–407. https://doi.org/10.1146/annurev-cellbio-092910-154148

5. Harmansa S et al (2015) Dpp spreading is required for medial but not for lateral wing disc growth. Nature 527(7578):317–322. https://doi.org/10.1038/nature15712

6. Müller P et al (2012) Differential diffusivity of Nodal and Lefty underlies a reaction-diffusion patterning system. Science 336 (6082):721–724. https://doi.org/10.1126/science.1221920

7. Rogers KW, Müller P (2018) Nodal and BMP dispersal during early zebrafish development. Developmental Biology pii:S0012-1606(17) 30925-9. http://doi.org/10.1016/j.ydbio.2018.04.002

8. Zinski J et al (2017) Systems biology derived source-sink mechanism of BMP gradient formation. eLife 6:e22199. https://doi.org/10.7554/eLife.22199

9. Pomreinke AP et al (2017) Dynamics of BMP signaling and distribution during zebrafish dorsal-ventral patterning. eLife 6:e25861. https://doi.org/10.7554/eLife.25861

10. Poo MM, Cone RA (1973) Lateral diffusion of rhodopsin in Necturus rods. Exp Eye Res 17 (6):503–510. https://doi.org/10.1016/0014-4835(73)90079-1

11. Liebman PA, Entine G (1974) Lateral diffusion of visual pigment in photoreceptor disk membranes. Science 185(4149):457–459. http://doi.org/10.1126/science.185.4149.457

12. Lorén N et al (2015) Fluorescence recovery after photobleaching in material and life sciences: putting theory into practice. Q Rev Biophys 48(3):323–387. https://doi.org/10.1017/S0033583515000013

13. Kicheva A et al (2007) Kinetics of morphogen gradient formation. Science 315 (5811):521–525. https://doi.org/10.1126/science.1135774

14. Gregor T et al (2007) Stability and nuclear dynamics of the Bicoid morphogen gradient. Cell 130(1):141–152. https://doi.org/10.1016/j.cell.2007.05.026

15. Umulis DM, Othmer HG (2012) The importance of geometry in mathematical models of developing systems. Curr Opin Genet Dev 22 (6):547–552. https://doi.org/10.1016/j.gde.2012.09.007

16. Bläßle A et al (2018) Quantitative diffusion measurements using the open-source software PyFRAP. Nature Communications 9(1):1582. http://doi.org/10.1038/s41467-018-03975-6

17. Nüsslein-Volhard C, Dahm R (2002) Zebrafish: a practical approach. The practical approach series, vol 261, 1st edn. Oxford University Press, Oxford

18. Westerfield M (2007) The zebrafish book: a guide for the laboratory use of zebrafish (Danio rerio). University of Oregon Press, Oregon

19. Schindelin J et al (2012) Fiji: an open-source platform for biological-image analysis. Nat Methods 9(7):676–682. https://doi.org/10.1038/nmeth.2019

20. Rogers KW et al (2015) Measuring protein stability in living zebrafish embryos using Fluorescence Decay After Photoconversion (FDAP). J Vis Exp 95:e52266. https://doi.org/10.3791/52266

21. Kimmel CB et al (1995) Stages of embryonic development of the zebrafish. Dev Dyn 203 (3):253–310. http://doi.org/10.1002/aja.1002030302

22. Bläßle A, Müller P (2015) PyFDAP: automated analysis of Fluorescence Decay After

Photoconversion (FDAP) experiments. Bioinformatics 31(6):972–974. https://doi.org/10.1093/bioinformatics/btu735

23. Xu Q (1999) Microinjection into zebrafish embryos. In: Guille M (ed) Molecular methods in developmental biology: xenopus and zebrafish. Humana Press, Totowa, NJ, pp 125–132.

https://doi.org/10.1385/1-59259-678-9:125

24. Rosen JN, Sweeney MF, Mably JD (2009) Microinjection of zebrafish embryos to analyze gene function. J Vis Exp 25:e1115. https://doi.org/10.3791/1115

Chapter 7

Generation of Ectopic Morphogen Gradients in the Zebrafish Blastula

Maraysa de Olivera-Melo, Peng-Fei Xu, Nathalie Houssin, Bernard Thisse, and Christine Thisse

Abstract

In the zebrafish embryo, cells of the early blastula animal pole are all equivalent and are fully pluripotent until the midblastula transition that occurs at the tenth cell cycle (512 to 1K cells). This naive territory of the embryo is therefore perfectly suited to assay for morphogen activity. Here we describe different methods to generate ectopic morphogen gradients, either in vivo at the animal pole of the embryo, or in vitro in animal pole explants or in aggregates of animal pole blastomeres (also named embryoid bodies). These methods include injection of mRNA coding for growth factor(s) into animal pole blastomere(s), transplantation of growth factor(s) secreting cells, implantation of beads coated with purified protein(s), and various combinations of these different approaches. Our comparative study reveals that all these methods allow to generate morphogen gradient(s) that are able to induce, both in vivo and in vitro, the formation of a well-patterned embryonic axis.

Key words Morphogen, Gradient, Zebrafish, Blastula, Injection, Cell transplantation, Bead implantation, Embryoid bodies

1 Introduction

The first 2½ h of zebrafish embryo development is characterized by a rapid succession of cell divisions. During this phase, the zygotic genome is mostly silent and the development depends almost exclusively on maternal products synthesized during oogenesis and stored in the egg. With the exception of a small number of dorsomarginal blastomeres, all cells of the embryo are equivalent and pluripotent until the midblastula transition that starts around cycle 10 (the 512 to the 1K-cell stages; 2.75–3 h post fertilization) [1]. The first evidence of morphogen signaling that will pattern the embryo is the accumulation of β-catenin in the nuclei of a small cluster of marginal blastomeres [2]. This stimulation of the canonical β-catenin signaling pathway is the result of the activity of Wnt8a, the dorsal determinant of the zebrafish embryo, whose mRNA is

Julien Dubrulle (ed.), *Morphogen Gradients: Methods and Protocols*, Methods in Molecular Biology, vol. 1863,
https://doi.org/10.1007/978-1-4939-8772-6_7, © Springer Science+Business Media, LLC, part of Springer Nature 2018

stored at the vegetal pole of the oocyte. During early cleavage stages, Wnt8a mRNA migrates asymmetrically toward the animal pole and activates the canonical Wnt/β-catenin signaling pathway in marginal blastomeres, breaking the initial radial symmetry along the vegetal–animal axis, and defining the dorsal side of the embryo [3]. Therefore, until the midblastula transition, blastomeres that do not display accumulation of beta-catenin in their nucleus are both pluripotent and naive and share properties with embryonic stem cells. This is in clear contrast with cells of the animal pole of amphibian blastula that display β-catenin in their nucleus with decreasing amounts from dorsal to ventral. Whereas Xenopus animal caps have been widely used to assay the molecular activity of a variety of factors [4], they are not as equivalent and naive as the animal pole cells of the zebrafish blastula.

Taken advantage of the pluripotency of these cells and of their easy accessibility for injection, the animal pole of the zebrafish blastula has been a model of choice to assay for the activity of morphogens. Morphogen gradients have been generated either by grafting cells expressing secreted growth factor(s) [5] or by injecting an animal pole blastomere at the 128–256-cell stage with in vitro synthesized mRNAs coding for these growth factors [6–10]. Generation of two opposing morphogen gradients either at the animal pole of the embryo or in animal pole explants cultured in vitro allowed to demonstrate that Bone Morphogenetic Protein (BMP) and Nodal signaling are sufficient to induce the molecular and cellular mechanisms required to organize the uncommitted cells of the animal pole into a well-developed embryo [10]. More recently, we found that generation of BMP and Nodal morphogen gradients in aggregates (referred as embryoid bodies—EBs) of dissociated then reassociated animal pole blastomeres can also induce embryonic development and generate an embryonic axis (unpublished observation).

Here we will describe the different methods allowing the generation of morphogen gradients at the animal pole of the embryo, in animal pole explants and in embryoid bodies made from animal pole blastomeres by (1) injection of mRNA coding for growth factors into blastomere(s) of blastula animal pole, (2) transplantation of cells expressing these growth factors, (3) implantation of beads coated with purified proteins or (4) use of a combination of these different approaches to assess the effect of these morphogens on the organization of a vertebrate embryo.

2   Materials

2.1   Embryo Preparation

1. 1× Danieau buffer: 58 mM NaCl, 0.7 mM KCl, 0.4 mM MgSO$_4$, 0.6 mM Ca(NO$_3$)$_2$, 5 mM HEPES, pH 7.6 supplemented with 100 U/mL of Penicillin/Streptomycin.

2. Agarose (molecular biology grade).

3. 1% w/v Pronase from Streptomyces griseus in 0.3× Danieau buffer and incubated for 2 h at 37 °C before storage at −20 °C.

4. 35 and 60 mm petri dishes.

5. Pasteur capillary glass pipettes.

6. Pipette pump pipettor.

2.2 Microinjection of RNA

1. Capped RNA synthesized in vitro using an mMessage mMachine kit (Invitrogen Ambion).

2. mRNA injection mix: in vitro synthesized sense mRNA in 0.1 M KCl, 0.1% phenol red and 0.2% w/v of fluorescein dextran (10,000 MW) or of rhodamine dextran (10,000 MW, neutral).

3. 100 mm petri dishes.

4. Microscope slides.

5. Injection needles made from borosilicate thin wall capillaries with an internal filament that allows easy back filling. Needles are prepared [11] using a needle puller (Flaming/Brown Pipette Puller Model P-97) with parameters set to generate extralong tapers (10–15 mm) (Fig. 1a).

2.3 Animal Pole Explants and Cell Transplantation

1. 27G½ sterile needles and 1 mL syringes.

2. Transplantation needles made from borosilicate thin wall capillaries *without* internal filament prepared using a needle puller. The tips are broken off at an angle using a handheld razor blade at a position along the tapper that corresponds to an internal diameter (around 20–30 μm at blastula stage) wider that the diameter of the cell to transplant (*see* **Note 1**).

3. Custom-made plastic mold to generate pockets of agarose 1 mm deep and 0.7 mm wide (Fig. 2a, b).

4. 100 mm petri dishes.

2.4 Protein Coated Beads

1. Heparin agarose beads 30–40 μm.

2. Cross-linked polyglucan microspheres 10–15 μm.

3. Polybeads, polystyrene 6 μm.

4. Purified recombinant proteins.

5. Dulbecco's phosphate buffered saline (DPBS).

6. Bovine serum albumin, FITC conjugate.

2.5 Aggregates of Dissociated Animal Pole Blastomeres

1. Calcium-free Ringer's solution: 116 mM NaCl, 2.9 mM KCl, and 5 mM HEPES pH 7.2 (sterilized by 0.22 μm filtration).

2. 27G½ sterile needles and 1 mL syringes.

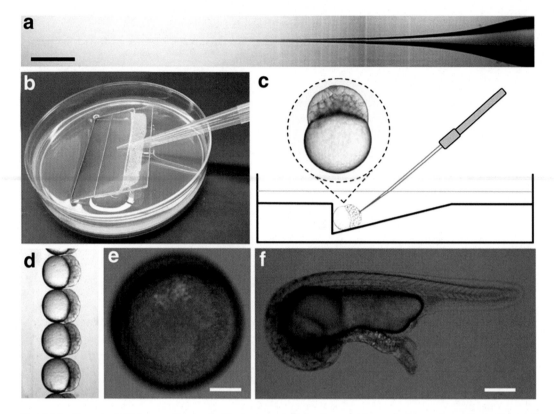

Fig. 1 Injection of mRNA in an animal pole blastomere. (**a**) Injection needle made from a borosilicate thin wall capillary with an internal filament that allows easy back filling. Needles are prepared using a needle puller in order to get extralong tapers. (**b**) Agarose coated plate with a groove that maintains the embryos in the correct orientation for injection. A microscope slide is placed before agarose solidifies and is maintained with an angle using a pipette tip. (**c**) Scheme showing how to position the embryo for injection in the groove made in the agarose. The injected animal pole blastomere (red) is filled with a mix of mRNA and phenol red. (**d**) Alignment in the groove of the embryos to inject. (**e**) Visualization at the onset of gastrulation (shield stage) of the clone derived from the injected animal pole blastomere (injection of a mix of BMP2b and Nodal related 2 (Ndr2) mRNAs together with fluorescein dextran). Embryo is in animal pole view. (**f**) Embryo injected with a 25:1 BMP2b–Ndr2 RNA ratio showing development of a secondary tail 24 h after injection. Scale bars (**a**) 1.5 mm, (**e**) 150 μm, (**f**) 300 μm

3. Zebrafish cell culture medium: Dulbecco's Modified Eagle Medium with Nutrient Mixture F-12 (DMEM/F-12) Media and 15 mM Hepes, 0.05 mM $CaCl_2$, 1 mM Sodium Pyruvate, 50 μg/mL Gentamycin, 1× antibiotic-antimycotic and 5% KnockOut Serum Replacement (KO-SR).

4. Glass beads 3 mm in diameter.

2.6 Equipment

1. 250 mL glass beakers.

2. Stereomicroscope with bottom lighting and a magnification up to 75× (e.g., zoom 5× and eyepiece magnification 15×).

3. Fluorescence microscope or fluorescence stereomicroscope.

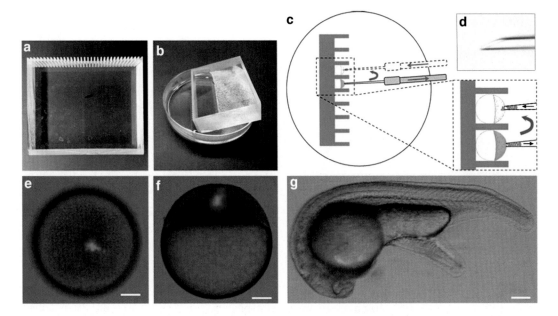

Fig. 2 Transplantation of cells from the blastula animal pole. (**a**) Plastic mold used to generate pockets of agarose 1 mm deep and 0.7 mm wide. (**b**) Transplantation plate made of a 100 mm petri dish coated with 2% agarose. After solidification, the plastic mold is placed with an angle on top of the agarose layer. Another layer of 2% agarose, 2–3 mm thick is poured. (**c**) A pocket containing a donor embryo (previously injected at the one-cell stage with morphogen(s) coding mRNA labeled with fluorescein dextran, green) and a pocket containing a wild-type host embryo both at the sphere stage. A group of cells from the donor is aspirated by applying a negative pressure. The needle is moved with the micromanipulator from the donor embryo to the host embryo. The needle is pushed through the EVL of the recipient and cells are delivered to the host by applying a positive pressure in the needle. (**d**) Detail of the sharp edge of the needle, prepared with a micropipette grinder. (**e**, **f**) Visualization of the cells transplanted from a donor injected with a mix of BMP2b and Nodal mRNAs together with fluorescein Dextran in animal pole view (**e**) and in lateral view (**f**). (**g**) Host embryo transplanted with cells from a donor previously injected with a 25:1 BMP2b–Nodal related 2 ratio showing development of a secondary tail 24 h after injection. Scale bars (**e**, **f**) 130 μm (**g**) 260 μm

4. Needle puller (e.g., Flaming/Brown Pipette Puller Model P-97).

5. Electronically regulated air-pressure microinjector (e.g., Eppendorf Femtojet).

6. Micromanipulator.

7. Micropipette grinder (e.g., Narishige EG-401).

8. Needle holder and tubing to connect to the air-pressure microinjector.

9. Incubator or warm room at 28.5 °C.

10. Dumont #5 fine forceps.

11. For cell transplantation or bead implantation: hydraulic manual microinjector (e.g., Eppendorf, Cell Tram vario).

3 Methods

3.1 Preparation of Embryos for Injection

1. Prepare agarose-coated plates by pouring the minimal amount of warm 2% agarose solution prepared in sterile water that will cover the bottom of a 60 mm petri dish. Once the agarose gel is solidified cover the petri dish with its lid and store the agarose-coated plates before use either at room temperature (for a couple of days) or at 4 °C in a close container or in a plastic bag (for several weeks).

2. Embryos are collected and placed in a 30 mm petri dish with just sufficient water from the fish system to cover them. Dechorionation is performed by adding 400 μL of Pronase solution (1% w/v) and by gently rotating the petri dish, looking under a stereomicroscope for the first embryos to quit the chorion. Pronase treatment is terminated by pouring the embryos in the pronase solution into a beaker containing 200 mL of water from the fish system. Let the embryos sink to the bottom of the beaker, pour the water out (just leave enough water to prevent the embryos to enter in contact with air). Rinse twice by pouring gently 200 mL of water from the fish system and letting the embryos sink at the bottom of the beaker. Then using a Pasteur capillary glass pipette, transfer the embryos to an agarose coated 60 mm petri dish filled with 0.3× Danieau buffer and let them develop until the 64-cell stage (*see* **Note 2**).

3. Prepare an agarose-coated plate in which a groove is made to maintain the embryos in proper orientation for injection. First, coat the bottom of a petri dish (100 mm in diameter) with a thin layer of 2% agarose prepared with sterile water. Let it solidify then poor an additional layer of agarose, 3 mm thick and place a microscope slide maintained with an angle using a pipette tip. Using two slides glued together using a cyanoacrylate based glue (super glue) (as shown in Fig. 1b) instead of a single slide allow creating two parallel grooves in the agarose plate. This is sufficient for injection of up to 100 embryos per dish. Let the agarose solidify and remove the slide. This creates a ramp with a gentle slope on one side and an abrupt edge on the other side (Fig. 1c).

4. Cover the agarose plate with 0.3× Danieau buffer.

5. Transfer the dechorionated embryos when they reach the 32–64-cell stage with a Pasteur glass pipette and orient them in such a way that they lay on their side with the yolk against the agarose edge (Fig. 1c). Let them be in contact with each other (Fig. 1d). This helps maintaining the correct orientation and prevents the embryos to move during injection.

3.2 Injection of RNA in Animal Pole Blastomeres

1. Prepare injection mix containing the morphogen(s) of interest (*see* **Note 3**).

2. Spin the mRNA injection mix to pellet any particles that could clog the needle.

3. Backload an injection needle by adding 2 μL of mRNA injection mix at the back tip of the needle that is maintained in a vertical orientation. The mix will move down to the needle tip in about 1–2 min thanks to the glass filament present in the capillary inner wall.

4. When the solution reaches the tip of the needle, place the needle into the needle holder.

5. Place the needle holder on the micromanipulator.

6. Connect the tubing to the microinjector.

7. Break the tip of the needle, in the 0.3× Danieau buffer, with a Dumont #5 forceps.

8. Apply air pressure to check for proper opening and for removing any air bubbles present in the needle in between the tip and the mRNA injection mix.

9. Under the stereomicroscope and using the micromanipulator, carefully move the needle close to the animal pole of an embryo and gently push the tip of needle into a blastomere (*see* **Note 4**). Apply air pressure until the blastomere is filled (about 15 pL) with the phenol red labeled mRNA injection mix (Fig. 1c).

   To generate two morphogen gradients, two different blastomeres should be injected. After injection of a first mRNA injection mix containing a first morphogen, take another needle and inject the mRNA injection mix for the second morphogen in a second blastomere (*see* **Note 5**).

10. Transfer injected embryos to an agarose-coated 60 mm petri dish filled with 0.3× Danieau buffer (with no more than 50 embryos per dish) and let them develop at 28.5 °C in an incubator or a warm room.

11. After a couple of hours visualize fluorescence to check for the presence of clones at the animal pole (Fig. 1e) and after a day of development, look for the effect of morphogen gradient(s) on development of ectopic structures (Fig. 1f).

3.3 Cell Transplantation

1. Prepare a transplantation plate made of a 100 mm petri dish coated with 2% agarose. After solidification, place the plastic mold (Fig. 2a) with an angle on top of the agarose layer (*see* Fig. 2b) and pour another layer of 2% agarose, 2–3 mm thick. Let the agarose solidify. Remove the plastic mold and cover with 1× Danieau buffer.

2. Inject 0.5–1 nL of in vitro synthesized mRNA coding for the morphogen of interest into the yolk of dechorionated embryos at the 1–4-cell stage in a 60 mm agarose coated petri dish in 0.3× Danieau buffer.

3. Let the injected embryos develop at 28.5 °C.

4. Fill the transplantation needle and tubing with mineral oil (Fisher Scientific, O121-1) and connect to the hydraulic manual microinjector. Make sure that there are no air bubbles between the microinjector and the tip of the needle.

5. Place the needle holder on the micromanipulator.

6. When embryos reach the blastula stage (1K-cell to the high stage), transfer them to the transplantation agarose plate.

7. Place one embryo per agarose pocket with their animal pole facing the opening as shown in Fig. 2c. Alternate pockets of donor embryos (injected with a morphogen coding mRNA) and pockets containing wild-type host embryos.

8. Place the tip of the needle in 1× Danieau buffer and apply a negative pressure to fill about a third of the needle.

9. With the micromanipulator, move the tip of the needle toward the animal pole of a donor embryo. Gently push the tip of the needle through the enveloping layer (EVL). When the needle enters in the deep cell layer, aspirate a group of cells (around 30 cells) by applying a negative pressure.

10. Slowly release the negative pressure until cells steadily hold in the needle.

11. Use the micromanipulator to gently move the needle from the donor embryo to the host embryo.

12. Slowly move down the cells in the needle close to its tip. Push the needle through the EVL.

13. Make sure that the tip of the needle is in the deep cell layer close to the surface of the embryo and carefully transplant a group of donor cells to the host by applying a positive pressure.

14. Move the needle out of the host embryo.

15. Repeat the procedure from **steps 8–14** until all host embryos have been transplanted with morphogen expressing cells.

16. After cell transplantation, remove the host embryos and place them into a 60-mm agarose coated petri dish filled with 0.3× Danieau buffer and let them develop at 28.5 °C.

17. At the desired time, look at the position of the fluorescently labeled clones of morphogen secreting cells (Fig. 2e, f).

18. Assess morphologically (Fig. 2g) and/or by in situ hybridization resulting phenotypes on embryo morphogenesis.

3.4 Beads Implantation

1. Transfer 50 μL of beads in suspension to a 1.5 mL Eppendorf microtube.

2. Centrifuge beads for 5 min at 15,000 × g in a tabletop centrifuge. Remove supernatant and rinse beads with 1× DPBS. Repeat twice.

3. Centrifuge for 5 min at 15,000 × g in a tabletop centrifuge. Remove the supernatant and add the solution containing the morphogen of interest (200 ng/μL) diluted in 1× DPBS at the desired concentration together with Albumin from Bovine Serum (BSA), FITC conjugated (100 ng/μL) that will allow visualization of the beads after implantation.

4. Pipet up-and-down to resuspend the bead pellet in the protein solution (*see* **Note 6**).

5. Incubate for 12 h at 4 °C to let the protein coating the beads.

6. Prepare a transplantation plate (as described in Subheading 3.3).

7. Place a small amount of morphogen-BSA-FITC coated beads in one pocket and embryos at the 1K-cell to the high stages in other pockets as described for cell transplantation (Subheading 3.3, **step 7**).

8. Collect a small number (10–20) of small beads (<10 μm in diameter) and implant them at the animal pole of host embryos as described in Subheading 3.3 for cell transplantation. For large beads (e.g., heparin-agarose beads of 30–40 μm in diameter) use a transplantation needle open without an angle. Maintain the bead(s) at the tip of the needle by applying a negative pressure. Then force the bead(s) through the EVL into the deep cell layer of the host embryo (Fig. 3a).

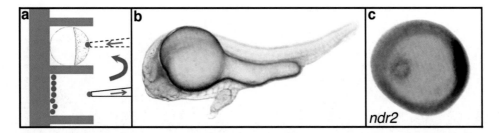

Fig. 3 Bead implantation. (**a**) Beads are placed in one pocket, and then implanted at the animal pole of a host embryo. Large beads are maintained at the tip of the needle by applying a negative pressure in the needle. The bead is then forced through the EVL into the deep cell layer of the host embryo. (**b**) Embryo 24 h after implantation of a bead coated with mouse BMP4 protein showing development of a secondary embryonic axis. (**c**) In situ hybridization showing endogenous expression of *nodal related 2* (*ndr2*) at the embryonic margin as well as an ectopic expression at the site of implantation of the bead. Embryos are in lateral view (**b**) and in animal pole view (**c**)

9. After bead(s) implantation, transfer embryos into a 60-mm agarose coated petri dish with 0.3× Danieau buffer.

10. Look for the position of the implanted bead(s) with a fluorescence microscope or stereomicroscope. Assess the effect of morphogen on embryo morphogenesis (Fig. 3b) or on gene expression (Fig. 3c) (*see* **Note 7**).

3.5 Morphogen Gradients in Aggregates of Dissociated–Reassociated Animal Pole Blastomeres

3.5.1 Formation of Aggregates (Embryoid Bodies—EBs) of Animal Pole Blastomeres

1. Prepare agarose microwells. First, coat a 60 mm petri dish with a thin layer of 2% agarose prepared in water. When solidified, pour another layer of 2% agarose 1.5–2 mm thick. Before solidification, add glass beads (3 mm in diameter) as shown in Fig. 4. When solidified remove the glass beads with a forceps and cover the plate with 10 mL of zebrafish cell culture medium.

2. Dechorionate embryos as described in Subheading 3.1, **step 1**.

3. When embryos reach the 1K-cell stage, place them into a 60-mm agarose coated petri dish covered with Ringer's solution without $CaCl_2$.

4. Using a 27G½ sterile needle on top of a 1 mL syringe, cut the animal pole of embryos to make an explant (the upper third of the embryo) and place this explant in another 60-mm agarose coated petri dish cover with Ringer's solution without $CaCl_2$ (*see* **Note 8**).

5. Due to the lack of calcium, explants progressively dissociate. This process can be sped up by gently pipetting explants up-and-down in the Ringer's solution without $CaCl_2$.

6. Slowly rotate the plate to accumulate dissociated cells in the center.

7. Using a 27G½ needle and a Pasteur capillary pipette, remove all nondissociated group of cells.

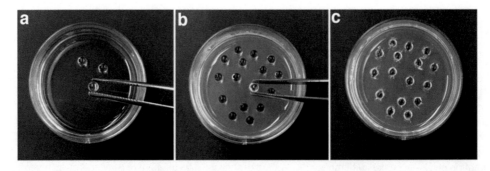

Fig. 4 Preparation of agarose microwells for facilitating formation of aggregates of animal pole blastomeres. (a) Glass beads are placed with a forceps in the second layer of agarose before its solidification. (b) After solidification of agarose glass beads are removed with a forceps. (c) Agarose microwells ready to be used

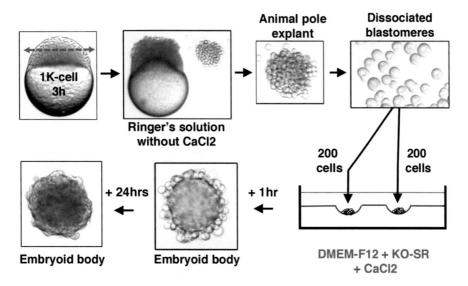

Fig. 5 Formation of Embryoid Bodies (EBs) made of animal pole blastomeres. The animal pole of an embryo at the 1K-cell stage is severed with a sharp sterile needle. This explant is placed into a fresh agarose coated dish containing Ringer's solution without $CaCl_2$. Absence of calcium results in the dissociation of the explant into individual blastomeres. Dissociated blastomeres are gently pipetted using a Pasteur capillary pipette and dispensed into agarose microwells (200 cells per well) filled with zebrafish cell culture medium. Embryoid bodies are formed in about 1 h and can be cultured for several days in that medium

8. Carefully collect the dissociated blastomeres with a Pasteur capillary pipette.

9. Dispense about 200 cells in each agarose microwell prepared in Subheading 3.5.1 (*see* **Note 9**).

10. EBs are formed in about 1 h (Fig. 5).

3.5.2 Generating Morphogen Gradients in EBs Following Transplantation of Morphogen Expressing Cells

1. Inject mRNA coding for the morphogen of interest together with fluorescein or rhodamine dextran into the yolk of dechorionated embryos at the 1–4 cell stage (*see* **Note 10**).

2. When they reach the high stage, place the injected embryos into a transplantation plate filled with zebrafish cell culture medium. Do the same for EBs, alternating one pocket with an injected embryo and one pocket with an EB.

3. When they reach the sphere stage, transplant cells of the injected embryos into the EBs (Fig. 6) as described in Subheading 3.3.

4. Using a Pasteur capillary pipette, transfer the transplanted EBs in a new agarose microwell petri dish filled with zebrafish cell culture medium and incubate them at 28.5 °C until analysis.

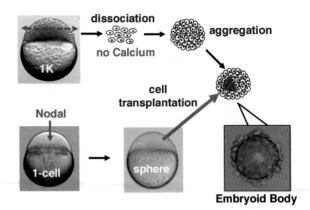

Fig. 6 Formation of a morphogen gradient in an Embryoid Body by transplantation of animal pole cells. At the one-cell stage (bottom left) an embryo is injected with a mix of Nodal mRNA and rhodamine dextran. At the 1K-cell stage, animal pole of uninjected embryos are severed (top left), cells are dissociated and then aggregated to generate an EB. When injected embryos reach the sphere stage their animal pole cells expressing Nodal are transplanted into the EB (bottom right)

3.5.3 Formation of Morphogen Gradients in EBs by Merging Aggregates of Animal Pole Blastomeres

1. Inject mRNA coding for the morphogen of interest together with fluorescein or rhodamine dextran into the yolk of dechorionated embryos at the 1–4 cell stage (*see* **Note 10**).

2. At the 1K-cell stage, explant animal poles of injected embryos, dissociate the blastomeres then aggregate them in an agarose microwell petri dish filled with zebrafish cell culture medium.

3. Proceed similarly with noninjected embryos.

4. After about 30 min, EBs made with blastomeres of injected embryos are collected and added (one per well) in wells containing EBs made of noninjected embryos.

5. When two EBs enter in contact they merge into a larger EB (Fig. 7). This process takes 1–2 h.

3.5.4 Generating Morphogen Gradients Using a Combination of Injection, Cell Transplantation, and Merging of Animal Pole Explants

1. Inject mRNA coding for BMP (BMP2b) together with fluorescein dextran in dechorionated embryos at the 1–4 cell stage (*see* **Note 10**).

2. Inject mRNA coding for Nodal (Nodal related 2—Ndr2) together with rhodamine dextran in dechorionated embryos at the 64-cell stage in two blastomeres located halfway in between the margin and the animal pole (Fig. 8) (*see* **Note 10**).

3. When embryos reach the 512-cell stage, animal poles of BMP and Nodal injected embryos are explanted. Merging of the two animal pole explants is obtained by apposing their two freshly cut surfaces (Fig. 8, [10]). After a few minutes, when the two explants are fused into a single mass of cells, transfer them into

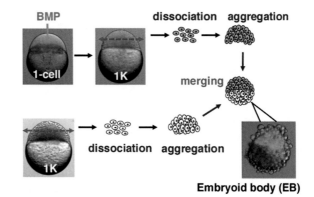

Fig. 7 Formation of a morphogen gradient in an Embryoid Body by merging two aggregates of animal pole blastomeres. At the one-cell stage, an embryo is injected with a mix of BMP2b mRNA and fluorescein dextran. At the 1K-cell stage, animal poles of injected and uninjected embryos are severed, cells are dissociated and then aggregated to generate two EBs, one of them expressing BMP2b. The two EBs are placed in contact at the bottom of an agarose well where they merge into a larger EB (bottom right). This merging process takes 1–2 h

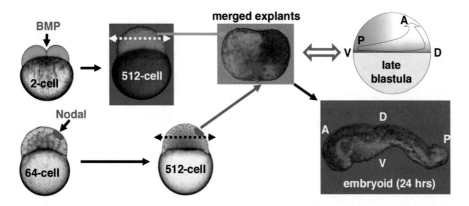

Fig. 8 Combination of mRNA injections and merging of animal pole explants to generate intersecting gradients of BMP and Nodal. In a wild-type embryo (top right), at late blastula stage, BMP signaling (green) is restricted to the ventral side (V) while Nodal (red) is expressed all around the margin with more expression on the dorsal side (D). Arrow indicates the posterior (P) to anterior (A) axis of the zebrafish blastula. To reproduce this pattern, an mRNA coding for BMP2b together with fluorescein dextran is injected in an embryo at the 1–4 cell stage (top left). mRNA coding for Nodal related 2 (Ndr2) together with rhodamine dextran is injected in another embryo at the 64-cell stage in two blastomeres located halfway in between the margin and the animal pole (bottom left). At the 512-cell stage, animal poles of BMP and Nodal injected embryos are explanted and merged in such a way that the clone of cells expressing Nodal is apposed to the BMP expressing explant. After 24 h this results in embryoids (bottom right) displaying a clear anterior–posterior (A-P) and dorsal–ventral (D-V) axes

another petri dish filled with fresh zebrafish cell culture medium and incubate them at 28.5 °C.

4. Observe embryos after 24 h under a fluorescence microscope. Analysis of the tissues and organs generated can be performed using in situ hybridization or immunohistochemistry for specific molecular markers.

3.5.5 Generating Morphogen Gradients Using a Combination of mRNA Injections and Merging of EBs

1. Inject mRNA coding for BMP together with fluorescein dextran into the yolk of dechorionated embryos at the one-cell stage (*see* **Note 10**).

2. Inject mRNA coding for Nodal together with rhodamine dextran into the yolk of dechorionated embryos at the one-cell stage (*see* **Note 10**).

3. When BMP injected embryos and WT noninjected embryos reach the high stage, explant their animal pole and dissociate their blastomeres in a Ringer's solution without CaCl$_2$. Then aggregate the isolated blastomeres to generate BMP and WT EBs (Fig. 9a).

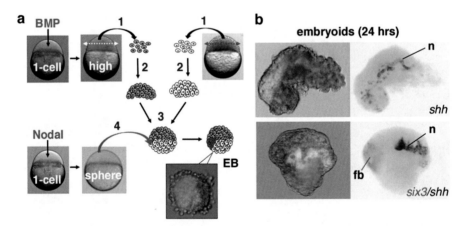

Fig. 9 Combination of mRNA injections and merging of animal pole embryoid bodies to generate intersecting gradients of BMP and Nodal. (**a**) The spatial organization of cells secreting BMP and Nodal morphogens in wild-type blastula can be reproduced in EBs made of animal pole blastomeres. Embryos are injected at the one-cell stage with BMP (BMP2b) and fluorescein dextran (top left) or with Nodal (Ndr2) and rhodamine dextran (bottom left). At the high stage, the animal poles of a BMP injected embryo and of a noninjected wild-type embryo are severed, cells are dissociated (1) and then reassociated (2) to generate two aggregates that are merged into a single EB (3). When the Nodal injected embryo reaches the sphere stage, animal pole cells of this embryo are transplanted (4) into the EB in between the BMP2b expressing and the WT cells. (**b**) After 24 h in culture, EBs instructed with BMP2b and Nodal have developed an embryonic axis with a notochord (n) expressing *sonic hedgehog* (*shh*) and that display a clear anterior–posterior axis characterized by the expression of the forebrain (fb) marker *six3* (red) at one end and the expression of *shh* (black) in the notochord at the other end. Left panels: position of cells expressing BMP2b (green) and Ndr2 (red). Right panels: in situ hybridization for *shh* and *six3*

4. When cell have aggregated, merge BMP expressing and WT EBs to form a larger EB that will express BMP in half of its cells.

5. Using a fluorescence dissecting microscope, transplant cells from Nodal injected embryos that have reached sphere stage into the merged BMP/WT EBs in a domain adjacent to the BMP expressing cells (step 4, Fig. 9a).

6. After a few minutes transfer these grafted EBs into another agarose microwell petri dish filled with zebrafish cell culture medium. Incubate at 28.5 °C until analysis either by live imaging under fluorescence microscope or using molecular markers (as illustrated in Fig. 9b).

4   Notes

1. An optimized needle with clean edges and precise tip angle (Fig. 2d) can be obtained by using a micropipette grinder.

2. Dechorionated embryos are very fragile and should be manipulated gently. They should not contact plastic surfaces or water--air interface.

3. Quantity of mRNA injected depends on the morphogen used and has to be experimentally defined. In our experiments, 0.5–10 pg Nodal related 2 mRNA and 10–100 pg of BMP2b mRNA were used to generate morphogen gradients at the animal pole. Because Ndr2 and BMP2b regulate their expression through positive autoregulatory loops, when a morphogen is tested alone, the precise amount of mRNA injected is not critical as long this is within the indicated range. However, when injected together in the same blatomere the BMP–Nodal ratio is a key factor to development of embryonic structures along the anterior–posterior axis: a high ratio (25:1) promotes formation of the tail whereas a 1:1 ratio induces formation of posterior head structures [9]. When injected in two separate blastomeres, the amount of BMP mRNA injected should be at least 10× higher than the amount of Nodal to allow, in the domains where BMP and Nodal gradient intersect, reproducing of the progressively decreasing ratio of activity from ventral to dorsal that characterizes the organizing activity of the blastula embryonic margin [9, 10, 12].

4. A freshly open needle enters very easily in blastomeres. However, after injection of several embryos the tip of the needle may be coated with membrane debris, making the needle sliding on the surface of the blastomere instead of entering in it. This can be fixed either by gently cleaning the tip of the needle with a forceps (with gentle movements from back to front near the tip of needle) or by breaking the needle a few microns back in

order to generate a new sharp tip. Because needles are prepared with extralong tappers it is possible to break the tip several times before the opening is too big. If this happens to be the case, change for a new injection needle.

5. The blastomere injected with the mRNA coding for the first morphogen is traced thanks to the phenol red contained in the mRNA injection mix. This color fades away in about half an hour. Therefore, it is important to inject the second blastomere soon after the first one. This is accomplished by limiting the number of embryos to inject (less than 50). Because a double injection is time consuming, there is one cell division in average between the first and the second injection. If the embryos are injected firstly at the 128-cell stage, the second injection will be performed at the 256-cell stage.

6. For control, use beads coated only with BSA, FITC conjugate.

7. It is essential to use exclusively proteins produced in *Danio rerio*. The use of heterospecific morphogens may result in artifactual results as illustrated in Fig. 3 for heparin agarose beads coated with mouse BMP4. Stimulation of animal pole cells of zebrafish blastula with zebrafish BMP4 (by injection of BMP4 mRNA or transplantation of zebrafish BMP4 secreting cells) does not induce formation of ectopic structures. On the other hand, implantation of beads coated with mouse BMP4 or injection of mouse BMP4 mRNA results in formation of ectopic embryonic axis that lacks axial structures (Fig. 3b). In zebrafish, this effect is only observed when both BMP and Nodal signaling pathways are concomitantly stimulated at the animal pole ([8, 9], Fig. 1f). In the mouse embryo, BMP4 secreted by the extraembryonic ectoderm promotes expression of Nodal in the adjacent embryonic cells [13]. We found that stimulation of the animal pole of the zebrafish blastula with mouse BMP4 results in ectopic expression of zebrafish *nodal related 2* (*ndr2—cyclops*) (Fig. 3c) and therefore to the concomitant activation of the BMP and Nodal signaling pathway.

8. Avoid taking cells close to the margin where maternal Wnt8a, transcribed from mRNA that have been transported from the vegetal pole during cleavage stages, activates the β-catenin signaling pathway [3].

9. Minimize the amount of Ringer's solution during the transfer of cells to avoid the dilution of the zebrafish culture medium.

10. For BMP and Nodal morphogens, 500 pg of BMP2b mRNA or 50 pg of Ndr2 mRNA is injected at the 1–4 cell stage.

Acknowledgments

This work was supported by funds from University of Virginia (B.T., C.T.), March of Dimes (1-FY15-298 to B.T.), Jefferson Trust (FAAJ3199 to C.T.), Conselho Nacional de Desenvolvimento Científico e Tecnológico, Brazil (grant number 200535/2014-5 to M. de O-M.).

References

1. Kane DA, Kimmel CB (1993) The zebrafish midblastula transition. Development 119:447–456

2. Schneider S, Steinbeisser H, Warga RM et al (1996) Beta-catenin translocation into nuclei demarcates the dorsalizing centers in frog and fish embryos. Mech Dev 57:191–198

3. Lu FI, Thisse C, Thisse B (2011) Identification and mechanism of regulation of the zebrafish dorsal determinant. Proc Natl Acad Sci U S A 108:15876–15880

4. Green J (1999) The animal cap assay. Methods Mol Biol 127:1–13

5. Kudoh T, Concha ML, Houart C et al (2004) Combinatorial Fgf and Bmp signalling patterns the gastrula ectoderm into prospective neural and epidermal domains. Development 131:3581–3592

6. Thisse B, Wright CV, Thisse C (2000) Activin- and Nodal-related factors control anteroposterior patterning of the zebrafish embryo. Nature 403:425–428

7. Chen Y, Schier AF (2001) The zebrafish Nodal signal Squint functions as a morphogen. Nature 411:607–610

8. Agathon A, Thisse C, Thisse B (2003) The molecular nature of the zebrafish tail organizer. Nature 424:448–452

9. Fauny JD, Thisse B, Thisse C (2009) The entire zebrafish blastula-gastrula margin acts as an organizer dependent on the ratio of Nodal to BMP activity. Development 136:3811–3819

10. Xu PF, Houssin N, Ferri-Lagneau KF, Thisse B, Thisse C (2014) Construction of a vertebrate embryo from two opposing morphogen gradients. Science 344:87–89

11. Dean DA (2006) Preparation (pulling) of needles for gene delivery by microinjection. CSH Protoc. https://doi.org/10.1101/pdb.prot4651

12. Thisse B, Thisse C (2015) Formation of the vertebrate embryo: moving beyond the Spemann organizer. Semin Cell Dev Biol 42:94–102

13. Shen MM (2007) Nodal signaling: developmental roles and regulation. Development 134:1023–1034

Chapter 8

Polarizing Region Tissue Grafting in the Chick Embryo Limb Bud

Holly Stainton and Matthew Towers

Abstract

The polarizing region of the developing limb bud is an important organizing center that is involved in anteroposterior (thumb to little finger) patterning and has three main functions that are now considered to depend on the secreted protein Sonic hedgehog (Shh). These are (1) specifying anteroposterior positional values by autocrine and graded paracrine signaling; (2) promoting growth in adjacent mesenchyme; (3) maintaining the distal epithelium that is essential for limb outgrowth by induction of a factor in adjacent mesenchyme. The polarizing region was identified using classical tissue grafting techniques in chicken embryos. Here we describe this procedure using tissue from transgenic Green Fluorescent Protein-expressing chicken embryos that allows the long-term fate of the polarizing region to be determined. This technique provides a highly useful and effective method to understand how the polarizing region patterns the limb and has implications for other organizing centers.

Key words Polarizing region, Chick, Limb, Morphogen, Sonic hedgehog, Grafting, Embryonic development

1 Introduction

In the early limb bud, molecules from distinct signaling centers act together to specify pattern and promote growth [1]. One important signaling center is the polarizing region (or Zone of Polarizing Activity), a group of morphologically indistinguishable mesenchyme cells located in the posterior part of the limb bud (Fig. 1a).

John Saunders discovered the polarizing region in classical grafting experiments that he performed on chick embryos in the 1960s, in which he transplanted tissue from the posterior margin of the wing bud to the anterior margin of the wing bud of a host embryo [2]; depicted in Fig. 1a, c (reviewed in Tickle [3]). This resulted in mirror image duplications of the normal pattern of chick wing digits, 1, 2, and 3, to obtain patterns such as 3-2-1-1-2-3 (Fig. 1b, note until recently, chick wing digits were numbered 2, 3, and 4). Figure 1c shows the procedure in which a polarizing region

Julien Dubrulle (ed.), *Morphogen Gradients: Methods and Protocols*, Methods in Molecular Biology, vol. 1863, https://doi.org/10.1007/978-1-4939-8772-6_8, © Springer Science+Business Media, LLC, part of Springer Nature 2018

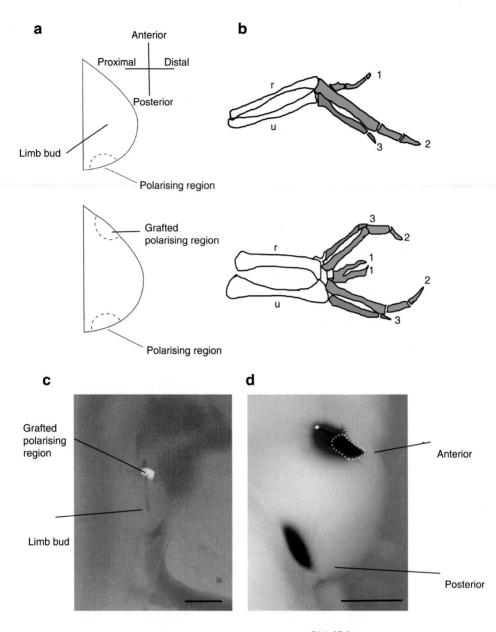

Shh/Gfp

Fig. 1 The chick limb polarizing region. (**a**) Schematic of the chick limb bud highlighting the polarizing region in the posterior portion of the limb bud, and below, a schematic illustrating the location of a polarizing region grafted anteriorly. (**b**) Schematic of the normal pattern of digits (1, 2, and 3 running from anterior to posterior; r = radius, u = ulna) of the chick wing and below a full duplication of the digit pattern (3-2-1-1-2-3), which is achieved when polarizing region grafts are performed to the anterior margin. (**c**) Picture taken *in ovo* after a polarizing region from a GFP-expressing chick wing bud (green tissue) was grafted to the anterior margin of a host wing bud. Scale bar represents 500 μm. (**d**) To confirm proficient grafting technique, *Shh* and *Gfp* expression can be determined by double whole-mount in situ hybridization after 24 h. The dark purple color indicates *Shh* expression in the distal part of the graft outlined by white dashed lines; orange indicates *Gfp* expression throughout the graft as it extends along the anterior margin of the wing bud. Note also *Shh* expression in host polarizing region. Scale bars represent 500 μm

has been excised from a Hamburger and Hamilton (HH) stage 20/21 (approximately 3.5 days post incubation [4]) transgenic chicken that constitutively expresses Green Fluorescent Protein (GFP) and grafted to a stage-matched wild type host [5, 6].

In 1993, the Tabin lab discovered that the product of the *Sonic hedgehog* (*Shh*) gene encodes the polarizing region signal. Cells expressing *Shh*, or beads soaked in recombinant Shh protein, produced similar duplications of the digits when grafted to the anterior margins of chick wing buds as polarizing region grafts [7–9]. Following a graft of GFP-expressing polarizing region tissue made to a wild type wing bud, *Shh* expression can be observed using whole-mount in situ hybridization in the distal part of the graft (dark purple in Fig. 1d), and *Gfp* expression can be observed in the whole graft (orange in Fig. 1d, note expression of *Gfp* in the distal region of the graft is masked by *Shh* signal). When left to develop until day 10, the GFP-expressing cells form a distinct stripe along the margin of the additional digit 3, but do not contribute to the digit skeleton (Fig. 2a) [5].

Many experiments have revealed that Shh displays many characteristics of the polarizing region signal identified in early studies (reviewed in Tickle and Towers [3]). However, grafts of the polarizing region—particularly ones expressing GFP—continue to be an important technique for studying limb development. For instance, they have revealed the long-term fate of the polarizing region [5, 10] and helped determine that the duration of *Shh* expression is intrinsically regulated [11]. Here, we describe the methodology behind making polarizing region grafts using tissue from GFP-expressing chick embryos. In doing so, we describe some experiments that led to our current understanding of how the polarizing region patterns the limb bud.

Lewis Wolpert described the effects that polarizing region grafts have on anteroposterior limb pattern to help formulate his concept of positional information, in which cells acquire a positional value and use it to instruct their differentiation into appropriately positioned structures (e.g., a thumb at the anterior side of the hand) [12, 13]. The polarizing region signal was predicted to act in a long-range graded paracrine fashion: low levels of the polarizing signal specifying the identities of the most-anterior digit (digit 1), and increasing levels specifying the identities of more-posterior digits (digits 2 and 3). In support of this hypothesis, as few as nine polarizing region cells implanted to the anterior part of a wing bud induced the formation of an additional digit 1, and digits with an increasing posterior character were produced when more cells were grafted [14]. The extent of digit duplication was also shown to depend on the duration of polarizing region signal. A polarizing region left in place for 15 h produced an additional digit 1, for 20 h, additional digits 1 and 2, and for more than 24 h, digits 1, 2, and 3 [15]. Similar digit patterns can

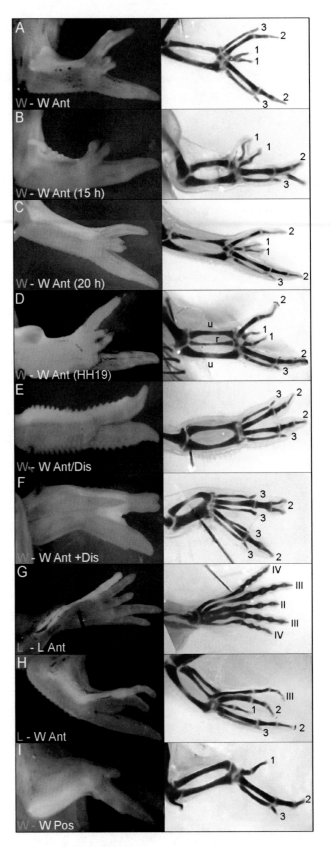

Fig. 2 GFP-expressing polarizing region grafts to wild type hosts. (**a–e**) Wing bud polarizing region grafted to the anterior margin of another wing bud (*see* Fig. 1a).

be produced when the distal half of the graft (the region now known to express *Shh*—*see* Fig. 1d) is excised at 15 h (Fig. 2b) and at 20 h Fig. 2c). In both cases, the remaining GFP-labeled cells of the graft contribute to soft tissues along the anterior margin of the wing.

Grafts of the polarizing region also thicken the overlying apical ectodermal ridge and provided additional evidence of a maintenance factor expressed in adjacent mesenchyme identified in earlier experiments [16]. The apical ectodermal ridge is required for the outgrowth of the limb and for the laying down of the structures along the proximodistal axis, i.e., upper arm to the digit tips [17]. The interdependence of anteroposterior patterning and proximodistal patterning was observed in experiments in which polarizing region grafts were made to host embryos of different stages [18]. Grafts made at the stage at which the forearm is being laid down (HH19—approximately 3 days post incubation [17]) can result in two ulnae forming that are separated by a single radius (Fig. 2d).

Polarizing region grafts made to embryos, which were then X-irradiated 2 h later to inhibit growth, resulted in duplications of the digit patterns in which digit 1 was absent—including patterns such as 3-2-2-3 [19]. This result was explained by the proposal that polarizing region signaling promotes mesenchyme expansion, and that this is required for the full specification of anteroposterior positional values by the graded signal [13]. This result can be

Fig. 2 (continued) (**a**) Full duplication of the three digits (3-2-1-1-2-3). (**b**) Duplication of digit 1 [1-1-2-3] following excision of the distal half of polarizing region graft (*see* Fig. 1d) at 15 h post-grafting. (**c**) Duplication of digits 2 and 1 [2-1-1-2-3] following excision of distal half of polarizing region graft at 20 h post-grafting. (**d**) HH20/21 graft made to early HH19 host wing resulting in duplication of the ulna and digits 1 and 2. (**e**) Graft placed closer to host polarizing region (distal to normal grafting position) resulting in duplication of digits 2 and 3. (**f**) Two wing bud polarizing region grafts—one made to the anterior margin and the other to the distal apex of another wing bud—resulting in a 3, 2, 3 pattern between grafted polarizing regions, and a 3, 2, 3 pattern between grafted and host polarizing region. (**g**) Polarizing region from a leg bud grafted to the anterior margin of the host wild type leg, resulting in a duplicated digit IV that arises from the polarizing region—complete pattern is IV-III-II-III-IV. (**h**) Leg polarizing region grafted into the anterior margin of a wild type wing bud, resulting in a digit III arising from the grafted polarizing region—complete pattern is III-2-1-2-3. (**i**) Wing bud polarizing region grafted in place of a host polarizing region gives rise to a band of posterior soft tissue along digit 3. Note in all cases (except in (**d**)), host and donor embryos are HH20/21. W = wing, L = Leg, Ant = Anterior, Dis = Distal; Green text indicates graft from GFP-expressing donor, white text indicates wild type host. Scale bar represents 500 μm

mimicked if one grafts a polarizing region immediately adjacent and distal to the normal grafting position at the anterior margin: the closer proximity of the host and donor polarizing regions results in the concentration of morphogen being unable to fall enough to specify anterior positional values (Fig. 2e).

Two duplicate patterns of 3-2-3 can also be made in the same wing by grafting two polarizing regions—one to the anterior margin and another to the distal apex (Fig. 2f). This experiment ruled out a model involving short-range signaling and intercalation that challenged a long-range graded signal (see Wolpert and Hornbruch [20] for more detail, also Iten et al. [21]), and again, demonstrates the requirement for sufficient tissue between the polarizing regions to accommodate the full range of anteroposterior positional values. Note that the graft made to the apex of the wing remains in a distal position.

Additional experiments revealed that grafts of the chick leg bud polarizing region are also able to produce mirror-image duplications of the digit pattern when grafted to the anterior margin of a recipient leg bud [22]. If this experiment is performed using a GFP-expressing leg polarizing region, it can be seen that the most-posterior digit IV arises from the graft [5] (Fig. 2g—note the chick leg has four digits numbered I, II, III, IV). This finding explained why a leg digit developed, in addition to a duplicated pattern of wing digits, when a leg polarizing region was grafted to the anterior margin of a recipient wing bud [22] (Fig. 2h—note leg digit forming from GFP-expressing graft). The finding that a digit arises from cells of the polarizing region helped reveal that the duration of autocrine Shh signaling specifies the identity of digit IV in the chick leg (see Towers et al. for additional detail [5]).

Almost all of the parameters of anteroposterior patterning have been determined by grafting polarizing regions to the anterior margins of recipient chick wing buds. However, grafting of GFP-expressing polarizing regions to the posterior margin of the wing bud in place of the host posterior polarizing region demonstrated that the polarizing region contributes to soft tissue along the posterior margin of digit 3 and not to the skeleton in normal development (Fig. 2i [5]). This observation supported an identification of chick wing digits of 1, 2, and 3, rather than 2, 3, and 4, and confirmed earlier predictions that the digits arise from cells adjacent to the polarizing region in response to paracrine Shh signaling (see Towers et al. for further detail [5]).

2   Materials

2.1   Dissecting Tools and Equipment

To perform polarizing region grafts you will need a dissection kit containing the following tools (shown in Fig. 3):

1. Two pairs of fine forceps (Fine Science Tools, Dumont #5)
2. Microdissecting spring scissors (Fine Science Tools).
3. A small spatula.
4. A sharpened tungsten needle.
5. Dissecting scissors.
6. Clear sellotape.
7. 25 μm diameter platinum wire (Goodfellow Metals).
8. A minimum of two insect pins.
9. Leica MZ165FC stereomicroscope with fluorescence.
10. Panasonic MIR-262-PE incubator set to 38 °C.

2.2   Reagents and Solutions

1. Sterile ice-cold phosphate buffered saline (1×PBS) in a 5 cm diameter petri dish.
2. 1.5% agarose made up in 1×PBS in a 5 cm diameter petri dish.
3. Sterile Dulbecco's modified Eagle's medium (DMEM, without phenol red) containing 1% penicillin/streptomycin to be added to the petri dish containing 1.5% agarose.
4. 90% ethanol in a petri dish.
5. Wild type and GFP-expressing chicken embryos, staged using the Hamburger Hamilton system (Hamburger and Hamilton [4]). Note, GFP-expressing chickens are produced by the Roslin Institute, Edinburgh, UK [6].

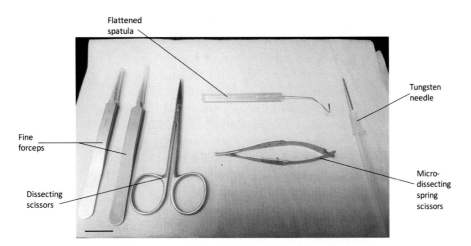

Fig. 3 Dissection kit used for polarizing region grafts. Scale bar represents 2 cm

3   Methods

1. In a petri dish containing 90% ethanol, cut the 25 μm platinum wire into small lengths (approximately 400 μm) using micro-dissection scissors. Bend the wire into "L" shaped pins with two pairs of forceps (*see* **Notes 1** and **2**).

2. Cut a hole in the shell of a wild type host chick egg (windowed with sellotape) using scissors then remove the vitelline membrane and amnion above the wing bud of the HH20/21 embryo using forceps. Cut through the block of tissue at the anterior part of the wing bud (or posterior if removing the host polarizing region) that you wish to excise using a sharpened tungsten needle—This block should be approximately 150×150 μm (*see* **Note 3**). Use forceps to completely remove the block of tissue if necessary. The cut should not be touching the flank of the embryo.

3. Remove the donor embryo HH20/21 (GFP-expressing) from the egg. To do this, cut through the vitelline and chorionic membranes around the embryo and remove with forceps or a spatula. Place the embryo in a petri dish containing PBS, carefully remove the membranes, internal organs and head using forceps, and then transfer the embryo to the agarose dish containing DMEM.

4. Pin the donor embryo out using insect pins so that the embryo is viewed with the dorsal side facing upwards. Make two cuts in the posterior part of the limb bud where the polarizing region is located using dissecting scissors or sharpened tungsten needles (*see* Fig. 4). Transfer the "L" shaped pin to the agarose dish. Then using forceps, grasp the hooked part of the pin in order to push it into the tissue to be grafted. Finally, make a final cut in the limb bud so that the graft attached to the pin can be removed from the tissue (*see* Fig. 4). The graft size should be approximately 150×150 μm.

5. Pick up the polarizing region affixed to the wire pin using forceps and place onto a spatula to transfer it to the host egg. Maneuver the graft with the "L" shaped part of the pin using forceps and push it into the anterior margin of the host wing bud (or the posterior margin if replacing the host polarizing region). The wire pin should be kept in the limb bud to hold the graft in place while the tissue heals (*see* also **Notes 4** and **5**). Note that grafts should be made in contact with the host apical ectodermal ridge (*see* **Note 6**).

6. Seal egg with sellotape and return it to the incubator.

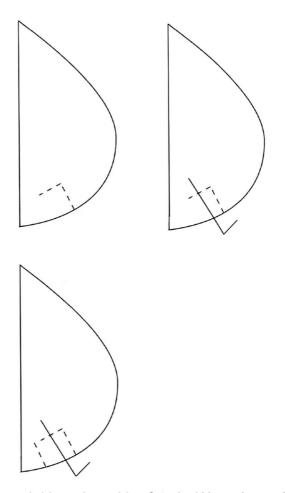

Fig. 4 Donor polarizing region excision. Cuts should be made according to the dashed lines. The wire pin should be inserted after two cuts have been made, prior to the final cut. The polarizing region can then be removed and held on the "L" shaped wire pin. To move the graft on the pin use the flattened spatula (*see* Fig. 3)

4 Notes

1. Grafts take approximately 15 min to complete when proficient.

2. Ensure that all dissecting tools are sterile and clean them often with a 70% ethanol solution to avoid infection.

3. Sharpen tungsten needles regularly to ensure accurate cutting of tissue.

4. To prevent infection when leaving embryos with grafts to develop until day 10 for cartilage staining, add approximately 50 μl of 1% penicillin/streptomycin in sterile DMEM to the egg.

5. To make sure that the correct region of cells has been grafted, detection of polarizing region cells can be determined in the host wing bud by analyzing *Shh* expression using whole-mount in situ hybridization (Fig. 1d). Maps of polarizing activity have also been made that can help with locating the polarizing region at different stages [23].

6. Polarizing region grafts can also be made under the apical ectodermal ridge. *See* tickle, 2008 for full procedure [24].

Acknowledgments

MT is funded by the Wellcome Trust; HS by the Wellcome Trust and the University of Sheffield. We thank Cheryll Tickle for comments on the manuscript and Helen Sang and Adrian Sherman (Roslin Institute, Edinburgh, UK) for providing GFP-expressing chicken eggs.

References

1. Towers M, Tickle C (2009) Growing models of vertebrate limb development. Development 136:179–190

2. Saunders J, Gasseling M (1968) Ectodermal mesenchymal interactions in the origin of limb symmetry. In: Fleischmajer RŽ, Billingham RE, Epithelial mesenchymal interactions, pp 78–97

3. Tickle C, Towers M (2017) Sonic hedgehog signaling in limb development. Front Cell Dev Biol 5:14

4. Hamburger V, Hamilton HL (1951) A series of normal stages in the development of the chick embryo. Dev Dyn 195:231–272

5. Towers M, Signolet J, Sherman A, Sang H, Tickle C (2011) Insights into bird wing evolution and digit specification from polarizing region fate maps. Nat Commun 2:426

6. McGrew MJ, Sherman A, Ellard FM et al (2004) Efficient production of germline transgenic chickens using lentiviral vectors. EMBO Rep 5:728–733

7. Riddle RD, Johnson RL, Laufer E, Tabin C (1993) Sonic hedgehog mediates the polarizing activity of the ZPA. Cell 75(7):1401–1416

8. Lopez-Martinez A, Chang DT, Chiang C et al (1995) Limb-patterning activity and restricted posterior localization of the amino-terminal product of Sonic hedgehog cleavage. Curr Biolland 5(7):791–796

9. Yang Y, Drossopoulou G, Chuang PT et al (1997) Relationship between dose, distance and time in Sonic hedgehog-mediated regulation of anteroposterior polarity in the chick limb. Development 124:4393–4404

10. Pickering J, Towers M (2016) Inhibition of Shh signalling in the chick wing gives insights into digit patterning and evolution. Development 143:3514–3521

11. Chinnaiya K, Tickle C, Towers M (2014) Sonic hedgehog-expressing cells in the developing limb measure time by an intrinsic cell cycle clock. Nat Commun 5:4230

12. Wolpert L (1969) Positional information and the spatial pattern of cellular differentiation. J Theor Biol 25:1–47

13. Tickle C, Summerbell D, Wolpert L (1975) Positional signalling and specification of digits in chick limb morphogenesis. Nature 254:199–202

14. Tickle C (1981) The number of polarizing region cells required to specify additional digits in the developing chick wing. Nature 289:295–298

15. Smith JC (1980) The time required for positional signalling in the chick wing bud. J Embryol Exp Morphol 60:321–328

16. Zwilling E, Hansborough LA (1956) Interaction between limb bud ectoderm and mesoderm in the chick embryo. III. Experiments with polydactylous limbs. J Exp Zool 132:219–239

17. Saunders J (1948) The proximo-distal sequence of origin of the parts of the chick

wing and the role of the ectoderm. J Exp Zool Part A: Ecol Integr Physiol 108:363–403

18. Summerbell D (1974) Interaction between the proximo-distal and antero-posterior co-ordinates of positional value during the specification of positional information in the early development of the chick limb-bud. J Embryol Exp Morphol 1:227–237

19. Smith JC, Wolpert L (1981) Pattern formation along the anteroposterior axis of the chick wing: the increase in width following a polarizing region graft and the effect of X-irradiation. J Embryol Exp Morphol 63:127–144

20. Wolpert L, Hornbruch A (1981) Positional signalling along the anteroposterior axis of the chick wing. The effect of multiple polarising region grafts. J Embryol Exp Morphol 63:145–159

21. Iten LE, Murphy DJ, Javois LC (1981) Wing buds with three ZPAs. J Exp Zool Part A: Ecol Integr Physiol 215:103–106

22. Summerbell D, Tickle C (1977) Pattern formation along the antero-posterior axis of the chick limb bud. In: Ede DA, Hinchliffe JR, Balls M (eds) M Vertebrate limb development and somite morphogenesis. Cambridge University Press, Cambridge, pp 41–53

23. MacCabe JA, Saunders JW Jr, Pickett M (1973) The control of the anteroposterior and dorsoventral axes in embryonic chick limbs constructed of dissociated and reaggregated limb-bud mesoderm. Dev Biol 31:323–335

24. Tickle C (2008) Grafting of apical ridge and polarizing region. Methods Mol Biol 461:313–323

Chapter 9

Analysis of PLETHORA Gradient Formation by Secreted Peptides During Root Development

Hidefumi Shinohara and Yoshikatsu Matsubayashi

Abstract

The formation of longitudinal zonation patterns is important for normal root development and is regulated by the transcription factor PLETHORA (PLT). PLT proteins form a concentration gradient, and PLT protein levels determine root zonation. A peptide hormone root meristem growth factor (RGF) and its receptors (RGFRs) act as key regulators of root development by regulating the PLT protein expression pattern. Here, we describe a method for monitoring PLT protein gradient patterns in *Arabidopsis* with exogenous RGF treatment.

Key words Plant, Root development, Root apical meristem, PLETHORA, Root meristem growth factor, Gradient, Receptor kinases, Ligand–receptor pairs

1 Introduction

During plant root growth, dividing cells in the root apical area must coordinate transitions from division to expansion and differentiation; thus, plants generate three distinct developmental zones: the meristem, the elongation zone, and the differentiation zone [1]. The meristem at the root tip, which is called the root apical meristem (RAM, Fig. 1a), is important for maintaining root development and is further divided into two regions: one region with slowly dividing stem cells and another with quickly dividing transit-amplifying cells. This longitudinal zonation pattern is important for normal root development and is regulated by PLETHORA (PLT). PLT is a transcription factor that is unique to plants and is known as a master regulator of root development [2–4]. PLT genes are only expressed in the stem cell region (Fig. 1b), whereas PLT proteins form a concentration gradient extending from the stem cell area into the elongation zone, across the region with rapidly dividing transit-amplifying cells (Fig. 1c). PLT protein levels determine root zonation; in particular, high PLT levels maintain stem cell identity,

Julien Dubrulle (ed.), *Morphogen Gradients: Methods and Protocols*, Methods in Molecular Biology, vol. 1863,
https://doi.org/10.1007/978-1-4939-8772-6_9, © Springer Science+Business Media, LLC, part of Springer Nature 2018

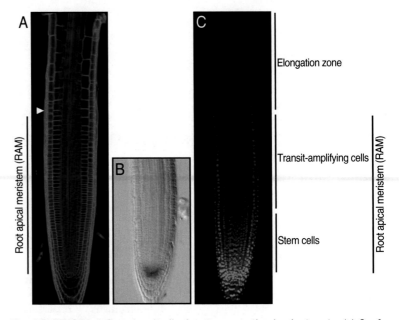

Fig. 1 PLETHORA defines longitudinal pattern zonation in plant roots. (**a**) Confocal image of the root apical meristem of a wild-type *Arabidopsis* seedling stained with propidium iodide (PI). White arrowheads indicate the root meristem boundary. (**b**) Whole-mount in situ hybridization with a *PLT2* antisense probe in the root meristem of a wild-type *Arabidopsis* seedling. (**c**) Confocal image of the root meristem of a wild-type *Arabidopsis* seedling expressing PLT2-GFP

intermediate PLT levels promote cell division, and low PLT levels permit cell elongation (Fig. 1c).

A peptide hormone, root meristem growth factor (RGF), is specifically expressed around the stem cell area and exhibits a concentration gradient in the RAM [5]. RGF is a 13-amino acid sulfated peptide hormone (Fig. 2a). Exogenous treatment of RGF peptide resulted in RAM enlargement (Fig. 2c, d), indicating that RGF positively regulates meristematic activity. RGF receptors have also been identified [6]. Three leucine-rich repeat receptor kinases (LRR-RKs) directly interact with RGF peptides in *Arabidopsis* (Fig. 2b). These three LRR-RKs, RGF receptor 1 (RGFR1), RGFR2, and RGFR3, are expressed in the meristem. A triple *rgfr* mutant was insensitive to RGF (Fig. 2e, f) and displayed a short root phenotype with a reduced number of meristematic cells (Fig. 2g).

The PLT protein gradient is decreased in both *rgf* and *rgfr* mutants [5, 6], indicating that RGFRs mediate the transformation of an RGF peptide gradient into a PLT protein gradient in the RAM and thereby act as key regulators of root meristem development. Here, we present a method for monitoring PLT protein gradient patterns in *Arabidopsis*. We introduce procedures for (1) expressing GFP-fused PLT2 (PLT2-GFP) in wild-type

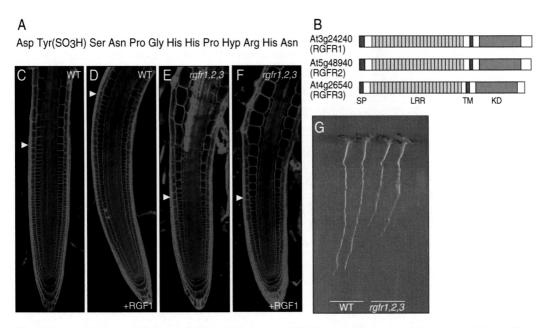

Fig. 2 Root meristem growth factor (RGF) and its receptors (RGFRs). (**a**) Sequence of the mature RGF1 peptide. (**b**) Schematic structures of RGF receptors. (**c–f**) Confocal images of the root apical meristem of wild type and the *rgfr* triple mutant stained with PI treated with 100 nM RGF1. White arrowheads indicate the root meristem boundary. (**g**) Root lengths of wild type and *rgfr* triple mutant seedlings

Arabidopsis and *Arabidopsis* with mutations in RGF signaling; (2) visualizing PLT2-GFP expression in the root tip; (3) detecting changes in the expression pattern of PLT2-GFP in response to treatment with exogenous RGF; and (4) performing time-course analysis in response to RGF application. These procedures show that RGF peptides regulate root development by stabilizing PLT protein expression patterns and mediate patterning of the root meristem.

2   Materials

2.1 Preparation of PLT2-GFP-Expressing Arabidopsis with Mutations in RGF Signaling

1. *Arabidopsis* accession Columbia (Col).

2. *Arabidopsis rgf* triple mutant derived from the T-DNA insertion lines (with a Col background) *rgf1-1* (SALK_132484), *rgf2-1* (SALK_145834), and *rgf3-1* (SALK_053439) (*see* **Note 1**).

3. *Arabidopsis rgfr* triple mutant derived from the T-DNA insertion lines (with a Col background) *rgfr1-1* (SALK_040393), *rgfr2-1* (SALK_096206) and *rgfr3-1* (SALK_053167) (*see* **Note 1**).

4. Gamborg B5 salt mixture (Wako, 399-00621).

5. Vitamin mixture: Add 25 mg of nicotinic acid, 25 mg of pyridoxine hydrochloride, 5 mg of thiamine hydrochloride, 5 g of myo-inositol, and 100 mg of glycine to a 500 mL glass beaker. Add 500 mL of water and mix. Store at −20 °C.

6. Sucrose.

7. Sterile dish.

8. Gamborg B5 medium: Add 1 bag of B5 salt mixture, 10 mL of vitamin mixture, and 10 g of sucrose to a 1 L glass beaker. Add water to a volume of 900 mL. Mix and adjust pH to 5.8 with aqueous KOH. Add water to a volume of 1 L, then autoclave.

9. Gamborg B5 plate: 0.7% W/V agar in Gamborg B5 medium with appropriate antibiotics (*see* **Note 2**).

10. Appropriate soil for *Arabidopsis* growth.

11. Luria–Bertani (LB) broth: 2% W/V tryptone, 1% W/V yeast extract, 2% W/V NaCl in H_2O.

12. LB plate: 1.5% agar in LB broth with appropriate antibiotics (*see* **Note 3**).

13. *Agrobacterium* strain C58C1 transformed with a binary vector containing a PLT2-GFP construct under the control of the PLT2 promoter (colonies on LB agar plates) (*see* **Note 4**).

14. Growth chamber set at 22 °C with continual light.

15. Silwet L-77.

16. Incubator shaker set at 30 °C.

2.2 Visualization of PLT2-GFP Expression Patterns in Wild-Type Plants and Plants with rgf and rgfr Mutations

1. Sterile B5 medium (*see* **Note 5**).

2. Sterile 24-well microplate.

3. Sterilized seeds of wild type, *rgf* triple mutant, and *rgfr* triple mutant *Arabidopsis* (Col) expressing PLT2-GFP.

4. Refrigerator set at 4 °C.

5. Growth chamber set at 22 °C with continual light.

6. Microforceps.

7. Slide glass.

8. Cover glass.

9. Kimwipes.

10. Confocal laser microscope (e.g., Olympus FV300).

2.3 Visualization of PLT2-GFP Expression Patterns with Exogenous RGF Treatment

1. Sterile B5 medium (*see* **Note 5**).

2. Sterile 24-well microplate.

3. RGF1 peptide solution: 400 µM RGF1 peptide in water (stock solution) (*see* **Notes 6** and **7**).

4. Sterilized seeds of wild type, rgf triple mutant, and rgfr triple mutant Arabidopsis (Col) expressing PLT2-GFP.

5. Growth chamber set at 22 °C with continual light.

6. Microforceps.

7. Slide glass.

8. Cover glass.

9. Confocal laser microscope (e.g., Olympus FV300).

2.4 Time-Course Analysis of PLT2-GFP Expression Patterns After Treatment with RGF

1. Sterile B5 medium (*see* **Note 5**).

2. Sterile 24-well microplate.

3. Seeds of *rgf* triple mutant *Arabidopsis* expressing PLT2-GFP.

4. Gamborg B5 plate (1.5% agar with a thickness of 5 mm that contains 100 nM of RGF peptide).

5. Growth chamber set at 22 °C with continual light.

6. Microforceps.

7. Slide glass.

8. Cover glass.

9. Humidity-maintaining box: a plastic petri dish with wet Kimwipes to prevent samples from drying up (Fig. 4a).

10. Confocal laser microscope (e.g., Olympus FV-300).

3 Methods

3.1 Preparation of PLT2-GFP-Expressing Arabidopsis in RGF Signaling Mutants

1. Put surface-sterilized *Arabidopsis* seeds on a B5 plate.

2. Vernalize seeds at 4 °C for 3 days.

3. Transfer plates to a growth chamber set at 22 °C with continual light.

4. After plants have bolted, place them into the soil in appropriate pods. Grow in a growth chamber set at 22 °C with continual light until flowering (*see* **Note 8**).

5. Streak *Agrobacterium* carrying a binary vector containing the PLT2-GFP construct onto LB medium containing appropriate antibiotics (*see* **Note 3**).

6. Pick colonies off the plate and transfer them into 100 mL of LB medium containing appropriate antibiotics for the plasmids that have been utilized. Culture for 16 h at 30 °C.

7. Collect *Agrobacterium* cells by centrifugation at 4000 × g for 15 min at room temperature, and resuspend cells in 50 mL of 5% sucrose.

8. Add Silwet L-77 to a concentration of 0.05% (vol/vol) and mix well.

9. Pour the *Agrobacterium* solution into an appropriately sized beaker.

10. Invert plants and dip the flowering parts of the plants into *Agrobacterium* solution for 2 to 3 s with gentle agitation (*see* **Note 9**).

11. Set dipped plants in a growth chamber set at 22 °C with continual light.

12. Grow plants up and harvest dry seeds (*see* **Note 10**).

13. Pour B5 plates containing appropriate antibiotics (*see* **Note 2**).

14. Place surface-sterilized transformant seeds onto B5 selection plates.

15. Vernalize seeds at 4 °C for 3 days.

16. Transfer plates to a growth chamber set at 22 °C with continual light. Grow until transformants can be readily distinguished as seedlings with healthy green cotyledons and roots that grow into the selection plates (*see* **Note 11**).

17. Transplant putative transformants (T1 generation) to B5 plates containing carbenicillin and cefotaxime and grow in a growth chamber set at 22 °C with continual light (*see* **Note 12**).

18. After bolting, replace T1 plants in the soil and grow in a growth chamber set at 22 °C.

19. Grow plants up for seeds (the T2 generation). Use T2 plants for the following experiments (*see* **Note 13**).

3.2 Visualization of PLT2-GFP Expression Patterns in Wild-Type Plants, rgf Mutants, and rgfr Mutants

1. Dispense 1 mL of sterile B5 medium into each well of a microplate.

2. Sow surface-sterilized *Arabidopsis* seeds directly into B5 medium in each well and keep the plate in the dark for 3 days at 4 °C (*see* **Note 14**).

3. Transfer microplates to a growth chamber set at 22 °C with continual light.

4. Incubate for 4 to 6 days.

5. Place seedlings carefully on a slide glass and cut the root tip using microtweezers (*see* **Note 15**).

6. Put the cut root tip on the slide glass with 2–3 drops of liquid B5 medium.

7. Put the cover glass on the root tip and use Kimwipes to remove unnecessary medium protruding from between the slide glass and cover glass (*see* **Note 16**).

8. Monitor PLT2-GFP fluorescence under a confocal microscope (exciters: 488 nm; emitter: 525 nm) (Fig. 3).

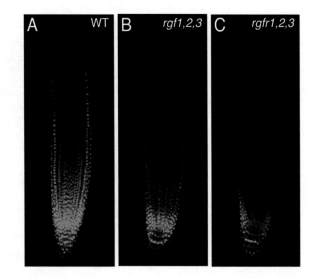

Fig. 3 PLT2 protein expression patterns. (**a**) Wild type. (**b**) *rgf* triple mutant. (**c**) *rgfr* triple mutant. The PLT2-GFP expression domain was strikingly diminished in the *rgfr* triple mutant compared with wild type. A similar reduction in the PLT2-GFP expression domain was also observed in the rgf triple mutant, albeit to a milder degree than in the rgfr triple mutant

3.3   Visualization of PLT2-GFP Expression Patterns After Exogenous RGF Treatment

1. Dispense 1 mL of sterile B5 medium into each well of a microplate.

2. Sow surface-sterilized *Arabidopsis* seeds directly into B5 medium in each well and keep the plate in the dark for 3 days at 4 °C.

3. Transfer microplates to a growth chamber set at 22 °C with continual light.

4. Incubate for 4–6 days.

5. Remove medium from each well and add new sterile B5 medium containing 100 nM RGF1.

6. Replace microplates in the growth chamber.

7. After plates have been incubated for 24 h, monitor PLT2-GFP expression patterns using the same procedure described in **steps 5–8** of Subheading 3.2 (Fig. 4).

3.4   Time-Course Analysis of PLT2-GFP Expression Patterns After Treatment with RGF

1. Dispense 1 mL of sterile B5 medium into each well of a microplate.

2. Sow surface-sterilized *Arabidopsis* seeds directly into B5 medium in each well and keep the microplate in the dark for 2 days at 4 °C. Transfer microplates to a growth chamber set at 22 °C with continual light.

3. Incubate 4 to 6 days.

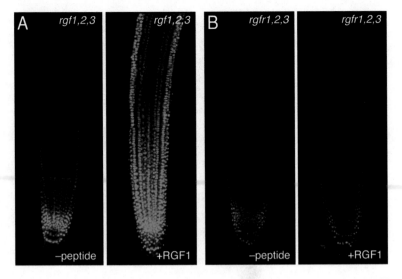

Fig. 4 PLT2 expression patterns in *rgf* and *rgfr* mutants after treatment with RGF1. PLT2 protein expression patterns after treatment with 100 nM RGF1 in the *rgf* triple mutant (**a**) and *rgfr* triple mutant (**b**). Exogenous application of RGF1 to the *rgf* triple mutant caused drastic shootward enlargement of PLT2-GFP expression. In contrast, the *rgfr* triple mutant was substantially less sensitive to RGF1

4. Cut a B5 agar plate containing 100 nM RGF1 to the size of the cover glass, and place the agar piece on the slide glass (Fig. 5a).

5. Replace seedlings carefully on the agar piece on a slide glass. Specifically, place the entire root on the agar, and place the hypocotyl and cotyledon off of the agar piece (Fig. 5a).

6. Put a cover glass that covers the whole root onto the agar piece.

7. Place a slide glass in a humidity-maintaining box and incubate in a growth chamber.

8. After slides have been incubated for 0 h, 1 h, 2 h, 4 h, 6 h, 12 h, and 24 h, monitor PLT2-GFP fluorescence under a confocal microscope (exciters: 488 nm; emitter: 525 nm) (Fig. 5b).

4   Notes

1. Seeds homozygous for each of the three *rgf* or *rgfr* mutants have previously been reported [5].

2. The antibiotics carbenicillin and cefotaxime are suitable to prevent bacterial contamination. Another suitable antibiotic for the positive selection of transformants should be added. In our case, we add hygromycin to B5 plates to select transformants containing T-DNA derived from the pBI101Hm binary vector.

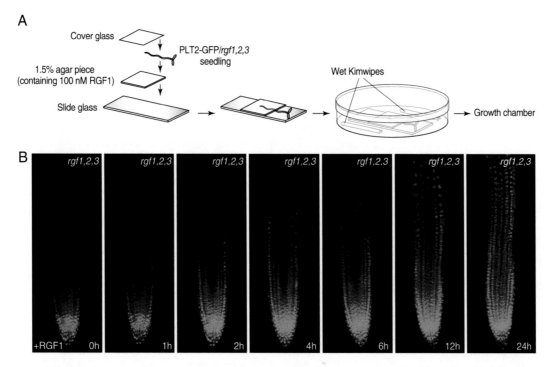

Fig. 5 Time-course visualization of PLT2 expression patterns after treatment with RGF1. (**a**) Schematic illustration of the time-course experiment. (**b**) PLT2-GFP expression patterns in the *rgf* triple mutant 0 h to 24 h after treatment with RGF1 peptide

3. Gentamicin and rifampicin are selective antibiotics for *Agrobacterium* strain C58C1 (pMP90). Another suitable antibiotic for selecting *Agrobacterium* containing a binary vector expressing PLT2-GFP should be also added. In our case, we add hygromycin to LB plates to select *Agrobacterium* containing the pBI101Hm binary vector.

4. Information regarding the construction of PLT2-GFP has previously been described [5].

5. At times, certain components of the medium sterilized via autoclaving may precipitate during seedling growth. These precipitants often interfere with visualization of the root tip. Filter sterilization is recommended for this experiment.

6. The strategy for the synthesis of RGF peptides has previously been described [5].

7. The peptide stock solution should be sterilized by filtration.

8. Under our conditions, growing *Arabidopsis* seedlings on a plate until bolting is helpful for the healthy growth of *Arabidopsis*.

9. A pipette can be used to drop solution onto some axillary floral buds that are too short to submerge into the solution.

10. Approximately 60 days are required to harvest seeds after *Agrobacterium* treatment.

11. During this step, fluorescence of GFP derived from transformed PLT2-GFP is detectable in root tips under a fluorescence microscope.

12. Antibiotics are essential for the positive selection of transformants but interfere with the growth of plants. We recommend removing antibiotics from the growth medium after selection.

13. Overall, 1/2 and 1/4 of the T2 plants will be heterozygous and homozygous, respectively, for the insert expressing PLT2-GFP. These lines are suitable for the following experiments. The remaining 1/4 of T2 plants will be wild type and should be excluded from the following experiments.

14. The use of six to eight seeds from each well is suitable for the following experiments.

15. A size smaller than approximately 1 mm of the root tip is suitable for detecting PLT2 expression patterns for microscopic analysis.

16. Gently tap the edge of the cover glass with Kimwipes and remove the liquid.

Acknowledgments

This research was supported by a Grants-in-Aid for Scientific Research on Innovative Areas (15H05957 to Y.M., 26113520 and 16H01234 to H.S.), a Grant-in-Aid for Scientific Research (S) (25221105 to Y.M.), and a Grant-in-Aid for Young Scientists (B) (25840111 to H.S.).

References

1. Dolan L, Janmaat K, Willemsen V, Linstead P, Poethig S, Roberts K, Scheres B (1993) Cellular organisation of the *Arabidopsis thaliana* root. Development 119(1):71–84

2. Aida M, Beis D, Heidstra R, Willemsen V, Blilou I, Galinha C, Nussaume L, Noh Y, Amasino R, Scheres B (2004) The PLETHORA genes mediate patterning of the *Arabidopsis* root stem cell niche. Cell 119:109–120

3. Galinha C, Hofhuis H, Luijten M, Willemsen V, Blilou I, Heidstra R, Scheres B (2007) PLETHORA proteins as dose-dependent master regulators of *Arabidopsis* root development. Nature 449:1053–1057

4. Mähönen A, Tusscher K, Siligato R, Smetana O, Díaz-Triviño S, Salojärvi J, Wachsman G, Prasad K, Heidstra R, Scheres B (2014) PLETHORA gradient formation mechanism separates auxin responses. Nature 515:125–129

5. Matsuzaki Y, Ogawa-Ohnishi M, Mori A, Matsubayashi Y (2010) Secreted peptide signals required for maintenance of root stem cell niche in *Arabidopsis*. Science 329:1065–1067

6. Shinohara H, Mori A, Yasue N, Sumida K, Matsubayashi Y (2016) Identification of three LRR-RKs involved in perception of root meristem growth factor in *Arabidopsis*. Proc Natl Acad Sci U S A 113:3897–3902

Chapter 10

Live Imaging of mRNA Transcription in Drosophila Embryos

Carmina Angelica Perez-Romero, Huy Tran, Mathieu Coppey, Aleksandra M. Walczak, Cécile Fradin, and Nathalie Dostatni

Abstract

Live imaging has been used in recent years for the understanding of dynamic processes in biology, such as embryo development. This was made possible by a combination of advancements in microscopy, leading to improved signal-to-noise ratios and better spatial and temporal resolutions, and by the development of new fluorescence markers, allowing for the quantification of protein expression and transcriptional dynamics in vivo. Here we describe a general protocol, which can be used in standard confocal microscopes to image early *Drosophila melanogaster* embryos, in order to learn about the transcriptional dynamics of a fluorescently labeled RNA.

Key words MS2 system, Live imaging, Confocal microscopy, RNA, Embryo

1 Introduction

The Drosophila embryo is a model system to study how cell identity is established at the right place and time during development. It was demonstrated using genetics that most of the patterning along the anteroposterior (AP) axis is set up by the expression of a set of gap genes under the control of morphogen gradients (for a review see [1]).

Quantification of gene expression during development was obtained so far from the detection of ribonucleic acid (RNA) on fixed material using fluorescence in situ hybridization (FISH) [2]. However, the rapid development of the embryo and frequent interruptions of gene expression during mitoses result in complex dynamics that cannot be inferred from static observations. Recently, the possibility to fluorescently tag RNA in living cells using well-characterized RNA targets (motifs) for specific RNA binding proteins fused to fluorescent domains provided access to

Cécile Fradin and Nathalie Dostatni contributed equally to this work.

Julien Dubrulle (ed.), *Morphogen Gradients: Methods and Protocols*, Methods in Molecular Biology, vol. 1863, https://doi.org/10.1007/978-1-4939-8772-6_10, © Springer Science+Business Media, LLC, part of Springer Nature 2018

the dynamics of transcription in real time. These developments combined with advances in confocal microscopy (which allowed reducing signal-to-noise ratio and improving spatiotemporal resolution) and in genome editing to tag the RNA of interest with motifs [3] give access to time dependent measurements of mRNA transcription in living organisms during their development.

Although several systems allowing to visualize RNA have been described [3–6], the MS2 system is probably the most popular one [7–9]. It relies on the MS2 bacteriophage RNA loops that can be added to a target RNA, and its core protein (MCP) that can be fused to a fluorescent protein like green fluorescent protein (GFP) allowing the visualization of the expression of the target RNA and quantify it [4]. Recently, the MS2 system has been used to tag different genes in fruit fly development, which has helped elucidate how gene circuits interact with each other in a quantitative manner [8, 10–15]. For example, we used this system to study the transcriptional response downstream of the Bicoid gradient, through the expression of its target gene *hunchback* [11, 12, 16]. The MS2 system has also been combined with another RNA tagging system (PP7) to tag two different RNA with different fluorescent proteins (GFP and RFP—red fluorescent protein) in the same embryo [17]. In our analyses, MS2 signals were very difficult to observe with epifluorescence and we had to use confocal microscopy to perform optical sectioning such that we could avoid the fluorescence signal coming from the whole thickness of the embryo. With the early fly embryo, our acquisition only involves the first few microns of the periphery of the embryo and thus does not require 2-photon microscopy. Yet, this type of microscopy can be used to image transcription in tissues that require deeper imaging. Finally, light sheet microscopy can also be used even though the technical bottleneck here will be image analysis on the collected data.

Here we describe a protocol to image transcription in living embryos using the MS2 system and standard laser scanning confocal microscopy.

2   Materials

2.1   Genetic Material

The system requires the simultaneous expression of three transgenes:

1. The MS2 reporter transgene: it includes the promoter of interest upstream of an MS2 reporter cassette (in general 24 MS2 loops) fused to the sequence of the iRFP (infrared fluorescent protein) or any coding sequence for a protein that you can easily detect (to insure that the tagged RNA is eventually translated). Importantly, the position of these two sequences relative to each other and to the promoter depends on the

question you ask. Briefly, a stronger MS2 signal will be obtained if the MS2 cassette is placed just downstream of the promoter (best signal-to-noise ratio) but faster dynamics of promoter bursting will be detected when the MS2 cassette is placed further away from the promoter (reviewed in [15]). Also, when designing the MS2 reporter make sure that cryptic binding sites for important transcription factors in your system are not localized in the spacers or stem loop sequences of the MS2 cassette. Finally, it is also possible to tag the 5' UTR, 3' UTR or introns of endogenous genes with the MS2 cassette by using genome editing approaches.

2. The MCP-GFP transgene: it expresses the MCP-GFP fusion protein at relatively low levels for increased signal-to-noise ratio. In the very early fruit fly embryo (syncytial), it was shown that an MCP-GFP without a nuclear localization signal (ΔNLS) improved detection of the specific transcription signal and reduced nuclear GFP aggregates at nuclear cycle 14 [12]. In the fly, there are available UAS-MCP-GFP lines [18] that can be combined with appropriate Gal4 drivers.

3. A transgene allowing the detection of nuclei: this reporter is required to identify the subcellular localization of the MS2 signal. These are for instance nucleoporin fused to RFP (Nup-RFP) to label the nuclear envelop (Bloomington # 35517) [19] or histone fused to RFP to label chromatin (Bloomington # 23650) [20]. In the early embryo, the signal detected from the His-RFP transgene is very strong and regenerates during each nuclear cycle, allowing for easier segmentation during image analysis. Even though it is slightly toxic for the embryo, we prefer it over other nuclear envelop markers such as maternally expressed Nup-RFP, whose signal is weaker and fades away during development.

The analyzed embryos result in general from a cross between females expressing the MCP-ΔNLS-GFP and His-RFP transgenes with males carrying the MS2 reporter. Once crossed, the resulting embryos will express maternally the MCP-GFP and His-RFP proteins and carry only a single MS2 locus, which helps for image analysis (*see* Chap. 11, this book). Embryos are then collected, removed from their chorion and imaged as follows using a confocal Laser Scanning Microscope 780 from Zeiss that enables the visualization of the reporter transgene in the green detection channel and the visualization of the histones in the red detection channel.

2.2 Sample Material

1. Appropriate fly stocks and crosses.

2. Paintbrush.

3. Embryo collection plates (ECP): Space nontreated petri dishes out, so that when mix is ready you can plate it rapidly. Weight

22 g of sucrose, and 14 g of Bacto agar. Add to 300 ml water and boil (using a microwave) until properly mixed and clear. While cool enough to touch, but still clear add 10 ml of pure ethanol 100% and 5 ml of glacial acetic acid. Add 100 ml of grape juice and mix well. Add mix to the petri dishes and let dry. Store in a closed plastic box with a little water to keep the humidity at 4 °C.

4. 10% acetic acid solution: 10% V/V glacial acetic acid in water.

5. Yeast paste: Gradually mix active dry yeast and water until they have a texture similar to cake frosting (sticky but not runny), this will favor embryo laying. Activated yeast will tend to expand so allow space for this to happen and break the air bubbles. The yeast paste can be kept fresh for a few days covered with parafilm and stored at 4 °C.

6. Spatula.

7. Embryo collection cage: To make homemade egg collection chamber you can use a narrow *Drosophila* vial. Carefully cut the bottom part of the vial, making a cylinder. On one end of the cylinder, add a cotton plug to allow for aeration of the chamber yet stop the flies from escaping. Flip the flies into the egg collection chamber and add the embryo collection plate on the other end, trapping the flies in the chamber. Secure the embryo collection plate to the egg collection chamber with tape.

8. Double-sided tape.

9. N-heptane glue: Cut small pieces of double-sided tape and put those inside a small glass bottle (use around 30 cm of double sided tape). Add 10 ml of *n*-heptane in the bottle to cover the tape (under a fume hood). Let the mixture incubate overnight. The heptane should dissolve the glue from the double-sided tape, which will make the heptane sticky. The mixture is good to use until most of the heptane-glue has evaporated. Remember that heptane evaporates fast and is flammable.

10. Fine point precision tweezers.

11. Dissecting microscope.

12. Dissecting needle.

13. Microscope glass coverslip (0.17 mm thickness).

14. Microscope slide.

15. 10 S Voltalef oil.

16. Confocal microscope (in our setup: Laser Scanning Zeiss, LSM 780 integrated with 488 and 585 laser lines with GaASp detection).

17. Heating and cooling chamber.

18. Immersion oil 518 F (23 °C).

19. Software for imaging acquisition.

20. Phosphate buffer saline (PBS, pH 7.4).

21. Fluorescein solution: Dissolve 1 g of fluorescein into 1.5 ml phosphate-buffer saline (PBS, pH 7.4) and vortex the solution. A fluorescein precipitate (red powder) should remain in the solution, indicating that the solution is saturated (this ensures that the concentration of dissolved fluorescein in the stock solution is always exactly the same, even if solvent evaporates). If kept in the dark at room temperature, this saturated fluorescein solution will be stable for at least 2 months. This solution can then be diluted to use for fluorescence standards.

3   Methods

3.1   Collecting Embryos

Virgin females carrying the MCP-GFPΔNLS and His-RFP are crossed with male flies carrying the MS2 reporter. The crosses must be performed 1 day in advance in culture vials to ensure that all females have been fertilized. Since live imaging experiments are usually carried out with a single embryo at a time, small numbers of embryos are sufficient to perform the experiment. However, it is recommended to have at least 10 females to carry out the study. Crosses will have a good embryo yield for about 4 days. Crosses are done in a normal feeding tube (Fig. 1a), and then flies are put to lay embryos on a collection plate (Fig. 1b).

3.1.1   Embryo Laying

1. Take 10 μl of 10% acetic acid solution and spread evenly with a paintbrush on an ECP.

2. With the end tip of the paint brush make a gentle grid of grooves on the ECP agar where flies will preferentially lay their eggs.

3. With a spatula add a small amount of yeast paste to the center of the ECP (Fig. 1a).

4. Take out crosses from narrow *Drosophila* vials and transfer to an embryo collection cage.

5. Rapidly cover the end of the embryo collection cage with the prepared ECP.

6. Safely secure the ECP and eggs collection cage together with tape (Fig. 1b).

7. Put back embryo collection cage into the incubator (*see* **Note 1**).

3.1.2   Chorion Removal and Preparation of the Coverslip for Imaging

Drosophila eggs are around 500 μm in length. They are white and can be seen with the naked eye on the contrasting background of ECPs. However, for chorion removal and preparation of the coverslip higher magnification is needed. Therefore, this procedure

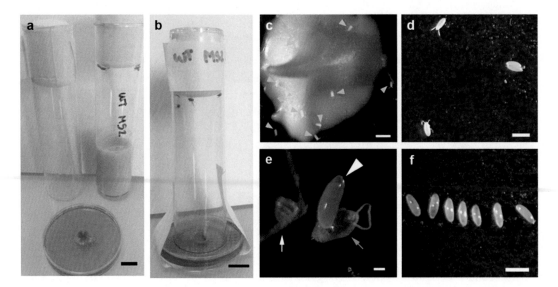

Fig. 1 *Drosophila* embryo collection and chorion removal. (**a**) Setup needed for making an egg laying chamber: an embryo collection plate prepared with 10% acetic acid, yeast paste and groves; an empty feeding tube closed with a plug; and an overnight cross of flies needed for laying. (**b**) Final setup of egg laying chamber once ready should be placed at 25 °C. (**c**) The embryo collection cage is removed from the incubator after the appropriate laying time, and eggs (marked by arrows) that are laid around and on top of the yeast paste (big white blob) are transferred to a slide (**d**) with a double-sided tape for removal of the chorion. (**e**) Embryos are rolled on double sided tape (white arrow) to be released from the chorion (orange arrow). Embryos at nc ~ 9 without a chorion (white arrow head) are transferred to a prepared coverslip with heptane glue. (**f**) The embryos are aligned and a drop of 10S oil is added before imaging. Scale bars: 5 mm (**a** and **b**), 1 mm (**c**), 500 μm (**d** and **f**), and 100 μm (**e**)

should preferably be carried out under a dissecting scope. A microscope with a low magnification objective with 2× or 5× objective can also be used.

1. Take 10 μl of heptane glue with a pipette and spread it over a microscope coverslip making fine lines, let dry (*see* **Note 2**).

2. Take the ECP from the collection cage by tapping gently the inverted cage and rapidly covering the opened cage end with another fresh ECP.

3. Using a precision tweezer or a double sided tape handled with tweezers, take the embryos from the ECP with care and transfer them to the double sided tape slide (Fig. 1c, d, *see* **Note 3**).

4. Remove the chorion of each embryo by gently touching its body, and rolling it over the double sided tape (Fig. 1d, e, *see* **Note 4**). Removal of the chorion by hand avoids the use of bleach (which prevents the embryo from sticking properly on the coverslip).

5. Once the chorion is removed, transfer the embryo to the coverslip, placing it on top of the glue line.

6. Place the embryo in the desired orientation for imaging (Fig. 1f).

7. Using the length of a dissecting needle gently tap the embryo in place to flatten it (*see* **Note 5**).

8. Cover the embryos with a generous drop of 10S Oil (*see* **Note 6**).

9. Secure coverslip into mounting device for imaging.

3.2 Setting and Optimizing Acquisition Parameters for Live Imaging

Here we describe a general protocol for adjusting the acquisition parameters of a Laser Scanning Microscope to obtain optimal spatial and temporal resolution for the imaging of live *Drosophila* embryos (or, in general, large-scale dynamic systems). This protocol should be repeated at the beginning of each different imaging series (e.g., each time a new type of embryo is being imaged). We used a Zeiss LSM 780 microscope and associated Zen Black software to develop this protocol; however, it can easily be adapted to any modern commercial confocal microscope. Tables 1 and 2 contain a brief

Table 1
Primary imaging parameters

Parameter	Symbol	Optimal value	Typical value
Magnification	M	Intermediate (allowing to image sufficiently large sample areas with an appropriate spatial resolution)	40×
Numerical aperture	NA	High (to achieve appropriate spatial resolution)	1.4
Excitation wavelength	λ	Intermediate (low λ generally result in higher signal levels but also more photodamage and more autofluorescence).	488 nm, 561 nm
Excitation power	P	Low (to avoid photobleaching)	3%–3.2% of laser power
Confocal pinhole diameter	D	~2wM = 1 Airy unit (AU)	31.5 μm = 0.99 AU
Detector gain	ɣ	Intermediate (to achieve both a linear response and sufficient signal level)	Master gain =700 Digital gain = 2
Pixel separation	d	~w	0.15 μm
Image separation	d_Z	~5w	0.5 μm
Pixel dwell time	δ	Short (to allow fast imaging)	0.5 μs
ROI width, height, depth	L_X, L_Y, L_Z		240 μm, 100 μm, 10 μm

Table 2
Derived imaging parameters

Parameter	Relation to other parameters	Typical value
Point-spread function radius	$w \approx \lambda/(2\mathrm{NA})$	488 nm/561 nm 0.17 μm/0.20 μm
Point-spread function half-height	$w_z = 5w$	1 μm
Pixels per line	$n_X = L_X/d$	1200
Lines per image	$n_Y = L_Y/d$	512
Images per z-stack	$n_Z = L_Z/d_Z$	20–30
Image acquisition time	$r > n_X\, n_Y\, \delta$	0.460 ms
Z-stack acquisition time (scan rate)	$R > n_X\, n_Y\, n_Z\, \delta$	10–20 s

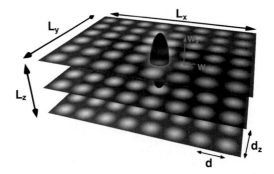

Fig. 2 Schematic representation of the confocal imaging process. The sketch shows the point-spread function at the laser focus (green revolution ellipsoid, with radius w and half-height w_z) as it scans through every pixel (each pixel is represented by a blue disk) in the ROI (width L_X, height L_Y, depth L_Z). The distance between two pixels is d. In this particular example, the confocal imaging results in a z-stack of $n_z = 3$ images, each with pixel dimensions $n_x = 8$ and $n_y = 6$. Also, in this example the images are undersampled ($d > w$)

description of the imaging parameters to be optimized, and give their typical value in our experiments as a reference. Figure 2 gives an illustration of the confocal imaging scheme and a graphic representation of some of these imaging parameters.

1. Select an objective with moderate magnification (e.g., 40×) and high numerical aperture (NA), which usually requires water or oil immersion. For example, the images shown here were acquired with a plan-apochromat 40×/1.4 NA oil immersion objective.

2. Activate the necessary laser lines and select appropriate filter cubes for the fluorophores to be imaged (in our case: GFP, maximally excited at 488 nm and visualized around 520 nm,

and RFP, maximally excited at 586 nm and visualized around 620 nm). The emission filters should have a bandpass as large as possible to maximize fluorescence signal collection, while avoiding cross talk between the two channels.

3. Set the pinhole diameter to 1 Airy unit for an optimal trade-off between resolution (both in the focal plane and along the optical axis) and signal collection. If necessary, the pinhole diameter can later be increased above 1 Airy unit to increase signal collection efficiency (and thereby possibly scan rate), at the cost of spatial resolution and sectioning power (and vice versa).

4. Place a representative sample (of the type that will be studied) on the microscope stage. Quickly select a laser power and detector gain that allows detecting the signal of interest (these two parameters will be adjusted more precisely at a later step).

5. Choose the dimensions of the region of interest (ROI) that is the size of the region in the sample that will be imaged. In general, this will need to be chosen according to your experimental question. Remember, however, that a large ROI can only be imaged at the expanse of spatial and/or temporal resolution.

6. Next choose the number of pixels per line and per row, or, alternatively, the pixel separation (Fig. 2). The Nyquist criterion can be used to determine the optimal pixel separation (*see* **Note 7**). As a rule, the more pixels the better the spatial resolution but the longer the acquisition time. Optimizing pixel separation is thus important, as it will greatly influence scan rate.

7. Choose the span (depth) and increment (separation between two consecutive images) of the z-stack for volumetric acquisition (Fig. 2). Optimizing the distance between images in the z-stack will influence the number of stacks needed for imaging a ROI with a particular depth, and thus affect time resolution. The optimal distance between images in the stack can also be estimated using the Nyquist criterion (*see* **Notes 7** and **8**).

8. The last adjustments to be made concern the pixel dwell time (or alternatively the scan acquisition time, as both are directly related). The pixel dwell time can be adjusted according to your experimental question (so that the dynamics of the process under study can be captured). Shorter pixel dwell times allow to capture faster processes and to minimize photobleaching; however, this is at the cost of signal collection.

9. To optimize the levels of fluorescence signal detected, after a pixel dwell time has been chosen, adjust the laser power and detector gain to achieve a sufficient signal-to-noise ratio to

visualize the process of interest (as estimated by eye from the acquired images). This has to be done separately for both channels. Laser power and detector gain must be adjusted together, until the best compromise is obtained. Remember that the laser power should be kept sufficiently low to minimize phototoxicity (short wavelength light is the most damaging to organisms) and photobleaching, while high detector gain can result in nonlinear detector response.

10. If time resolution is of great importance, after all the other settings have been optimized for this purpose, a bidirectional scanning mode can be selected. This can greatly improve the time resolution; however, it can also cause image artifacts when using high magnification objectives (*see* **Note 9**). Once all the imaging parameters have been optimized, remember to save them (e.g., as an experimental setup in the Zen Black software) in order to be able to use them again in subsequent imaging sessions.

11. As laser power and microscope optical alignment may vary with time, affecting the amount of fluorescence signal collected from embryos, we recommend using a control fluorescence slide to keep a record of the intensity of the signal that can be obtained in this particular instrumental configuration (*see* **Note 10**). This can be used as a benchmark in subsequent experiments.

Once the imaging parameters have been optimized for a particular set of experiments, they should be saved and used consistently in subsequent experiments of the same type, allowing reproducible experiments to be used for quantitative analysis.

3.3 Live Imaging of Drosophila Embryos

Here we describe the basic procedure to record a 3D movie of a live embryo. This procedure generates a series of two-color image stacks. Each image stack is the 3D confocal image of the same region (the region of interest, or ROI) of the embryo at a different time (as illustrated in Fig. 2). Along with the 3D movie, we recommend performing a tile scan of the whole embryo, which will allow for the measurement of the embryo dimensions (antero-posterior and dorso-ventral axes) and exact localization of the ROI within the embryo. Finally, we explain how to use maximum projection to produce a 2D movie that can be then exported as an .avi file and used for quick visualization.

1. Set the microscope incubator chamber to 23 °C, and allow to equilibrate for 30 min (*see* **Note 11**).

2. Turn on the lasers to be used for excitation, and allow their output intensity to equilibrate if necessary.

3. Select the correct objective and carefully add a drop of water or immersion oil if required.

4. Load the confocal imaging parameters saved at the end of the setup procedure described in the previous section, and record an image of a fluorescence calibration slide (*see* **Note 10**). Check that the signal obtained from this slide is similar to that obtained in previous experiments. A signal that is significantly lower (or higher) than usual indicates that there is an issue with the alignment or the settings of the microscope (e.g., incorrect laser intensity, incorrect fluorescence filters, incorrect detector gain, misaligned confocal pinhole), something that should be dealt with before further imaging. Small daily variations in signal, on the other hand, are to be expected and no cause for concern. The intensity recorded for the calibration slide can then be used to normalize the intensity of the images acquired during this session, allowing comparison between the data obtained in different sessions.

5. Place the coverslip with the embryos (prepared as explained in Subheading 3.1) onto the microscope stage. Turn on the bright field (transmitted) illumination and move the region of the sample containing the embryos (that should be visible to the naked eye) in the field of view.

6. Adjust the objective focus until the embryos come into focus.

7. Still using bright field imaging, you can quickly observe each embryo in the sample to determine which ones are in the desired developmental stage (*see* **Note 12**).

8. Once an embryo in the correct developmental stage has been spotted, you can rotate the coverslip to roughly orient the embryo as desired relative to the scanning axis.

9. Start live confocal acquisition to check for the presence of a fluorescence signal in the chosen embryo and confirm its developmental stage (*see* **Note 13**).

10. At this point, the image of the embryo may be digitally rotated to position it properly within the computer screen (e.g., with the antero-posterior axis oriented horizontally).

11. Move the stage to place the region of interest in the center of the field of view.

12. Take a full sagittal plane image of the embryo by acquiring three images (tiles) that can be assembled to form an image of the whole embryo (Fig. 3a). This will allow the measurement of the embryo dimensions, as well as for the exact location of the ROI within the embryo (e.g., along the antero–posterior axis). This is important for quantification and image analysis purposes. This tile scan can also be performed at the end of the experiment (*see* **Note 14**).

13. Choose the ROI (region of interest, Fig. 3) that encompasses the area of interest in the embryo (*see* **Note 15**).

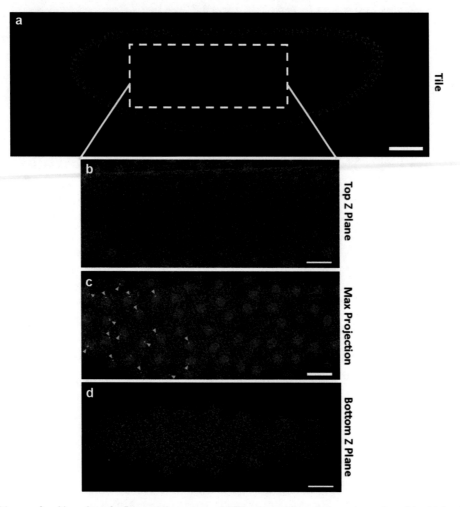

Fig. 3 Live confocal imaging of a Drosophila embryo. (**a**) Tile view of the whole embryo at nc 14, obtained from three adjacent tiles, auto-stitched at the end of the acquisition. The ROI within the embryo is indicated by the dashed line rectangle. Scale bar: 50 μm. **b–d**. Snapshots of different views of the ROI at nc 11. (**b**): Uppermost image in the z-stack, above the layer of nuclei. (**d**): Bottom image in the z-stack, below the layer of nuclei. (**c**): Maximum projection of the 23 images in the z-stack (i.e., all the images acquired between planes (**b**) and (**d**), each separated by a distance $d_z = 0.5$ μm). Active transcription sites are indicated by white arrowheads. Scale bar: 20 μm (**b–d**)

14. Define the depth of the region to be imaged, by choosing the position of the top and bottom planes in the z-stack (Fig. 3b, d, *see* **Note 16**). The number of images in the z-stack will then be calculated according to the optimal image separation determined during the optimization step (Subheading 3.2).

15. Select continuous acquisition, or a very high amount of imaging cycles. In this way the acquisition can be terminated based on the completion of the biological process under study.

16. Finally, start the acquisition and let the embryo develop without disturbing (*see* **Note 17**).

17. When appropriate, stop the acquisition and save the resulting data (*see* **Note 18**).

18. One can perform a fast image processing of the acquired 3D movie, using maximum projection, in order to generate a 2D movie with a single image per time point (Fig. 3c). This simplified version of the data can then be saved in .avi or .mov format, making it easy to disseminate or to share with collaborators. It also allows for a rapid visualization of the studied process.

    Movies of transcriptional dynamics obtained with flies carrying a MS2 reporter and expressing MCP-GFPΔNLS, as well as a nuclear marker such as His-RFP, can then be analyzed using the *LiveFly* toolbox described in Chap. 11.

4 Notes

1. Regular 12 h day–night incubation cycles are important to have consistent growth and a good yield in embryo laying. Since egg laying occurs preferentially during sunset and dark periods, it can be useful to set up the incubator on an inverted day–night schedule to optimize live imaging experiments during the working hours of the laboratory.

2. Since the heptane evaporates quickly, using too much of heptane glue will create an uneven surface of glue for the embryos to lay. It is thus important to take a small quantity of heptane glue and spread it rapidly in a stripe like manner over the coverslip. The more even the surface, the easier it is to position the embryos and flatten them, which can influence both the quality of imaging and the development of the embryo.

3. The embryos with their chorions will easily stick to one another and to the inside or outside part of the tweezers, making the transfer easier.

4. The chorion will stick to the double sided tape. To release the embryo from the chorion, roll it gently until the chorion breaks. Once the embryo is partially released, move slightly the chorion to release the embryo completely, and using tweezers or a piece of double sided tape handled with tweezers transfer the embryo without chorion to the coverslip. Removal of the chorion by hand should be a gentle process, and it may take time and practice before one is able to do it properly. Also, the younger the embryos, the more fragile they are and the harder it is to remove their chorion without damage.

5. Since the embryo is slightly curved, focusing on its central part may result in both its extremities being slightly out of focus, making them hard to image with high enough resolution. For studies relying on proper imaging of both the central part of the embryos and their anterior and/or posterior end, it is thus important to flatten the embryos. We have found it hard to flatten embryos without affecting their viability and expression. However, the simple step of tapping an embryo with the length (not the tip!) of a dissecting needle helps not only to glue the embryo firmly to the coverslip but also to flatten its ends.

6. The 10S Oil is very viscous, so it is hard to pipette and handle. However, it is important to cover the embryos properly and shortly after removal of their chorion to prevent them from drying. This oil is oxygen permeable such that embryos can continue to develop while under the microscope.

7. The pixel size (sometime referred to as sampling rate), d, is related to the width of the ROI (L_X) and to the number of pixels per line (n_X) through $d = L_X/n_X$. The pixel size can usually either be set directly in the software, or controlled by varying L_X (in the Zeiss LSM, this is done by changing the so-called zoom factor) and/or n_X (that is by changing the image pixel dimensions). Usually, one would use square pixels, with $L_Y/n_Y = L_X/n_X$. For the Nyquist criterion to be met, a minimum of two pixels must be used to image the length of an area with the dimension of the point spread function. In other words, the pixel size, d, should be no larger than the radius of the point-spread function, w (*see* Table 2). If $d > w$, undersampling occurs, which means that spatial information might be lost. If $d < w$, oversampling occurs, which means that one is sacrificing time resolution without any gain in spatial resolution.

8. The Nyquist criterion is also relevant to the choice of the distance between two images in a z-stack, d_Z. As the point-spread function is elongated along the optical axis, usually by a factor of about 5, the optimal image separation, d_Z, should be around $5d_X$. However, just as for the pixel size, one can choose to trade off spatial resolution (by choosing a larger d_Z) for temporal resolution.

9. To check that the bidirectional scan is not affecting your signal take a small five frame time series of your biological sample. Focus on an area with bright spots that remains immobile over time. Then select a single bright pixel, and check that it does not move along the scanning direction in subsequent frames, and that the image is not blurred. If problems with bidirectional scanning occur, they are usually easily seen from one frame to the next in your biological signal. If problems with

bidirectional acquisitions occur, return to the line scanning mode, which unfortunately has a much lower time resolution.

10. A saturated solution of fluorescein can be used to prepare slides with reproducible fluorophore concentration, for calibration purposes. When needed, dissolve a small amount of the saturated fluorescein solution in fresh PBS. (It is very important to always use the same dilution; it is also very important that the pH of the PBS used for dilution is exactly at 7.4, as fluorescein is a pH-sensitive dye.) Use this diluted solution to prepare a calibration slide (stick a coverslip on a microscope slide with heated parafilm spacers, fill the gap between slide and coverslip with the dye solution, then seal with melted wax or transparent nail polish to avoid evaporation). A fresh calibration slide (usually prepared on the same day) can be used at the beginning of each imaging session for reference: a similar signal obtained from the calibration slides before two different imaging sessions means that a quantitative comparison of the images obtained in both sessions can be made, while decrease in the signal measured from the calibration slide over time may indicate variations in the laser power output or misalignment of the optics (e.g., confocal pinhole alignment). If quantitative image analysis is to be done using the red channel, a similar fluorescence slide standard can be prepared, e.g., using rhodamine. Commercial fluorescence standards such as Argolight (Argo-HM) that work for most wavelengths and are photostable for years are also available, but at a cost.

11. It is important to equilibrate the temperature of the system since changes in temperature can cause focus drift and affect image quality. A temperature control chamber that allows not only heating but also cooling is recommended, since usually heating devices are optimized for 37 °C applications. However, live imaging of *Drosophila* embryos requires a temperature of 23 °C. Typical incubator systems usually have difficulty maintaining the system at this temperature during acquisition, due to the additional heat emanating from the lasers. Therefore, it is recommended to place the microscope in a room kept at a cool temperature, and/or an incubator system with both heating and cooling capabilities, with a feedback to keep the system at a stable 23 °C temperature during imaging, such as the one available from Tokai Hit (INUC-KPP Series).

12. Developmental features of the drosophila embryos can be used to determine their developmental stage (their exact nuclear cycle (nc)). The earliest signal that we have been able to detect with our MS2 system has been around nc 9 or 10. A useful marker for this stage is the emergence of the polar bodies at the posterior side of the embryo in early nc 9. On the other hand,

nc 14 is easily recognized by its very long duration and the size and amount of nuclei in the embryos.

13. Using a *Drosophila* strain expressing either a histone or a nucleoporin fused to RFP can help determine the developmental stage of the embryo during live imaging (by quickly inspecting the red channel), as well as the exact position of nuclei, which is important both during imaging to determine the optimal ROI, and during the segmentation step of image analysis (*see* Chap. 11). If an embryo is found to be at an earlier developmental stage than desired, one can wait for it to develop to the right stage. It is a good idea to keep track of a few embryos at a time in the sample to anticipate which ones might reach the correct stage to be imaged. This can easily be done using a multidimensional acquisition setup to save the position of different embryos. Once one of the embryos is at the desired developmental stage, the multidimensional acquisition can be canceled and the correct embryo selected for imaging.

14. The tile scan should cover the entire length of the embryo, including the ROI in the sagittal plane. If the embryo is horizontal, this can usually be achieved using three horizontal tiles. The tiles can be processed with auto-stitching, or manually stitched. It is then possible, for example, to measure the distance between the ROI and the anterior and posterior poles of the embryo. This might be important for quantification and image analysis when comparing several different embryos. The length of the embryo can also be measured using this tile scan.

15. Once a ROI has been created it can be saved and loaded, allowing to use ROI with the exact same size between experiments. The smaller the ROI, the faster the scan rate, so it is really important to make sure the imaging region is as small as possible.

16. When one is interested in imaging the cortical region of the embryo, where nuclei can be found after nc 8, we recommend to set the bottom and top image planes of the z-stack (rather than just setting the center one). The first (bottom) plane should be set above the lower membrane of the embryo and before the first nuclear signal is seen in the red channel. The last (top) plane should be set up a few microns above the first plane, after most nuclei have been observed (while moving the focus higher in the sample from the first plane). It is a good idea, however, to add one or a few planes up and down to account for focus drift during image acquisition. It is important also to use the same number of images per z-stack between acquisitions, since this will influence the scan rate, which should be

consistent between experiments if quantitative comparisons are to be made.

17. At the beginning of acquisition, make sure that the embryo displays the expected features, for example that foci corresponding to nascent RNA appear in the GFP channel, and that the embryo development appears normal, e.g., by checking the nuclear signal in the RFP channel. Sometimes the chosen embryo might not develop correctly or might not develop at all, in which case another embryo will have to be chosen, or a new slide will have to be prepared, and the procedure repeated.

18. The acquired image stacks can be saved in .czi or .lsm format if using the Zen software. Other formats, for example .tiff can be used with other imaging software. The *LiveFly* toolbox described in Chap. 11 is capable of opening any microscopy format compatible with the Bio-formats LOCI tools [21].

Acknowledgments

The authors thank Patricia Le Baccon and the Imaging Facility PICT-IBiSA of the Institut Curie. This work was supported by a PSL IDEX REFLEX Grant for Mesoscopic Biology (ND, AMW, MC), an Ontario Trillium Scholarship for International Students (CAPR), a Mitacs Global Link Scholarship (CAPR) and an Internal Curie Institute Scholarship (CAPR), ARC PJA20151203341 (ND), a Mayent Rothschild sabbatical Grant from the Curie Institute (CF) and an NSERC discovery grant RGPIN/06362-15 (CF), a Marie Curie MCCIG grant No. 303561 (AMW), ANR-11-LABX-0044 DEEP Labex (ND), ANR- 11-BSV2-0024 Axomorph (ND and AMW) and PSL ANR-10-IDEX-0001-02. Cécile Fradin and Nathalie Dostatni contributed equally to this work. The funders had no role in study design, data collection and analysis, decision to publish, or preparation of the manuscript.

References

1. Porcher A, Dostatni N (2010) The bicoid morphogen system. Curr Biol 20:R249–R254. https://doi.org/10.1016/j.cub.2010.01.026

2. Jensen E (2014) Technical review: in situ hybridization. Anat Rec 297:1349–1353. https://doi.org/10.1002/ar.22944

3. Urbanek MO, Galka-Marciniak P, Olejniczak M, Krzyzosiak WJ (2014) RNA imaging in living cells – methods and applications. RNA Biol 11:1083–1095. https://doi.org/10.4161/rna.35506

4. Bertrand E, Chartrand P, Schaefer M, Shenoy SM, Singer RH, Long RM (1998) Localization of ASH1 mRNA particles in living yeast. Mol Cell 2:437–445. https://doi.org/10.1016/S1097-2765(00)80143-4

5. Elf J, Li GW, Xie XS (2011) Probing transcription factor dynamics at the single-molecule level in a living cell. Science 316:1191–1194. https://doi.org/10.1126/science.1141967

6. Nelles DA, Fang MY, O'Connell MR, Xu JL, Markmiller SJ, Doudna JA, Yeo GW (2016) Programmable RNA tracking in live cells with

CRISPR/Cas9. Cell 165:488–496. https://doi.org/10.1016/j.cell.2016.02.054

7. Cusanelli E, Perez-Romero CA, Chartrand P (2013) Telomeric noncoding RNA TERRA is induced by telomere shortening to nucleate telomerase molecules at short telomeres. Mol Cell 51:780–791. https://doi.org/10.1016/j.molcel.2013.08.029

8. Bothma JP, Garcia HG, Ng S, Perry MW, Gregor T, Levine M (2015) Enhancer additivity and non-additivity are determined by enhancer strength in the Drosophila embryo. Elife 4. https://doi.org/10.7554/eLife.07956

9. Desponds J, Tran H, Ferraro T, Lucas T, Perez-Romero CA, Guillou A, Fradin C, Coppey M, Dostatni N, Walczak AM (2016) Precision of readout at the hunchback gene. PLoS Comput Biol 12(12): e1005256. https://doi.org/10.1371/journal.pcbi.1005256

10. Bothma JP, Garcia HG, Esposito E, Schlissel G, Gregor T, Levine M (2014) Dynamic regulation of eve stripe 2 expression reveals transcriptional bursts in living Drosophila embryos. Proc Natl Acad Sci U S A 111:10598–10603. https://doi.org/10.1073/pnas.1410022111

11. Lucas T, Ferraro T, Roelens B, De Las Heras Chanes J, Walczak AM, Coppey M, Dostatni N (2013) Live imaging of bicoid-dependent transcription in Drosophila embryos. Curr Biol 23:2135–2139. https://doi.org/10.1016/j.cub.2013.08.053

12. Garcia HG, Tikhonov M, Lin A, Gregor T (2013) Quantitative imaging of transcription in living Drosophila embryos links polymerase activity to patterning. Curr Biol 23:2140–2145. https://doi.org/10.1016/j.cub.2013.08.054

13. Lim B, Levine M, Yamakazi Y (2017) Transcriptional pre-patterning of Drosophila gastrulation. Curr Biol 27(2):286–290. https://doi.org/10.1016/j.cub.2016.11.047

14. Esposito E, Lim B, Guessous G, Falahati H, Levine M (2016) Mitosis-associated repression in development. Genes Dev. 30(13): 1503–1508. https://doi.org/10.1101/gad.281188.116

15. Ferraro T, Esposito E, Mancini L, Ng S, Lucas T, Coppey M, Dostatni N, Walczak AM, Levine M, Lagha M (2016) Transcriptional memory in the Drosophila embryo. Curr Biol 26:212–218. https://doi.org/10.1016/j.cub.2015.11.058

16. Ferraro T, Lucas T, Clémot M, De Las Heras Chanes J, Desponds J, Coppey M, Walczak AM, Dostatni N (2016) New methods to image transcription in living fly embryos: the insights so far, and the prospects. Wiley Interdiscip Rev Dev Biol 5:296–310. https://doi.org/10.1002/wdev.221

17. Fukaya T, Lim B, Levine M (2016) Enhancer control of transcriptional bursting. Cell 166:358–368. https://doi.org/10.1016/j.cell.2016.05.025

18. Forrest KM, Gavis ER (2003) Live imaging of endogenous RNA reveals a diffusion and entrapment mechanism for nanos mRNA localization in Drosophila. Curr Biol 13:1159–1168. https://doi.org/10.1016/S0960-9822(03)00451-2

19. Katsani KR, Karess RE, Dostatni N, Doye V (2008) In vivo dynamics of Drosophila nuclear envelope components. Mol Biol Cell 19:3652–3666. https://doi.org/10.1091/mbc.E07-11-1162

20. Pandey R, Heidmann S, Lehner CF (2005) Epithelial re-organization and dynamics of progression through mitosis in Drosophila separase complex mutants. J Cell Sci 118:733–742. https://doi.org/10.1242/jcs.01663

21. Linkert M, Rueden CT, Allan C, Burel J-M, Moore W, Patterson A, Loranger B, Moore J, Neves C, Macdonald D, Tarkowska A, Sticco C, Hill E, Rossner M, Eliceiri KW, Swedlow JR (2010) Metadata matters: access to image data in the real world. J Cell Biol 189:777–782. https://doi.org/10.1083/jcb.201004104

Chapter 11

LiveFly: A Toolbox for the Analysis of Transcription Dynamics in Live Drosophila Embryos

Huy Tran, Carmina Angelica Perez-Romero, Teresa Ferraro, Cécile Fradin, Nathalie Dostatni, Mathieu Coppey, and Aleksandra M. Walczak

Abstract

We present the *LiveFly* toolbox for quantitative analysis of transcription dynamics in live *Drosophila* embryos. The toolbox allows users to process two-color 3D confocal movies acquired using nuclei-labeling and the fluorescent RNA-tagging system described in the previous chapter and export the nuclei's position as a function of time, their lineages and the intensity traces of the active loci. The toolbox, which is tailored for the context of *Drosophila* early development, is semiautomatic, and requires minimal user intervention. It also includes a tool to combine data from multiple movies and visualize several features of the intensity traces and the expression pattern.

Key words Drosophila melanogaster, Real-time monitoring, Transcription dynamics, Image analysis, Early development

1 Introduction

The adaptation of the RNA fluorescent tagging system to fly embryos allows us to capture the temporal aspects of transcription [1, 2]. This system makes use of interactions between a specific Coat Protein (e.g., MCP) and an RNA stem loop generally derived from viral systems (e.g., MS2 stem loop) to monitor the dynamics of nascent RNA at the active transcription loci (see previous Chapter). These constructs, such as the MCP-GFP [1, 2] and PP7 [3] systems have been successfully implemented in the *Drosophila* embryo and used to reveal several important features of gene expression in the early phase of development [1–6].

Obtaining biological information from the unprecedented real-time observations of transcription dynamics using the MCP-MS2 system requires processing of the movies obtained from

---

Mathieu Coppey and Aleksandra M. Walczak contributed equally to this work.

Julien Dubrulle (ed.), *Morphogen Gradients: Methods and Protocols*, Methods in Molecular Biology, vol. 1863, https://doi.org/10.1007/978-1-4939-8772-6_11, © Springer Science+Business Media, LLC, part of Springer Nature 2018

microscopy. Many available tools [7, 8] were developed for bacterial colonies and are not well adapted for the analysis of fly images. They lack the flexibility of switching between a specialized automatic mode, which is essential for efficient processing, and manual intervention mode, which is often required to maximize the amount of samples acquired from each time-consuming experiment.

In this chapter, we present the *LiveFly* toolbox, which provides an interface for the analysis of the MCP-MS2 movies acquired in the early nuclear cycles, from nuclear cycle (nc) 10 to nc 14, using the confocal microscope. From the movie's two-colored channels, one for nuclear labeling and one for observations of nascent mRNA dynamics, the interface exports the nuclei's position, lineage and the intensity of the active loci as a function of time. The processing is semiautomatic: it requires user input only at the beginning of the process, but allows users to intervene for manual corrections if needed. It also includes a Visualizer to combine the data from multiple movies and visualize several features of the intensity traces and the pattern. The toolbox has been used to study the hunchback patterning system [9, 10], reducing processing time for each movie from days to hours.

The *LiveFly* toolbox has three modules as shown in Fig. 1, forming a pipeline for movie processing and analysis. We present the functions and guide the reader through each module in the following sections.

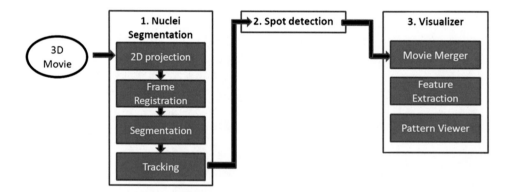

Fig. 1 Modules of the *LiveFly* toolbox. (1) The Nuclei segmentation module extracts from the 3D movie's nuclei channel (marked by either fluorescent histones or NUP) the temporal position and lineage of each nucleus. (2) The Spot detection module extracts the intensity trace of active transcription loci in each nucleus. (3) The Visualizer module allows users to manage the data from multiple movies and to visualize the pattern of several features of the transcription dynamics

2   Materials

2.1   System Requirements

1. MATLAB 2013 or above, with the Image Processing Toolbox.

2. The *LiveFly* toolbox: available for download from: https://github.com/huytran216/LiveFly_toolbox (*see* **Note 1**).

3. MATLAB toolbox for Bioformat: version 5.2.1 or higher, available for download from: https://www.openmicroscopy.org/bio-formats/downloads/. The Bioformat toolbox (zipped) is to be extracted to the *Tool* folder of the LiveFly's main folder.

2.2   Livefly Inputs

1. A time-lapse 3D movie captured with a confocal microscope. The compatible file formats are *.tif, *.czi, or *.lsm commonly used in time-lapse microscopy. The movie has two channels: the first red channel visualizes the nuclei marker (either NUP or Histone) and the second green channel visualizes the MCP-GFP/PP7 protein (*see* **Note 2**). The movies usually capture a thin layer of the embryo so that there are no overlapping nuclei on the same z-stack.

2. Input parameters: time resolution and XYZ pixel resolution, embryo orientation and the coordinates of the imaging region relative to the embryo anterior/posterior poles.

3. (Optional) List of nuclei IDs, time frames to be ignored for spot detection (*see* **Note 3** and Subheading 3.3).

3   Methods

3.1   Nuclei Segmentation

The segmentation module works with the "nuclear" signal of His-RFP or NUP-RFP acquired using the red channel. It should be noted that the fly nuclear signal is distinct from that of bacterial colonies: nuclei are generally homogeneous in shape and size and are far apart from one another. However, nuclei division is very synchronous and rapid (it takes ~30 s for the whole nuclei population to fully divide). Therefore, nuclear signal processing requires very robust nuclear tracking while the segmentation algorithm can be kept simple.

We opted to combine popular cell segmentation methods for static images [8, 11, 12] with frame-by-frame image registration for precise and efficient nuclei tracking. For fast computation, we perform nuclei segmentation on 2D images created by a maximum projection of the image stacks onto the Z-axis (optical axis).

3.1.1   Maximum Projection

This step creates a 2D movie from the histone/NUP channel of the original 3D movie for nuclei segmentation. The image in each frame of the 2D movie is the maximum projection of the corresponding 3D image. Added to this, between every two consecutive frames, an image registration, which records the displacement of image regions from one frame to another, is computed. This registration will be used to predict the movement of nuclei in the subsequent nuclei tracking and lineage construction.

1. In the LiveFly toolbox's main folder, open `Cell segmentation/LiveFly_CREATE_PROJECTION.m` with the MATLAB editor.

2. Specify the 3D movie file name and path (Line 4–6), the output 2D movie name, and the frame range of interest (Line 8–9) (*see* **Note 4**).

3. Run the script (Press *F5*).

This process will create two files in the input folder: a maximum projection file of the nuclei in TIF format (hereby referred to as `RED.tif`), and a MATLAB data file containing the frame registration information (named `RED_align.mat`).

3.1.2 Nuclei Segmentation

The nuclei segmentation process is performed on the Segmentation platform. The user interface layout is shown in Fig. 2. At any time,

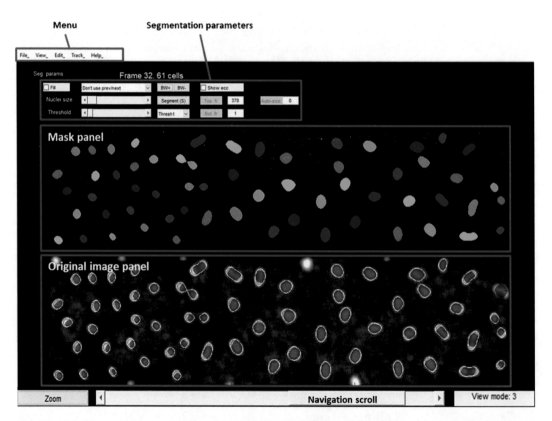

Fig. 2 Nuclei segmentation user interface. The interface allows users to observe the segmentation (performed automatically) result, illustrated in the mask panel and the original image panel, in each frame. Users can adjust the segmentation parameters or perform manual corrections through a range of tools accessible via the menu and hotkey

you can load and save the segmentation progress. The segmentation in each frame is automatic based on the specified parameters, followed by manual correction if necessary.

Step by step (*see* **Note 5**):

1. In the LiveFly toolbox's main folder, run `Cell segmentation/LiveFly_SEGMENTATION.m` to start the nuclei segmentation user interface.

2. Load the 2D maximum projection movie via `Menu> File> Load image...`.

3. Specify the average nuclear size and intensity threshold for nuclear segmentation. You will see the preview of the segmentation result in the mask panel.

4. Click on `Segment` for automatic segmentation of the current frame with the specified parameters. You can examine the segmentation result by cycling through the view modes using `View>Cycle` `View (X)` (*see* **Note 6**). The view mode 4 is recommended to check both the segmentation result and the overlap of nuclei masks between frames (which is critical in nuclei tracking).

5. Manually increase or decrease the size of all masks by one pixel by pressing `BW+/BW-` respectively (*see* **Note 7**).

6. Update the number of pixels needed to be added to the mask's size in the `Auto-size` section (use negative numbers to decrease the mask size). This will enable automatic mask size adjustment following the segmentation process. The `Auto-size` parameter is generally constant throughout the processing of the whole movie.

7. Select `Menu> Edit> Realign` for image re-registration if necessary. You can opt for automatic registration or manual registration using the MATLAB built-in tool (*see* **Note 8**).

8. Manually correct the masks if needed. You can add new nuclei masks or remove existing masks by accessing `Menu> Edit`.

9. Apply the automatic segmentation and mask size adjustment to the next frame by selecting `Menu> Edit> Propagate Next/Prev` (*see* **Note 9**). Continue this frame-by-frame automatic segmentation until parameter adjustments and manual corrections are required.

3.1.3 Nuclei Tracking

In the nuclei tracking process, the program uses the image registration to determine a nuclei's ancestor in the previous frame. Two nuclei in two consecutive frames are considered ancestor–descendant if their masks, realigned based on the image registration, overlap (*see* **Note 10**). If this nucleus has only one descendant in the next frame, this descendant is treated as the same nucleus. If it has more than one descendant in the next frame, the two most

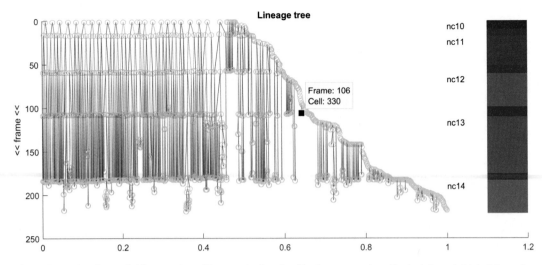

Fig. 3 Example of a nuclei lineage tree. The y-axis denotes the frame number. Each data point (circle) marks the birth of a new nucleus (either by division or by entering the captured image region). The data point's x-coordinate is set to the nucleus' ID. Mother and daughter nuclei are linked by solid lines. The textbox and the colored column on the right side specify the frames' nuclear cycle: Frames of the same nuclear cycle will have the same color. The tag specifying the selected frame and nucleus ID is accessed via MATLAB's data cursor

overlapping descendants are considered as the two daughters of that nucleus.

1. Press Menu> Track to begin nuclei tracking across frames.

2. Check the lineage tree by selecting Menu> Track> Plot Lineage Tree (as shown in Fig. 3) for abnormalities in the timing of nuclear divisions. You can identify the frames and nuclear ID with abnormalities by clicking on the lineage tree.

3. Most abnormalities are caused by imperfect image registration and nuclei masks being too small (there is no overlap between a nucleus and its intended ancestor). Examine the frames and nuclei with abnormalities by selecting Menu> Track> Check Track and perform manual corrections with the provided tool (add/remove/resize masks) on them if needed (*see* **Note 11**).

4. Once all the abnormalities are corrected, rerun the tracking process.

5. Reassign the nuclear cycle by selecting Menu> Track> Reassign nuclear cycle. You will be asked to specify the frame range for each nuclear cycle. The input will be used to determine which nuclear cycle each nucleus belongs to.

3.2 Spot Detection

The spot detection module extracts the fluorescent spot location and intensity from the 3D green channel. Spots are first identified using the threshold method (*see* below) and then fitted with a 2D Gaussian landscape, from which the spot intensity is extracted. Here, th(1) is the minimum absolute spot intensity and th(2) is the minimum absolute spot intensity above the local background level (determined using a median filter). These two values are used for the initial identification of the fluorescent spots. Only the brightest spot(s) located in the nucleus space are considered active transcription loci. This module will also combine the spot information and the nuclei segmentation result to create the movie's complete output file, which is used in subsequent analyses.

1. In *LiveFly* toolbox's main folder, open `Spot detection> main_Spot_detection.m` in the MATLAB editor. Set the raw 3D movie and 2D nuclei movie path (line 7–9).

2. Run the script. An input dialog will appear asking you to either enter the debug mode to determine `th` or proceed with the spot detection. Select "Yes" to enter the debug mode.

3. Select the frame in which you know that there is a spot in the next input dialog.

4. Zoom into the spot location and determine the threshold values, based on the two panels (*see* **Note 12**) (Fig. 4) using the built-in MATLAB Zoom In and data cursor.

5. Quit the script by pressing Ctrl+C.

6. Set the value of `th(1)` and `th(2)` in the editor. Set the frame range of interest by modifying both `it_start` and `it_end` (line 15–16). Run the script. Skip the debug mode this time.

7. The complete movie output file, containing frame-by-frame nuclei locations, lineages and its spot intensity, are stored in the subfolder `table_summary` of the 3D movie's directory (*see* **Note 13**). Check `Spot_detection/Header.txt` for the output file's header. A MATLAB file (in the format _config.m) specifying the movie's time and spatial resolutions, and the parameters for spot detection is also generated.

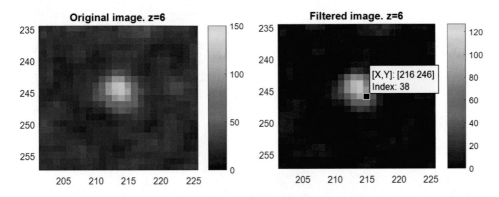

Fig. 4 Determining the threshold value for spot detection. Example of a zoomed-in spot in a specific z stack, from which the value of *th(1)* and *th(2)* from the original image panel (left) and median filtered image panel (right) are retrieved, respectively. The pixel intensities are scaled on a color map for easy spot identification. The data on the spot intensity (Index) is shown using the MATLAB built-in data cursor

8. Check the spot detection results by opening the newly generated movies, located in the subfolder <output_images> folder (*see* **Note 14**).

3.3 Postanalysis Correction

The exported output file is generally ready for subsequent analysis. However, due to many reasons, such as nuclei moving in and out of the captured region, it may be necessary to crop the movie frame, ignore specific nuclei, or a whole nuclear cycle.

1. Copy the file Data analysis/correction.m to the output file folder.

2. Specify the output file name and possible exceptions (e.g., bad nuclei, frames, nuclear cycles) in this file. See the comments in the correction file for details (*see* **Note 15**).

3.4 Pattern Visualizer

The pattern visualizer is a graphical user interface that aids users to merge data from multiple fly movies of the same strain and genetic construct and visualize them, either individually or together. The interface is shown Fig. 5 in with examples of data.

To access the visualizer, run Data analysis/ LiveFly_VISUALIZER.m.

3.4.1 Managing the Dataset

1. Create a new dataset by selecting Menu>File>Create dataset.

2. Select Add... to add a new movie to the dataset. A dialog will appear inquiring for the movie correction file.

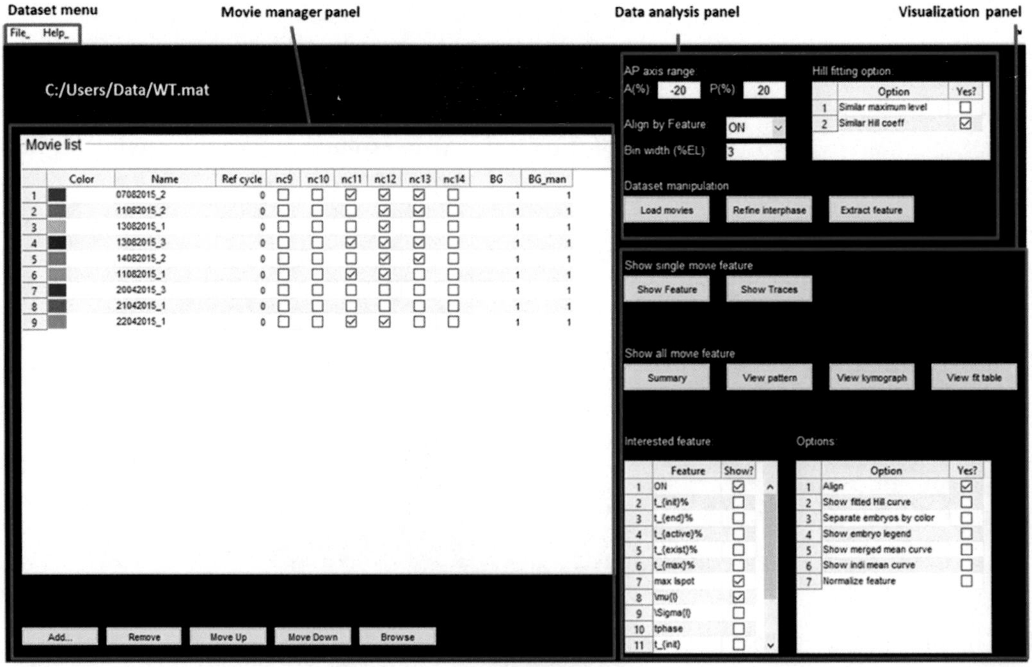

Fig. 5 Pattern visualizer. The interface allows the user to manage movies acquired with similar conditions, extract features from the trace intensity of active transcription loci and visualize their pattern in individual or merged embryos

3. Edit the movie list with Remove/Move Up/Move Down buttons.

4. Select the nuclear cycle of interest in each movie by editing the tick boxes.

5. Save/load the dataset any time using Menu>File (*see* **Note16**).

3.4.2 Initial Analysis

1. Click on Load movies to load data from all the movies.

   This step will read the movie output files, and extract information about each nucleus, including its position, nuclear cycle, lineage, and the intensity traces of active transcription loci. The exceptions specified in the correction file are taken into account. The program will extract several features of the traces, such as the total spot intensity (ΣI), the mean spot intensity (μI), time of first spot appearance (t_{init}), the total spot appearance duration (t_{active}), etc. *See* <Menu>Help>Show feature description> for the feature list and description.

2. Select a range of proper interphase duration for each nuclear cycle (*see* **Note 17**) by selecting Refine interphase.

3. Extract the feature pattern by clicking Extract feature.

   This step will fit the trace feature values along the embryo anterior–posterior (AP) axis in each movie with a Hill function, from which the maximum intensity level, the pattern's border position (coordinate of the half-maximum expression point) along the AP axis and the Hill coefficient are extracted. It is possible to impose the same Hill coefficient or maximum intensity level on all movies by selecting the options in Hill fitting option. Note that only patterns in the movies that are selected with tick boxes will be fitted.

4. Following this step, all the nuclei information and the Hill fitting results can be exported in an Excel file by selecting Menu>Export xls. It is recommended to save the dataset after this step to save time in future analyses.

3.4.3 Pattern Viewer

Once the movie data have been imported and processed, the trace feature patterns can be viewed in individual embryos (using Show single movie feature buttons) (Example in Fig. 6) or the movie patterns can be merged together (using Show all movie feature buttons) (Example in Fig. 7).

Edit the tick boxes list at the bottom of the interface to select the features of interest and determine how to visualize their patterns.

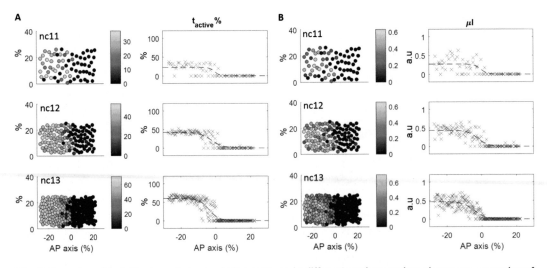

Fig. 6 Examples of individual movie transcription patterns in different nuclear cycles: shown are examples of patterns for the relative promoter active time (t_{active}, **a**) and the mean spot intensity (μI, **b**) for each nuclear cycle. The subplots on the left show the nuclei's 2D position (brighter nuclei corresponds to higher feature values, as in the color bars). The y-axis can be either side-by-side or ventral–dorsal axis, with unit in percentage of embryo length. The subplots on the right show the feature values (blue cross) along the AP axis along with the fitted Hill function (dashed). The x-axis describes the nuclei position along the embryo AP axis, with the middle point at position 0%

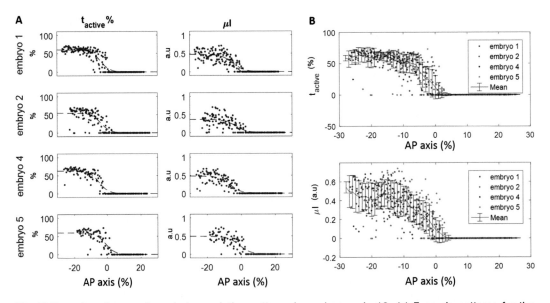

Fig. 7 Examples of merged movie transcription patterns in nuclear cycle 13: (**a**) Example patterns for the relative promoter active time (t_{active}) and the mean spot intensity (μI) for 4 selected movies, accessed by <Summary> button. (**b**) Merged patterns from 4 selected movies, accessed by the <View pattern> button. Also shown are the mean and standard deviation (solid black line) of the feature values at each position along the AP axis

4   Notes

1. The `README.md` file contains a link to a 3D sample movie and the microscopy configurations, which can be used to familiarize oneself with the whole process.

2. The color of the nuclei or MCP channel is not necessarily fixed. The nuclei channel can be specified in the variable `channel` (line 8).

3. The information on exceptions (e.g., bad nuclei, frames, nuclear cycles) is only needed after analysis of the raw movie and can be updated anytime.

4. It is possible to analyze only a subsection of the movie by specifying the frame range (`it_start`, `it_end` variables (line 35–36)). In general, incomplete nuclear cycles or the second half of nuclear cycle 14 (after cellularization) is ignored.

5. Once the basics of the segmentation process are understood, it is advisable to familiarize oneself with the hotkeys. The segmentation tool is designed so that most frames can be segmented with hotkeys alone. Actions involving pointers, such as mask adding/removing, are required infrequently during mitosis.

6. There are four view modes, accessible from the menu or the hotkey *X*.

    (a) Mode 1 shows the original nuclei image.

    (b) Mode 2 shows the original image with the overlays of the nuclei masks in red.

    (c) Mode 3 shows the original image with nuclei masks as wireframes. This is useful when visualizing densely populating masks (e.g., nuclear cycle 14).

    (d) Mode 4 shows in wireframe the nuclei mask plus the previous frame masks as red overlay. Mode 4 is useful to recognize new cells that appear during segmentation. If a nucleus is detected in the current frame but has no overlap with any nuclei in the previous frame, its mask will appear as a blue wireframe instead of white.

7. The nuclei real radius is generally a few pixels larger than the detected size. It is advisable to adjust the mask size to the real size to minimize errors in the subsequent spot detection process.

8. View mode 4 is useful to check for the displacement of the nuclei between frames. While the image registration should work properly throughout the cycle interphase, it may fail during mitosis due to rapid mitotic waves. In this case, manual registration, which is done by a few clicks, is needed.

9. This step can be done with the hotkey *Space*. Generally, the automatic segmentation parameters require few adjustments during the whole process. The whole nuclear cycle interphase can be analyzed by using *Space*, inspection, *Space*, inspection etc.

10. Given the movement of nuclei during the development phase, it is recommended that the movie time resolution is kept below 20 s per frame if the users want to track nuclei across mitosis. However, if users are only interested in specific nuclear cycles, the time resolution can be as slow as 40 s per frame.

11. Due to the fast movement of nuclei during mitosis, nuclei masks between frames may not overlap with one another leading to bad nuclei tracking. In this case, you should increase the mask size (using BW+) until they overlap. View Mode 4 is very useful for this correction.

12. The first spots that are visually detected are typically bright spots, and the threshold values should therefore be slightly lower than half of their peak intensity. Note that setting the value of `th(1)` and `th(2)` too low will increase the rate of false positive detections (there is no spot in the nucleus but the program detects that there is), resulting in detected spots with near-zero intensity. This does not affect the spot intensity traces drastically but it may be computationally more expensive, due to the high number of spots detected. If the thresholds are set too low, a warning about too many (over 500) detected spots will appear. When it happens, interrupt the program and increase `th(1)` and `th(2)` slightly.

13. The output file, organized by columns (features) and rows (time frame, nuclei ID), can be opened with any sheet editing program.

14. Open the movie `Check_image.tif` to see a frame-by-frame comparison between the maximum projection image with the segmented nuclei and fluorescent spots.

15. The correction file, though not strictly required for the analysis, is strongly recommended. If no exceptions are declared, only the movie output file name should be specified.

In this correction file, one can specify:
(a) The XY frame of interest (with `xlim_left, xlim_right, ylim_up, ylim_down` variables): this is useful to discard nuclei at the border of the capture region, which pop in and out frequently during the movie.

(b) Frames to be discarded (with `rm` variable): spots detected in these frames are ignored (nuclei are still segmented and tracked). Specifying these frames is useful if the MCP-GFP forms aggregates right before mitosis, which are sometimes falsely detected as active transcription loci.

(c) Whether the movie begins or ends in the middle of a nuclear interphase (with `leftcensored, rightcensored` variables). These incomplete nuclear cycles will be ignored.

16. The data file is saved in the MATLAB *.mat format. This file can directly be loaded into the MATLAB workspace instead of using the Visualizer.

17. This step will retain the nuclei that are detected within a certain time range. It is helpful to remove nuclei that move in and out of the captured image region.

Acknowledgments

This work was supported by a Marie Curie MCCIG grant No. 303561 (AMW), PSL IDEX REFLEX (ND, AMW, MC), ANR-11-LABX-0044 DEEP Labex (ND), ANR- 11-BSV2-0024 Axomorph (ND and AMW), and PSL ANR-10-IDEX-0001-02. The funders had no role in study design, data collection and analysis, decision to publish, or preparation of the manuscript.

References

1. Garcia HG, Tikhonov M, Lin A, Gregor T (2013) Quantitative imaging of transcription in living Drosophila embryos links polymerase activity to patterning. Curr Biol 23:2140–2145. https://doi.org/10.1016/j.cub.2013.08.054

2. Lucas T, Ferraro T, Roelens B et al (2013) Live imaging of bicoid-dependent transcription in Drosophila embryos. Curr Biol 23:2135–2139. https://doi.org/10.1016/j.cub.2013.08.053

3. Fukaya T, Lim B, Levine M (2016) Enhancer control of transcriptional bursting. Cell 166:358–368. https://doi.org/10.1016/j.cell.2016.05.025

4. Bothma JP, Garcia HG, Esposito E et al (2014) Dynamic regulation of eve stripe 2 expression reveals transcriptional bursts in living Drosophila embryos. Proc Natl Acad Sci U S A 111:10598–10603. https://doi.org/10.1073/pnas.1410022111

5. Desponds J, Tran H, Ferraro T et al (2016) Precision of readout at the hunchback gene: analyzing short transcription time traces in living fly embryos. PLOS Comput Biol 12: e1005256. https://doi.org/10.1371/journal.pcbi.1005256

6. Ferraro T, Esposito E, Mancini L et al (2016) Transcriptional memory in the Drosophila embryo. Curr Biol 26:212–218. https://doi.org/10.1002/wdev.221

7. Santinha J, Martins L, Häkkinen A et al (2016) iCellFusion: tool for fusion and analysis of live-cell images from time lapsemultimodal mycroscopy. In: Karâa WBA, Dey N (eds) Biomedical image analysis and mining techniques for improved health outcomes, IGI Global. doi: https://doi.org/10.4018/978-1-4666-8811-7

8. Young JW, Locke JCW, Altinok A et al (2011) Measuring single-cell gene expression dynamics in bacteria using fluorescence time-lapse microscopy. Nat Protoc 7:80. https://doi.org/10.1038/nprot.2011.432

9. Driever W, Nusslein-Volhard C (1988) A gradient of bicoid protein in Drosophila embryos. Cell 54:83–93. https://doi.org/10.1016/0092-8674(88)90183-3

10. Porcher A, Dostatni N (2010) The bicoid morphogen system. Curr Biol 20:R249–R254. https://doi.org/10.1016/j.cub.2010.01.026

11. Häkkinen A, Muthukrishnan AB, Mora A et al (2013) CellAging: a tool to study segregation and partitioning in division in cell lineages of Escherichia coli. Bioinformatics 29:1708–1709. https://doi.org/10.1093/bioinformatics/btt194

12. Otsu N (1979) A threshold selection method from gray-level histograms. IEEE Trans Syst Man Cybern 9(1):62–66

Part III

Modeling of Morphogen Behaviors and Activities

Chapter 12

Discrete-State Stochastic Modeling of Morphogen Gradient Formation

Hamid Teimouri and Anatoly B. Kolomeisky

Abstract

In biological development, positional information required for pattern formation is carried by the gradients of special signaling molecules, which are called morphogens. It is well known that the establishment of the morphogen gradients is a result of complex physical-chemical processes that involve diffusion, degradation of locally produced signaling molecules, and other biochemical reactions. Here we describe a recently developed discrete-state stochastic theoretical method to explain the formation of morphogen gradients in complex cellular environment.

Key words Morphogen gradient, Local accumulation time, Reaction–diffusion processes, Spatially varying degradation rate, Discrete-state stochastic modeling, Nonlinear degradation mechanism, Direct-delivery mechanism

1 Introduction

Several classes of biological signaling molecules, known as morphogens, play a critical role in tissue patterning during the embryonic development of multicell organisms [18, 54, 80, 81]. Analysis of the developmental processes in different systems suggested the existence of several universal mechanisms governing the establishment of signaling profiles and their activities [1, 11–13, 15, 51, 53, 55, 60, 64, 71, 72, 77, 82, 84]. Stimulated by these experimental observations, a large number of theoretical ideas on the mechanisms of formation of morphogen gradients have been proposed [2–7, 9, 12, 19, 20, 22–25, 28, 31, 33–38, 40, 43–46, 50, 51, 57, 60, 62, 64, 67, 69, 75, 76, 78, 79, 83]. Most of these studies suggest that the establishment of biological signaling profiles in development is a result of complex physical-chemical processes that include the localized production of morphogens that later can diffuse and be removed from the cellular medium by various types of biochemical transitions [12, 51, 60, 64, 79]. Based on some experimental observations [8, 17, 30, 39, 49, 66, 68], the

Julien Dubrulle (ed.), *Morphogen Gradients: Methods and Protocols*, Methods in Molecular Biology, vol. 1863, https://doi.org/10.1007/978-1-4939-8772-6_12, © Springer Science+Business Media, LLC, part of Springer Nature 2018

possibility of alternative mechanisms of the direct delivery of morphogens to the target cells via dynamic cellular extensions called cytonemes has also been discussed [42, 66, 75]. It was argued that in some situations the complex environment of the embryo systems might prevent the free diffusion from establishing the distinguishable morphogen gradients at different regions, implying a different mechanism of the biological signal transduction [42, 66].

Two main theoretical methods have been developed in clarifying the mechanisms of the formation of signaling profiles. One of them utilizes a continuum description, while another one is based on more general discrete-state stochastic analysis of the processes. In this chapter we present only the discrete-state stochastic formalism of morphogen gradient formation since it reflects better the discrete nature of involved biochemical processes [47].

2   Theoretical Method

The discrete version of the so-called synthesis-diffusion-degradation (SDD) model is shown in Fig. 1. It is assumed that each embryo cell is associated with a lattice site. In the simplest version of the model, the morphogens are produced only at the origin with a rate $Q_0 = Q$ ($L = 1$ case in Fig. 1). Signaling molecules can diffuse with a rate u along the lattice. At any position, the morphogen molecule can be degraded and removed from the system with a rate k. To simplify the calculations, a single-molecule view, according to which the concentration of morphogens at a given site is equivalent to the probability of finding the signaling molecule at this location, is adopted. The probability of finding the morphogen at site n at time t is characterized by a function $P_n(t)$. The time evolution of this probability is controlled by the following master equations:

$$\frac{dP_0(t)}{dt} = Q + uP_1(t) - (u + k)P_0(t), \tag{1}$$

for $n = 0$; and

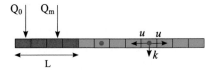

Fig. 1 A scheme for the one-dimensional discrete-state stochastic model for the formation of the morphogen gradient. Signaling molecules are produced at the sites $0 \leq m \leq L$ (shown in red) with rates Q_m. The case of $m = 0$ and $L = 1$ corresponds to the source localized at the origin. Morphogens might diffuse along the lattice to the neighboring sites with a rate u, or they might be degraded with a rate k. Adapted with permission from Ref. [74]

$$\frac{dP_n(t)}{dt} = u[P_{n-1}(t) + P_{n+1}(t)] - (2u + k)P_n(t), \qquad (2)$$

for $n > 0$. At large times, when $\frac{dP_n(t)}{dt} = 0$, these equations can be easily solved, producing an exponentially decaying concentration profile,

$$P_n^{(s)} = \frac{2Qx^n}{k + \sqrt{k^2 + 4uk}}, \qquad (3)$$

with

$$x = (2u + k - \sqrt{k^2 + 4uk})/(2u). \qquad (4)$$

It can be shown that in the continuum limit ($u \gg k$), Eq. 3 reduces to the following form:

$$C^{(s)}(x) = \frac{Q}{\sqrt{Dk}} \exp\left(-\frac{x}{\lambda}\right). \qquad (5)$$

One of the main functions of the morphogens is to transfer biological information to embryo cells. It is still debated how it happens at the molecular level, and if the transfer of information is taking place in the stationary state or before reaching the steady-state conditions [12]. Let us assume for simplicity that the signaling profiles should reach the stationary state before the full transferring of the information can take place. However, our analysis can be also generalized for the pre-steady-state coding possibilities. The important characteristics of the dynamics of morphogen gradients formation are times needed to achieve the steady-state concentration levels at specific spatial locations. These times are known as local accumulation times (LAT), and a theoretical framework for computing these quantities has been developed recently by Berezhkovski and coworkers [5]. They introduced a local relaxation functions $R(n;t)$, which can be written as

$$R(n;t) = \frac{P(n;t) - P^{(s)}(n)}{P(n;t = 0) - P^{(s)}(n)} = 1 - \frac{P(n;t)}{P^{(s)}(n)} \qquad (6)$$

The physical meaning of the local relaxation function is simple: it gives a measure of how close is the system to the steady-state conditions. It ranges from $R = 1$ at $t = 0$ to $R = 0$ when the stationary state at given location is achieved. Introducing the Laplace transform of the function, $\tilde{R}(n,s) = \int_0^\infty R(n;t)e^{-st}dt$, it can be shown that the LAT are given by

$$t(n) = -\int_0^\infty t \frac{\partial R(n;t)}{\partial t} dt = \int_0^\infty R(n,t)dt = \tilde{R}(n;s = 0) \qquad (7)$$

From this relation the explicit expression for the local accumulation times can be found:

$$t(n) = \frac{1}{\sqrt{k^2 + 4uk}} \left[\frac{2u + k + \sqrt{k^2 + 4uk}}{k + \sqrt{k^2 + 4uk}} + n \right] \qquad (8)$$

again, for $u \gg k$ it can be shown that

$$t(x) = \frac{1}{2k}\left(1 + \frac{x}{\lambda}\right). \qquad (9)$$

For bicoid morphogen gradient in fruit fly embryos, using the expression Eq. 9 along with the estimate of the decay length $\lambda \simeq 60$ μm and with a better estimate of the diffusion constant $D \simeq 1$ μm^2/s [13, 57], the time to reach the stationary state at the most distant boundary was evaluated to be less than 200 min, which is much closer (although still not perfect) to the experimental values (≃90 min). The difference between these theoretical predictions and experiments is probably due to not precise measurements of the diffusion constant and the decay length, as well as due to over-simplified theoretical assumptions of the strongly localized source region, as we discuss below in more detail [5]. Thus, the systematic approach to evaluate LAT as a measure of the dynamics of the formation of morphogen gradients was able to mostly resolve the apparent paradox of slow diffusion [5]. However, it also raised several questions. Equation 9 indicates a linear scaling as the distance from the source for the SDD model instead of the expected quadratic scaling for the unbiased random-walk motion since there are no apparent external driving forces in the system. It led to a conclusion that signaling profiles formed much faster than was previously estimated [5, 47, 70]. But the mechanism of this speedup was not clarified.

2.1 The Effect of Source Delocalization

Although the simplest SDD model was able to explain some aspects of the development of morphogen gradients, it was pointed out that many realistic features of the process that might strongly influence the dynamics are not taken into account [2, 3, 5, 6, 19, 47, 52, 74]. Experiments show that in many biological systems the production region of signaling molecules is not strongly localized as assumed in the simplest SDD model [12, 51, 52, 60, 64, 79]. Morphogens are protein molecules that must be first synthesized from the corresponding RNA molecules, but the distributions of the RNA species in various embryos are typically more diffuse [56]. For example, for the bicoid system it is known that the maternal RNA molecules can be found in the region of size 30–50 μm, which should be compared with the total length of embryo of ≃400 μm [60].

A more general theoretical method to evaluate the role of the source delocalization was introduced later by our group [74].

A discrete-state stochastic SDD model in one dimension with the extended source range, as illustrated in Fig. 1, was considered. We assumed that the signaling molecules are produced over the interval of length L with rates Q_m for $0 \leq m \leq L$: see Fig. 1. The total production rates is equal to $Q = \sum_{m=0}^{L} Q_m$. Because the production of morphogens at different sites is independent from each other, it was suggested that the general solution for the probability to find a signaling molecule at the site n at time t with a delocalized production region as specified in Fig. 1, $P(n, t)$, can be written as a sum of the probabilities $P(n, t;m)$ for the single localized sources at the sites m [74]. More specifically, the probability function $P(n, t;m)$ is governed by the following master equations:

$$\frac{dP(n,t;m)}{dt} = Q_m \delta_{m,n} + u[P(n-1,t;m) + P(n+1,t;m)] \\ - (2u+k)P(n,t;m), \tag{10}$$

for $n > 0$, and

$$\frac{dP(0,t;m)}{dt} = Q_0 \delta_{m,0} + uP(1,t;m) - (u+k)P(0,t;m). \tag{11}$$

for $n = 0$. In the steady-state limit, $t \to \infty$, these equations can be solved explicitly, yielding

$$P_1^{(s)}(n;m) = \frac{Q_m[(k+\sqrt{k^2+4uk})x^{m-n} + (-k+\sqrt{k^2+4uk})x^{n+m})]}{(k+\sqrt{k^2+4uk})\sqrt{k^2+4uk}} \tag{12}$$

for $0 \leq n \leq m$, and

$$P_2^{(s)}(n;m) = \frac{Q_m[(k+\sqrt{k^2+4uk})x^{n-m} + (-k+\sqrt{k^2+4uk})x^{n+m})]}{(k+\sqrt{k^2+4uk})\sqrt{k^2+4uk}}, \tag{13}$$

for $m \leq n$. The parameter x is given in Eq. 4. Using the superposition arguments presented above, it can be shown that

$$P(n,t) = \begin{cases} \sum_{m=0}^{n} P_2(n,t;m) + \sum_{m=n+1}^{L} P_1(n,t;m), & \text{for } 0 \leq n \leq L; \\ \sum_{m=0}^{L} P_2(n,t;m), & \text{for } L \leq n. \end{cases} \tag{14}$$

This method allowed us to analyze the formation of morphogen gradients for arbitrary length of the production region and for arbitrary production rates [74].

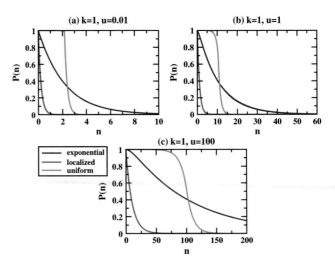

Fig. 2 Stationary-state density profiles for the formation of morphogen gradients with different production regions and for variable diffusion and degradation rates. Red curves describe the single-site localized source, green curves describe the uniform production over the finite interval, and blue curves describe the exponential production over the semi-infinite interval. Adapted with permission from Ref. [74]

To clarify the role of the source delocalization, the development of the morphogen gradients with uniformly distributed production over the finite interval and with the exponentially distributed production along the semi-infinite interval was compared with the formation of the signaling profile in the case of sharply localized source at the origin [74]. The corresponding density profiles are presented in Fig. 2, while the estimated LAT are given in Fig. 3. It was concluded that the extended sources delivered the signaling molecules much further in comparison with the single localized source. In addition, the delocalized sources were able to create sharp boundaries which are needed to controllably turning genes on and off as required. These systems were also generally reaching the stationary states faster.

2.2 The Formation of Morphogen Gradients in Two and Three Dimensions

Most of theoretical models applied for describing the establishment of the biological signaling profiles are essentially one-dimensional [2, 3, 5, 6, 47, 74]. However, a more realistic description of these processes should take into account a complex structure of the embryos [57, 67]. This led to extending the original SDD models [27, 38, 73].

A general multi-dimensional discrete-state stochastic models were investigated in full detail in our research group [73]. The d-dimensional system with the production at the origin, as shown in Fig. 4, was studied. In this system, each lattice cite is characterized by d coordinates, $\mathbf{n} = (n_1, n_2, \ldots, n_d)$. The source of signaling molecules is at the origin, $\mathbf{n}_0 = (0, 0, \ldots, 0)$, with the production

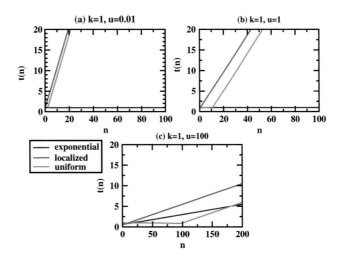

Fig. 3 Local accumulation times for the formation of morphogen gradients with different production regions and for variable diffusion and degradation rates. Red curves describe the single-site localized source, green curves describe the uniform production over the finite interval, and blue curves describe the exponential production over the semi-infinite interval. Adapted with permission from Ref. [74]

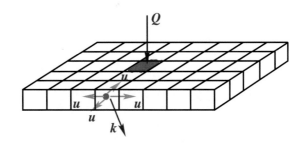

Fig. 4 A schematic view of the formation of morphogen gradients for $d = 2$ dimensions. The signaling molecules are created at the origin with the rate Q. They diffuse with the rate u in all directions without a bias. At each cell, morphogens can be degraded with a rate k. Adapted with permission from Ref. [73]

rate Q (*see* Fig. 4). Morphogens can diffuse to the nearest neighboring sites with the rate u, and the degradation rates at each site is equal to k (Fig. 4). To solve the problem, a function $P(n_1, n_2, \ldots, n_d; t)$, defined as the probability density at the cell $\mathbf{n} = (n_1, n_2, \ldots, n_d)$ at time t, was analyzed at all times using the following master equations [73]:

$$\frac{dP(n_1, n_2, \ldots, n_d; t)}{dt} = u \sum_{nn} P(n_1, n_2, \ldots, n_d; t) - (2ud + k) P(n_1, n_2, \ldots, n_d; t),$$ (15)

where Σ_{nn} corresponds to summing over all nearest neighbors, and

$$\sum_{nn} P(n_1, n_2, \ldots, n_d; t) = P(n_1 - 1, n_2, \ldots, n_d; t)$$

$$+ P(n_1 + 1, n_2, \ldots, n_d; t)$$
$$+ P(n_1, n_2 - 1, \ldots, n_d; t)$$
$$+ P(n_1, n_2 + 1, \ldots, n_d; t) + \cdots.$$

(16)

At the origin, the dynamics is slightly different due to morphogen production,

$$\frac{dP(0, 0, \ldots; t)}{dt} = Q + u \sum_{nn} P(0, 0, \ldots; t) - (2du + k) P(0, 0, \ldots; t).$$

(17)

At large times, the system achieves a stationary state with exponentially decaying signaling concentration profile,

$$P^{(s)}(n_1, n_2, \ldots, n_d) = \frac{2Q x^{|n_1| + |n_2| + \cdots + |n_d|}}{\sqrt{k^2 + 4duk}}$$

$$= \frac{2Q}{\sqrt{k^2 + 4duk}} \exp\left(\frac{-|n_1| - |n_2| - \cdots - |n_d|}{\lambda}\right),$$

(18)

where

$$x = (2du + k - \sqrt{k^2 + 4duk})/(2du), \quad \lambda = -1/\ln x. \quad (19)$$

It was shown also that the dynamics of approaching the stationary state is characterized well by the LAT [73]

$$t(n_1, n_2, \ldots, n_d) = \frac{(2du + k)}{(k^2 + 4duk)} + \frac{|n_1| + |n_2| + \cdots + |n_d|}{\sqrt{k^2 + 4duk}}.$$

(20)

The equivalent expression for the LAT at the distance r from the origin produces [73],

$$\tau(r) = \frac{(2du + k)}{(k^2 + 4duk)} + \left(\frac{\sqrt{d}}{\sqrt{k^2 + 4duk}}\right) r.$$

(21)

In the fast degradation limit, $k \gg u$, this equation gives

$$\tau(r) \simeq \frac{1}{k} + \frac{r\sqrt{d}}{k}.$$

(22)

In the continuum limit, $u \gg k$, it was found that there is no dependence on the dimensionality [73],

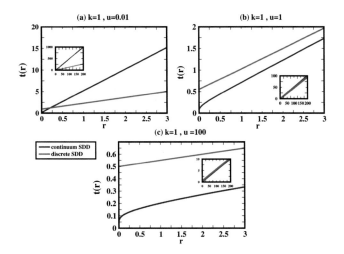

Fig. 5 Comparison of the LAT for the discrete-state stochastic models and for radially symmetric continuum models as a function of the distance from the source r in two dimensions. (**a**) Fast degradation limit, $k = 1$, $u = 0.01$; (**b**) comparable diffusion and degradation rates, $k = u = 1$; and (**c**) fast diffusion limit, $k = 1$, $u = 100$. Insets show the same plots for larger length scales. Adapted with permission from Ref. [73]

$$\tau(r) \simeq \frac{1}{2k} + \frac{r}{2\sqrt{uk}}. \tag{23}$$

To rationalize these deviations between various theoretical predictions, the LAT in two and three dimensions for both approaches have been compared. The results are presented in Figs. 5 and 6. One can see that the continuum limit of the discrete-state models and radially symmetric continuum models do not agree with each other, although for large distances ($r \gg 1$) the differences are getting smaller [73]. It was noticed also that the radially symmetric models predict that $\tau(r = 0) = 0$, while for the discrete-state case $\tau(r = 0) = 1/2k \neq 0$. But the relaxation times to the stationary profiles can never be zero because originally in the system there are no morphogens. Thus, the radially symmetric continuum models cannot properly describe the dynamics of the formation of morphogen gradients for $d > 1$ near the production region. The main reason for this is the assumption of spherically symmetric solutions of the corresponding reaction–diffusion equations at all length scales. Theoretical approach based on the discrete-state stochastic framework does not assume the spherical symmetry, and this led to the correct description of the dynamics at all scales and in all dimensions [73].

2.3 Nonlinear Degradation Mechanisms

Several experimental studies suggested that in some systems the development of the signaling profiles might be associated with more complex nonlinear degradation processes [14, 26, 37, 41]. In these situations, the presence of other morphogens can

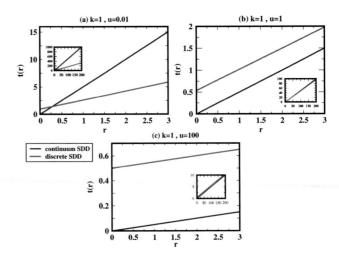

Fig. 6 Comparison of the local accumulation times for discrete-state stochastic models and for radially symmetric continuum models as a function of the distance from the source r in three dimensions. **(a)** Fast degradation limit, $k = 1$, $u = 0.01$; **(b)** comparable diffusion and degradation rates, $k = u = 1$; and **(c)** fast diffusion limit, $k = 1$, $u = 100$. Insets show the same plots for larger length scales. Adapted with permission from Ref. [73]

catalyze or inhibit the process of the removal from the medium, and this should affect the dynamics of the formation of morphogen gradients. In this case, the temporal evolution of the concentration profile can be written as [37]

$$\frac{\partial C(x,t)}{\partial t} = D\frac{\partial^2 C(x,t)}{\partial x^2} - kC(x,t)^m, \tag{24}$$

with $m \neq 1$. Using numerical solutions and mathematical bounds initial studies have shown that the dynamics of approaching to the stationary state significantly varies depending on the parameter m [37]. For $m = 0$ and $m = 1$ (linear degradation) LAT are linear functions of the distance from the source, but for $m = 2$, 3, and 4 the scaling changes from linear to quadratic. The explanations for these surprising observations were presented in the theoretical analysis that proposed to view the degradation process as an effective driving potential [9]. The degradation creates a gradient by removing molecules from the system, and this is equivalent to the action of the potential that drives the molecules away from the production region. Then the original reaction–diffusion model with the degradation can be well approximated as a biased-diffusion model without degradation [9]. This is illustrated in Fig. 7.

It was assumed that the equivalent biased-diffusion model has L ($L \to \infty$) sites [9]. The morphogens start the motion at $t = 0$ at the origin. The particles can move to the right (left) from the site n with the rate $g_n(r_n)$: *see* Fig. 7. When the particle reaches the site

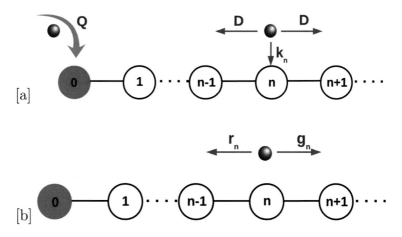

Fig. 7 Schematic view of equivalent models for the formation of morphogen gradients. (**a**) Synthesis-diffusion-degradation model; (**b**) biased-diffusion model. Adapted with permission from Ref. [9]

L it is instantaneously moved back to the origin, $n = 0$. The model is nonequilibrium, so that there is always a flux in the system in the direction away from the source. In the biased-diffusion model the probability to find the molecule at the site n at time t is given by a function $\Pi_n(t)$ [9]. The temporal evolution of these probabilities is described by the corresponding master equations [9],

$$\frac{d\Pi_n(t)}{dt} = r_{n+1}\Pi_{n+1}(t) + g_{n-1}\Pi_{n-1}(t) - (r_n + g_n)\Pi_n(t), \quad (25)$$

for $0 < n < L$, while for $n = 0$ and $n = L$ we have

$$\frac{d\Pi_0(t)}{dt} = J + r_1\Pi_1(t) - g_0\Pi_0(t), \quad (26)$$

$$\frac{d\Pi_L(t)}{dt} = g_{L-1}\Pi_{L-1}(t) - r_L\Pi_L(t) - J, \quad (27)$$

where J is the flux from the site L back to the origin $n = 0$. When the system achieves the stationary-state behavior, the flux through every site is equal to J.

Comparing the SDD model and the equivalent biased-diffusion model, it should be clear that the mapping between them is not exact [9]. It can be seen by noting that in the biased-diffusion model there is always a conservation of the probability, while in the SDD model the conservation is only achieved at the stationary-state limit. The relations between the parameters of both models can be made quantitative by using the following arguments. The diffusion rates g_n and r_n are related to each other via the effective potential as can be shown using the detailed balance-like arguments [9],

$$\frac{g_n}{r_{n+1}} = \exp\left(\frac{U_n^{eff} - U_{n+1}^{eff}}{k_B T}\right). \qquad (28)$$

The physical meaning of this expression is that the stronger the potential, the faster the motion in the positive direction, $g_n > r_{n+1}$. But to obtain the explicit formulas for transition rates a second condition is needed [9],

$$g_n + r_n = 2D + kC^{m-1}. \qquad (29)$$

This implies that the residence of each molecule at site n is identical in both models. Together, Eqs. 28 and 29 uniquely define the transition rates in the biased-diffusion model via parameters of the SDD model [9].

For linear degradation ($m = 1$) this approach leads to the following transition rates in the equivalent biased-diffusion model [9],

$$g_n = g = \frac{2D + k}{x + 1}, \quad r_n = r = x\frac{2D + k}{x + 1}, \qquad (30)$$

with $x = (2D + k - \sqrt{k^2 + 4kD})/2D$. The results for mean first-passage times from both models are given in Fig. 8. We conclude that the approximate mapping works quite well everywhere, but especially it is successful for the large degradation rates.

Extending this method to nonlinear degradation processes indicates that for $m \geq 2$ the steady-state profile is given by [9],

$$P_n^{(s)} \simeq \frac{1}{(1 + n/\lambda)^{\frac{2}{m-1}}}, \qquad (31)$$

where the parameter λ is defined as

$$\lambda = \frac{1}{m - 1}\left[\frac{(2D)^m(m + 1)}{kQ^{m-1}}\right]^{\frac{1}{m+1}}. \qquad (32)$$

This concentration profile corresponds to the logarithmic potential,

$$\frac{U_n^{eff}}{k_B T} \simeq -\frac{2}{m - 1}\ln(1 + n/\lambda). \qquad (33)$$

It can be shown that the mean first-passage times for equivalent biased-diffusion model at large distances from the source are equal to [9]

$$\tau_n \simeq \frac{(m - 1)}{(m + 1)}\frac{n^2}{2D}. \qquad (34)$$

This is an important result since it predicts a quadratic scaling for relaxation times with the nonlinear degradation, as illustrated also in Fig. 9.

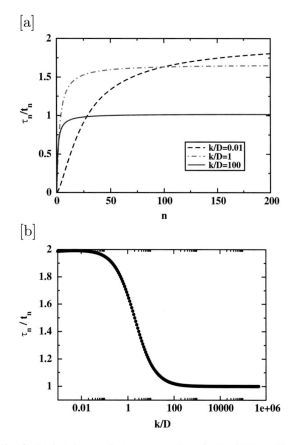

Fig. 8 Ratio of calculated mean first-passage times for the SDD model and for the equivalent biased-diffusion model. (**a**) The dependence on the distance from the source; (**b**) the dependence on the ratio of degradation rate over the diffusion rate. Distance from the source is set to $n = 10^4$. Adapted with permission from Ref. [9]

Theoretical calculations using the mapping of the SDD model to the equivalent biased-diffusion model clearly show different scaling behavior depending on the mechanisms of degradation. Linear scaling is observed for $m = 0$ or 1, while quadratic scaling is found for nonlinear degradations with $m \geq 2$ [9, 37]. The different dynamic behavior was explained using the concept of the effective potentials due to degradation [47]. Linear degradation corresponds to strong driving potentials, as shown in Fig. 10. In this case, there is a unique length scale λ across the whole system. This leads to effectively driven diffusion which has the expected linear scaling. The situation is different for all nonlinear degradation processes. The stationary state in this case can be described by a power-law concentration profiles, which do not possess unique length scales. As a result, the effective potential (logarithmic) is weak enough so that it cannot destroy the quadratic scaling of the unbiased diffusion. It might only affect the amplitude of the random-walk fluctuations for each signaling molecule, i.e., the effective mobility of the molecules is increased.

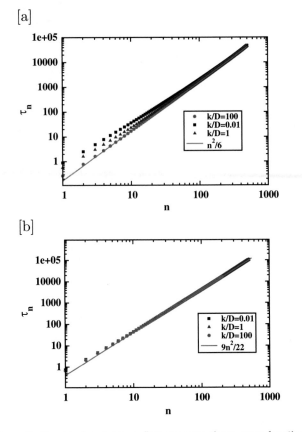

Fig. 9 Theoretically calculated mean first-passage times as a function of the distance from the source for different degrees of nonlinearity for the biased-diffusion model. (**a**) $m = 2$; (**b**) $m = 10$. Adapted with permission from Ref. [9]

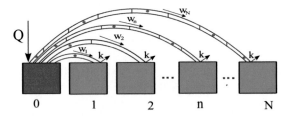

Fig. 10 A simplified view of the direct-delivery mechanism of signaling molecules via utilization of cytonemes. The red cell is the source. Green cells labeled as $n = 1, \ldots, N$ are target cells. Cytonemes are shown as tubular extensions from the source cell to the target cells. Morphogens are small red circles inside the cytonemes. Adapted with permission from Ref. [75]

2.4 Alternative Mechanisms: Direct Delivery of Morphogens

Recent experimental advances in studying the development processes in various systems revealed that there is a significant number of experimental observations that cannot be explained by the reaction–diffusion picture presented above [30, 32, 42, 63, 66]. In embryo systems with complex internal structures simple free diffusion might not be always very efficient in establishing the morphogen gradient [42, 66]. An alternative direct-delivery mechanism has been proposed [42, 48, 59, 65]. It was suggested that the signaling molecules can be transported to target cells via special cellular tubes, which are called cytonemes [32, 42, 48, 59, 65]. Cytonemes are dynamic extensions that cells can produce and retract very quickly with the help of actin filaments. Their length is varying from 1 to 100 μm with the diameter of less than 100 nm. Cytonemes have been recently observed in several biological systems, but their cellular functions are still unclear [8, 17, 30, 32, 39, 48, 66, 68]. It was proposed that morphogens can be transported by myosin motor proteins along the actin filaments inside the cytonemes directly from the source cells to the target cells, as shown schematically in Fig. 10 [32, 42, 68]. The direct-delivery mechanism thus avoids the problems where geometrically complex environment prevents the free diffusion to form the signaling profiles.

We developed a first quantitative physical-chemical method to describe the direct delivery via cytonemes [10, 75]. It is based on the model as illustrated in Fig. 10. The model postulates that there are $N + 1$ embryo cells in the system (shown as squares in Fig. 10). One of them (red square, $n = 0$) is a special source cell where the signaling molecules are produced with a rate Q. The source cell also generates N cytonemes that extend and attach to each of the target cell (green squares in Fig. 10). It was assumed that the cytonemes are already established at the beginning of the process and they are stable until the morphogen gradient is fully established. Signaling molecules (shown as small red circles in Fig. 10) are transported to the n-th target cell from the source cell with a rate w_n ($n = 1, 2, \ldots,$ N). When they reach their target cells, morphogens can be degraded with a rate k. It was argued that this is a minimal model that takes into account the most relevant processes such as the direct delivery via cytonemes and the degradation of morphogens.

This model was solved by analyzing the single-molecule probability density function $P_n(t)$ of finding the morphogen at the site n at time t. The temporal evolution of this probability function follows the set of master equations [75]

$$\frac{dP_0(t)}{dt} = Q - \sum_{n=1}^{N} w_n P_0(t), \qquad (35)$$

for $n = 0$, and

$$\frac{dP_n(t)}{dt} = w_n P_0(t) - k P_n(t) \qquad (36)$$

for $n > 0$. Assuming that initially there were no morphogens in the system, $P_n(t = 0) = 0$ for all n, these master equations can be solved exactly at all times, leading to

$$P_0(t) = \frac{Q}{\eta}[1 - e^{-\eta t}]; \tag{37}$$

$$P_n(t) = \left[\frac{Q w_n}{\eta(\eta - k)}\right]e^{-\eta t} - \left[\frac{Q w_n}{k(\eta - k)}\right]e^{-kt} + \frac{Q w_n}{\eta k}, \tag{38}$$

where $\eta = \sum_{n=1}^{N} w_n$ is defined as a total productions rate from the source cell to all target cells. These results imply that the concentration of signaling molecules at each cell is an exponentially decaying function of the time. It can be viewed as a result of balancing between two opposing processes: the direct delivery with the rate η and the removal with the rate k. In the stationary-state limit, $(t \to \infty)$, the density profiles reduce to,

$$P_n^{(s)} = \frac{Q}{k\eta} w_n, \quad P_0^{(s)} = \frac{Q}{\eta}. \tag{39}$$

The dynamics of approaching the stationary state can be again understood from analyzing the local relaxation function, $R_n(t) \equiv \frac{P_n(t) - P_n^{(s)}}{P_n(0) - P_n^{(s)}}$ [5]. Simple calculations yield the following expressions for LAT [75]:

$$\langle \tau_n \rangle = \frac{1}{k} + \frac{1}{\eta}, \quad \langle \tau_0 \rangle = \frac{1}{\eta}. \tag{40}$$

The model predicts that there is no dependence on the target cell position, n, in contrast to reaction–diffusion mechanisms of the formation of morphogen gradients. The relaxation dynamics to the stationary concentration profiles is identical for all target cells. The main reason for this is that the processes at all target cells are independent from each other, and the stationary state at each of them cannot be established until the steady-state behavior is observed in the source cell [75]. This can only happen simultaneously in all cells in the system (see Fig. 10).

Local relaxation functions have also been applied to obtain the second moment of LAT, which allowed us to evaluate the robustness of the direct-delivery mechanism [75]. It was shown that

$$\langle \tau_n^2 \rangle = \frac{2(\eta^2 + k\eta + k^2)}{k^2 \eta^2}, \quad \langle \tau_0^2 \rangle = \frac{2}{\eta^2}, \tag{41}$$

which are again independent of the position of the target cell, n. The normalized variance was then computed to be [75],

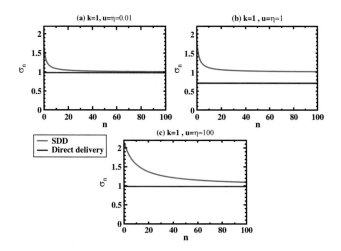

Fig. 11 Comparison of normalized variances as a function of the distance from the source for reaction–diffusion and direct-delivery mechanisms. Red lines correspond to the SDD model with a diffusion rate u. Blue lines describe the direct delivery via cytonemes with the total transportation rate η. Adapted with permission from Ref. [75]

$$\sigma_n = \left[\frac{\eta^2 + k^2}{\eta^2 + 2k\eta + k^2} \right]^{1/2}, \quad \sigma_0 = 1. \tag{42}$$

The normalized variances for the direct-delivery mechanisms are compared with the corresponding predictions from the reaction–diffusion processes in Fig. 11. One can see that σ_n is always less for the direct-delivery transport. This means that moving signaling molecules through cytonemes is a more robust mechanism of the formation of morphogen gradients because it is affected less by the stochastic noise [75]. In the reaction–diffusion mechanism signaling molecules can fluctuate spatially between different cells due to diffusion, but this option is not available for the direct-delivery mechanism. So the advantage of using the transport via cytonemes in creating signaling profiles is not only in overcoming the geometric constraints but also in reducing the influence of the stochastic noise [75].

To understand better how the direct-delivery process works, a more microscopic description of the transportation rates w_n was utilized for calculating the dynamic properties of the system [75]. In the first approach, it was suggested to use the fact that motor proteins drive the morphogens along the cytonemes. We argued that the rates should be related to the free energy difference of moving the signaling molecule from the source cell to the target cell [75],

$$w_n = \exp\left[-\frac{\Delta G(n)}{k_B T} \right], \tag{43}$$

where $\Delta G(n)$ is the energy required to displace the morphogen to the target cell n. One can assume that the length of the cytoneme to the target cell n, L_n is proportional to n, i.e., $L_n = An$, and the motor proteins spend energy ε (in units of $k_B T$) by moving every signaling molecule a distance l. Then the free energy difference can be written as

$$\Delta G(n) = \frac{L_n \varepsilon k_B T}{l} = \frac{An\varepsilon k_B T}{l} = \frac{nk_B T}{a}, \tag{44}$$

where $a = l/A\varepsilon$. The explicit expression for the transportation rate is given by $w_n = \exp\left[-\frac{n}{a}\right]$. This finally leads to the following expression for the stationary concentration profile [75]:

$$P_n^{(s)} = \frac{Q}{k\eta}\exp\left[-\frac{n}{a}\right], \tag{45}$$

where the total transportation rate η is equal to

$$\eta = \sum_{n=1}^{N}\exp\left[-\frac{n}{a}\right] = \frac{\exp\left[-\frac{1}{a}\right] - \exp\left[-\frac{(1+N)}{a}\right]}{1 - \exp\left[-\frac{1}{a}\right]}. \tag{46}$$

This model predicts the exponential decaying morphogen gradient (*see* Eq. 45), which is similar to predictions from the reaction–diffusion models [12, 51, 60, 79]. However, the difference is that the decay length in the direct-delivery mechanism, specified by the parameter a, is larger for more efficient motor proteins that spend less energy in driving the morphogens along the cytonemes. In the reaction–diffusion mechanism the decay length is controlled by the ratio of diffusion and degradation rates [5, 47]. Thus the energy dissipation in the transportation of signaling molecules through cytonemes is important for direct-delivery mechanism [75].

Because cytonemes are narrow cylindrical tubes, the transport of signaling molecules can be viewed as effectively one-dimensional, and this suggested intermolecular interactions, e.g., due to exclusion, might affect the dynamics [75]. This possibility was investigated using the concept of totally asymmetric exclusion processes (TASEP) [75]. TASEPs are nonequilibrium multi-particle models that were successfully utilized for uncovering the mechanisms of many complex biological processes [16]. It was proposed that each cytoneme can be viewed as 1D lattice on which morphogens move in the direction of the target cell. The problem of describing the dynamics of the formation of morphogen gradients in such system is identical to a set of open-boundary TASEP segments coupled at the source cell. Stationary-state fluxes for TASEP on finite lattice segments with an entrance rate α and an exit rate β are well known [21],

$$J(\alpha, \beta; n) = \frac{S_{n-1}(1/\beta) - S_{n-1}(1/\alpha)}{S_n(1/\beta) - S_n(1/\alpha)}, \tag{47}$$

where

$$S_n(y) = \sum_{i=0}^{n-1} \frac{(n-i)(n+i-1)!}{n!i!} y^{n-i+1}. \tag{48}$$

For the model presented in Fig. 11, the entrance and exit rates on each cytoneme are given by [75],

$$\alpha = Q/N, \quad \beta = k. \tag{49}$$

The transition rate from the source cell to the target cell n can be written as

$$w_n = J(Q/N, k; n). \tag{50}$$

The stationary-state profile in this system of interacting morphogens is equal to

$$P_n^{(s)} = \frac{Q}{k\eta} \frac{S_{n-1}(1/k) - S_{n-1}(N/Q)}{S_n(1/k) - S_n(N/Q)}. \tag{51}$$

Figure 12 illustrates the morphogen gradients for this system of interacting signaling molecules. The possibility of interactions between the morphogen molecules has a dramatic effect on the stationary profiles. While at the distances not far away from the source the effect is minimal, for larger distance the density profile saturates. But this leveling is not useful for the morphogen gradients because the information can be transferred efficiently only from strongly decaying profiles. It was suggested that these intermolecular interactions might present a problem for the direct-delivery mechanism on very large distances, but experimental tests of these predictions are needed because many other factors might change the outcome [75].

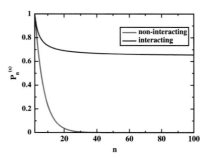

Fig. 12 Steady-state concentration profiles for interacting and non-interacting signaling molecules during the transportation along the cytonemes. Adapted with permission from Ref. [75]

3   Conclusions

To conclude, we presented a review of recent developments in theoretical understanding of the mechanisms that lead to the formation of biological signaling profiles. The dynamics of formation of morphogen gradients was analyzed first using the reaction–diffusion framework. This is assumed to be the main mechanisms for creating the concentration profiles of signaling molecules that can efficiently transfer the information in embryo systems. We discussed the critical role of the degradation processes, and it was argued that its action is similar to the driving potential that accelerates the dynamics of formation of morphogen gradients. Several other important aspects of the development of signaling profiles, including the effect of the size of the production region, dimensionality, nonlinearity in the degradation, and discreteness of these processes, have been thoroughly analyzed. We also discussed the alternative direct-delivery mechanisms in the establishment of morphogen gradients. The presented theoretical methods are applicable to a broad range of biological development phenomena, as well as for the cell signaling and tissue and organ formation processes.

Acknowledgements

A.B.K. acknowledges the support from the Center for Theoretical Biological Physics (NSF Grant PHY-1427654).

References

1. Alaynick WA, Jessell TM, Pfaff SL (2011) SnapShot: spinal cord development. Cell 146:178–178.e1

2. Berezhkovskii AM (2011) Renewal theory for single-molecule systems with multiple reaction channels. J Chem Phys 134:074114

3. Berezhkovskii AM, Shvartsman SY (2011) Physical interpretation of mean local accumulation time of morphogen gradient formation. J Chem Phys 135:154115

4. Berezhkovskii AM, Shvartsman SY (2013) Kinetics of receptor occupancy during morphogen gradient formation. J Chem Phys 138:244105

5. Berezhkovskii AM, Sample C, Shvartsman SY (2010) How long does it take to establish a morphogen gradient? Biophys J 99:L59–L61

6. Berezhkovskii AM, Sample C, Shvartsman SY (2011) Formation of morphogen gradients: local accumulation time. Phys Rev E 83:051906

7. Bergmann S, Sandler O, Sberro H, Shnider S, Schejter E, Shilo B-Z, Barkai N (2007) Pre-steady-state decoding of the bicoid morphogen gradient. PLoS Biol 5:e46

8. Bischoff M, Gradilla AC, Seijo I, Andrés G, Rodríguez-Navas C, González-Méndez L, Guerrero I (2013) Cytonemes are required for the establishment of a normal Hedgehog morphogen gradient in Drosophila epithelia. Nature Cell Biol 15:1269–1281

9. Bozorgui B, Teimouri H, Kolomeisky AB (2015) Theoretical analysis of degradation mechanisms in the formation of morphogen gradients. J Chem Phys 143:025102

10. Bressloff PC, Hyunjoong K (2018) Bidirectional transport model of morphogen gradient formation via cytonemes. Phys Biol 15:026010

11. Briscoe J (2009) Making a grade: Sonic Hedgehog signalling and the control of neural cell fate. EMBO J 28:457–465

12. Briscoe J, Small S (2015) Morphogen rules: design principles of gradient-mediated embryo patterning. Development 142:3996–4009

13. Castle BT, Howard SA, Odde DJ (2011) Assessment of transport mechanisms underlying the bicoid morphogen gradient. Cell Mol Bioeng 4:116–121

14. Chen Y, Struhl G (1996) Dual roles for patched in sequestering and transducing Hedgehog. Cell 87:553–563

15. Cheung D, Miles C, Kreitman M, Ma J (2014) Adaptation of the length scale and amplitude of the Bicoid gradient profile to achieve robust patterning in abnormally large Drosophila melanogaster embryos. Development 141:124–135

16. Chou T, Mallick K, Zia RKP (2011) Non-equilibrium statistical mechanics: from a paradigmatic model to biological transport. Rep Prog Phys 74:116601

17. Cohen M, Georgiou M, Stevenson NL, Miodownik M, Baum B (2010) Dynamic filopodia transmit intermittent Delta-Notch signaling to drive pattern refinement during lateral inhibition. Dev Cell 19:78–89

18. Crick FH (1970) Diffusion in embryogenesis. Nature 225:420–421

19. Dalessi S, Neves A, Bergmann S (2012) Modeling morphogen gradient formation from arbitrary realistically shaped sources. J Theor Biol 294:130–138

20. Deng J, Wang W, Lu LJ, Ma J (2010) A Two-dimensional simulation model of the bicoid gradient in Drosophila. PLoS Biol 5: e10275

21. Derrida B, Evans MR, Hakim V, Pasquier V (1993) Exact solution of a ID asymmetric exclusion model using a matrix formulation. J Phys A 26:1493–1517

22. Dessaud E, Yang LL, Hill K, Cox B, Ulloa F, Ribeiro A, Mynett A, Novitch BG, Briscoe J (2007) Interpretation of the sonic hedgehog morphogen gradient by a temporal adaptation mechanism. Nature 450:717–720

23. Dilao R, Muraro D (2010) mRNA diffusion explains protein gradients in Drosophila early development. J Theor Biol 264:847–853

24. Drocco JA, Grimm O, Tank DW, Wieschaus E (2011) Measurement and perturbation of morphogen lifetime: effects on gradient shape. Biophys J 101:1807–1815

25. Drocco JA, Wieschaus EF, Tank DW (2012) The synthesis-diffusion-degradation model explains Bicoid gradient formation in unfertilized eggs. Phys Biol 9:055004

26. Eldar A, Rosin D, Shilo B-Z, Barkai N (2003) Self-enhanced ligand degradation underlies robustness of morphogen gradients. Dev Cell 5:635–646

27. Ellery AJ, Simpson MJ, McCue SW (2013) Comment on local accumulation times for source, diffusion, and degradation models in two and three dimensions. J Chem Phys 139:017101

28. England JL, Cardy J (2005) Morphogen gradient from a noisy source. Phys Rev Lett 94:078101

29. Entchev EV, Schwabedissen A, Gonzales-Gaitan M (2000) Gradient formation of the TGF-beta homolog Dpp. Cell 103:981–991

30. Fairchild CL, Barna M (2014) Specialized filopodia: at the 'tip' of morphogen transport and vertebrate tissue patterning. Curr Opin Genet Devel 27:67–73

31. Fedotov S, Falconer S (2014) Nonlinear degradation-enhanced transport of morphogens performing subdiffusion. Phys Rev E 89:012107

32. Gradilla A-C, Guerrero I (2013) Cytoneme-mediated cell-to-cell signaling during development. Cell Tissue Res 352:59–66

33. Gregor T, Wieschaus EF, McGregor AP, Bialek W, Tank DW (2007) Stability and nuclear dynamics of the bicoid morphogen gradient. Cell 130:141–152

34. Grimm O, Coppy M, Wieschaus EF (2009) Modelling the bicoid gradient. Development 137:2253–2264

35. Gordon PV, Muratov CB (2012) Self-similarity and long-time behavior of solutions of the diffusion equation with nonlinear absorption and a boundary source. Netw Heterog Media 7:767–780

36. Gordon PV, Muratov CB (2015) Eventual self-similarity of solutions for the diffusion equation with nonlinear absorption and a point source. SIAM J Math Anal 47:2903–2916

37. Gordon PV, Sample C, Berezhkovskii AM, Muratov CB, Shvartsman SY (2011) Local kinetics of morphogen gradients. Proc Natl Acad Sci USA 108:6157–6162

38. Gordon PV, Muratov CB, Shvartsman SY (2013) Local accumulation times for source, diffusion, and degradation models in two and three dimensions. J Chem Phys 138:104121

39. Guerrero I, Kornberg TB (2014) Hedgehog and its circuitous journey from producing to target cells. Seminars Cell Dev Biol 33:52–62

40. Hecht I, Rappel W-J, Levine H (2009) Determining the scale of the Bicoid morphogen gradient. Proc Natl Acad Sci USA 106:1710–1715

41. Incardona JP, Lee JH, Robertson CP, Enga K, Kapur RP, Roelink H (2000) Receptor-mediated endocytosis of soluble and membrane-tethered Sonic hedgehog by Patched-1. Proc Natl Acad Sci USA 97:12044–12049

42. Kornberg TB (2012) The imperatives of context and contour for morphogen dispersion. Biophys J 103:2252–2256

43. Kerszberg M, Wolpert L (1998) Mechanisms for positional signalling by morphogen transport: a theoretical study. J Theor Biol 191:103–114

44. Kerszberg M, Wolpert L (2007) Specifying positional information in the embryo: looking beyond morphogens. Cell 130:205–209

45. Kicheva A, Pantazis P, Bollenbach T, Kalaidzidis Y, Bittig T, Jülicher F, Gonzales-Gaitan M (2007) Kinetics of morphogen gradient formation. Science 315:521–525

46. Kicheva A, Bollenbach T, Wartlick O, Jülicher F, Gonzalez-Gaitan M (2012) Investigating the principles of morphogen gradient formation: from tissues to cells. Curr Opin Gen Dev 22:527–532

47. Kolomeisky AB (2011) Formation of a morphogen gradient: acceleration by degradation. J Phys Chem Lett 2:1502–1505

48. Kornberg TB, Roy S (2014) Communicating by touch neurons are not alone. Trends Cell Biol 24:370–376

49. Kornberg TB, Roy S (2014) Cytonemes as specialized signaling filopodia. Development 141:729–736

50. Krotov D, Dubuis JO, Gregor T, Bialek W (2014) Morphogenesis at criticality. Proc Natl Acad Sci USA 111:3683–3688

51. Lander DA (2007) Morpheus unbound: reimagining the morphogen gradient. Cell 128:245–256

52. Lipshitz HD (2009) Follow the mRNA: a new model for Bicoid gradient formation. Nature Rev Mol Cell Biol 10:509–512

53. Little SC, Tkacik G, Kneeland TB, Wieschaus EF, Gregor T (2011) The formation of the bicoid morphogen gradient requires protein movement from anteriorly localized mRNA. PLoS Biol 9:e1000596

54. Lodish H, Berk A, Kaiser C, Krieger M, Scott MP, Bretscher A, Ploegh H, Matsudaira P (2007) Molecular cell biology, 6th edn. W.H. Freeman, New York

55. Martinez-Arias A, Stewart A (2002) Molecular principles of animal development. Oxford University Press, New York

56. Medioni C, Mowry K, Bess F (2012) Principles and roles of mRNA localization in animal development. Development 139:3263–3276

57. Mogilner A, Odde D (2011) Modeling cellular processes in 3D. Trends Cell Biol 21:692–700

58. Müller P, Rogers KW, Jordan BM, Lee JS, Robson D, Ramanathan S, Schier AF (2012) Differential diffusivity of Nodal and Lefty underlies a reaction-diffusion patterning system. Science 336:721–724

59. Müller P, Rogers KW, Yu SR, Brand M, Schier AF (2013) Morphogen transport. Development 140:1621–1638

60. Porcher A, Dostatni N (2010) The Bicoid morphogen system. Curr Biol 20:R249–R254

61. Redner S (2001) A guide to first-passage processes. Cambridge University Press, New York

62. Reingruber J, Holcman D (2014) Computational and mathematical methods for morphogenetic gradient analysis, boundary formation and axonal targeting. Seminars Cell Dev Biol 35:189–202

63. Richards DM, Saunders TE (2015) Spatiotemporal analysis of different mechanisms for interpreting morphogen gradients. Biophys J 108:2061–2073

64. Rogers KW, Schier AF (2011) Morphogen gradients: from generation to interpretation. Annu Rev Cell Dev Biol 27:377–407

65. Rørth P (2014) Reach out and touch someone. Science 343:848–849

66. Roy S, Kornberg TB (2015) Paracrine signaling mediated at cell-cell contacts. Bioessays 37:25–33

67. Sample C, Shvartsman SY (2010) Multiscale modeling of diffusion in the early Drosophila embryo. Proc Natl Acad Sci USA 107:10092–10096

68. Sanders TA, Llagostera E, Barna M (2013) Specialized filopodia direct long-range transport of SHH during vertebrate tissue patterning. Nature 497:628–632

69. Saunders T, Howard M (2009) When it pays to rush: interpreting morphogen gradients prior to steady-state. Phys Biol 6:046020

70. Sigaut L, Pearson JE, Colman-Lerner A, Dawson SP (2014) Messages do diffuse faster than messengers: Reconciling disparate estimates of the morphogen bicoid diffusion coefficient. PLoS Comp Biol 10:e1003629

71. Spirov A, Fahmy K, Schneider M, Frei E, Noll M, Baumgartner S (2009) Formation of

the bicoid morphogen gradient: an mRNA gradient dictates the protein gradient. Development 136:605–614

72. Tabata T, Takei Y (2004) Morphogens, their identification and regulation. Development 131:703–712

73. Teimouri H, Kolomeisky AB (2014) Development of morphogen gradient: the role of dimension and discreteness. J Chem Phys 140:085102

74. Teimouri H, Kolomeisky AB (2015) The role of source delocalization in the development of morphogen gradients. Phys Biol 12:026006

75. Teimouri H, Kolomeisky AB (2016) New model for understanding mechanisms of biological signaling: direct transport via cytonemes. J Phys Chem Lett 7:180–185

76. Teimouri H, Bozorgui B, Kolomeisky AB (2016) Development of morphogen gradients with spatially varying degradation rates. J Phys Chem B 120:2745–2750

77. Tompkins N, Li N, Girabawe C, Heymann M, Ermentrout GB, Epstein IR, Fraden S (2013) Testing Turing's theory of morphogenesis in chemical cells. Proc Natl Acad Sci USA 111:4397–4402

78. Tufcea DE, Francois P (2015) Critical timing without a timer for embryonic development. Biophys J 109:1724–1734

79. Wartlick O, Kicheva A, Gonzales-Gaitan M (2009) Morphogen gradient formation. Cold Spring Harb Perspect Biol 1:a001255

80. Wolpert L (1969) Positional information and the spatial pattern of cellular differentiation. J Theor Biol 25:1–47

81. Wolpert L (1998) Principles of development. Oxford University Press, New York

82. Yu SR, Burkhardt M, Nowak M, Ries J, Petrasek Z, Scholpp S, Schwille P, Brand M (2009) Fgf8 morphogen gradient forms by a source-sink mechanism with freely diffusing molecules. Nature 461:533–536

83. Yuste SB, Abad E, Lindenberg K (2010) Reaction-subdiffusion model of morphogen gradient formation. Phys Rev E 82:061123

84. Zhou S, Lo WC, Suhalim JL, Digman MA, Grattom E, Nie Q, Lander AD (2012) Free extracellular diffusion creates the Dpp morphogen gradient of the drosophila wing disc. Curr Biol 22:668–675

Chapter 13

Simulation of Morphogen and Tissue Dynamics

Michael D. Multerer, Lucas D. Wittwer, Anna Stopka, Diana Barac, Christine Lang, and Dagmar Iber

Abstract

Morphogenesis, the process by which an adult organism emerges from a single cell, has fascinated humans for a long time. Modeling this process can provide novel insights into development and the principles that orchestrate the developmental processes. This chapter focuses on the mathematical description and numerical simulation of developmental processes. In particular, we discuss the mathematical representation of morphogen and tissue dynamics on static and growing domains, as well as the corresponding tissue mechanics. In addition, we give an overview of numerical methods that are routinely used to solve the resulting systems of partial differential equations. These include the finite element method and the Lattice Boltzmann method for the discretization as well as the arbitrary Lagrangian-Eulerian method and the Diffuse-Domain method to numerically treat deforming domains.

Key words In silico morphogenesis, Morphogen dynamics, Tissue dynamics, Tissue mechanics

1 Introduction

During morphogenesis, the coordination of the processes that control size, shape, and pattern is essential to achieve stereotypic outcomes and comprehensive functionality of the developing organism. There are two main components contributing to the precisely orchestrated process of morphogenesis: morphogen dynamics and tissue dynamics. While signaling networks control cellular behavior, such as proliferation and differentiation, tissue dynamics in turn modulate diffusion, advection, and dilution, and affect the position of morphogen sources and sinks. Due to this interconnection, the regulation of those processes is very complex. Although a large amount of experimental data is available today, many of the underlying regulatory mechanisms are still unknown. In recent years, cross-validation of numerical simulations with experimental data has emerged as a powerful method to achieve an integrative understanding of the complex feedback structures underlying morphogenesis, see, e.g., [1–5].

Julien Dubrulle (ed.), *Morphogen Gradients: Methods and Protocols*, Methods in Molecular Biology, vol. 1863,
https://doi.org/10.1007/978-1-4939-8772-6_13, © Springer Science+Business Media, LLC, part of Springer Nature 2018

Simulating morphogenesis is challenging because of the multiscale nature of the process. The smallest regulatory agents, proteins, measure only a few nanometers in diameter, while animal cell diameters are typically at least a 1000-fold larger cp. [6, 7], and developing organs start as a small collection of cells, but rapidly develop into structures comprised of ten thousands of cells, cf. [8]. A similar multiscale nature also applies to the time scale. The basic patterning processes during morphogenesis typically proceed within days. Gestation itself may take days, weeks, or months-in some cases even years, see [9]. Intra-cellular signaling cascades, on the other hand, may be triggered within seconds, and mechanical equilibrium in tissues can be regained in less than a minute after a perturbation, see, e.g., [10]. The speed of protein turn-over, see [11], and of transport processes, see, e.g., [12], falls in between. Together, this results in the multiscale nature of the problem.

Given the multiscale nature of morphogenesis, combining signaling dynamics with tissue mechanics in the same computational framework is a challenging task. Where justified, models of morphogenesis approximate tissue as a continuous domain. In this case, patterning dynamics can be described by reaction-advection-diffusion models. Experiments have shown that a tissue can be well approximated by a viscous fluid over long time scales, i.e., several minutes to hours, and by an elastic material over short time scales, i.e., seconds to minutes, cf. [13]. Accordingly, tissue dynamics can be included by using the Navier-Stokes equation and/or continuum mechanics. In addition, cell-based simulation frameworks of varying resolution have been developed to incorporate the behavior of single cells. These models can be coupled with continuum descriptions where appropriate.

In this review, we provide an overview of approaches to describe, couple, and solve dynamical models that represent tissue mechanics and signaling networks. Subheading 2 deals with the mathematical representation of morphogen dynamics, tissue growth, and tissue mechanics. Subheading 3 covers cell-based simulation frameworks. Finally, Subheading 4 presents common numerical approaches to solve the respective models.

2   Mathematical Representation of Morphogen and Tissue Dynamics

2.1   Morphogen Dynamics

A fundamental question in biology is that of self-organization, or how the symmetry in a seemingly homogeneous system can be broken to give rise to stereotypical patterning and form. In 1952, Alan Turing first introduced the concept of a *morphogen* in his seminal paper "The Chemical Basis of Morphogenesis," cf. [14], in the context of self-organization and patterning. He hypothesized that a system of chemical substances "reacting together and diffusing through a tissue" was sufficient to explain the main phenomena of morphogenesis.

Morphogens can be transported from their source to target tissue in different ways. Transport mechanisms can be roughly divided into two categories: extracellular diffusion-based mechanisms and cell-based mechanisms [12]. In the first case, morphogens diffuse throughout the extracellular domain. Their movement can be purely random, or inhibited or enhanced by other molecules in the tissue. For example, a morphogen could bind to a receptor which would hinder its movement through the tissue. Morphogens can also be advected by tissue that is growing or moving. Cell-based transport mechanisms include transcytosis [15] and cytonemes [16].

According to the transcytosis model, morphogens are taken up into the cell by endocytosis and are then released by exocytosis, facilitating their entry into a neighboring cell [17]. In this way, morphogens can move through the tissue. The morphogen Decapentaplegic (Dpp) was proposed to spread by transcytosis in the *Drosophila* wing disc, see [18]. However, transport by transcytosis would be too slow to explain the kinetics of Dpp spreading, cp. [19], and further experiments refuted the transcytosis mechanism for Dpp transport in the wing disc, cf. [20].

According to the cytoneme model, cytonemes, i.e., filopodia-like cellular projections, emanate from target cells to contact morphogen producing cells and vice versa, cf. [16]. Morphogens are then transported along the cytonemes to the target cell. Several experimental studies support a role of cytonemes in morphogen transport across species, see [21–23]. However, so far, the transport kinetics and the mechanistic details of the intra-cellular transport are largely unknown, and a validated theoretical framework to describe cytoneme-based transport is still missing. Accordingly, the standard transport mechanism in computational studies still remains diffusion. In this book chapter, we will only consider diffusion- and advection-based transport mechanisms.

A key concept for morphogen-based patterning is Lewis Wolpert's *French Flag model*, cf. [24]. According to the French Flag model, morphogens diffuse from a source and form a gradient across a tissue such that cells close to the source experience the highest morphogen concentration, while cells further away experience lower concentrations (Fig. 1). To explain the emergence of patterns such as the digits in the limb, Wolpert proposed that the fate of a tissue segment depends on whether the local concentration is above or below a patterning threshold. Thus, the cells with the highest concentration of the morphogen (blue) differentiate into one type of tissue, cells with a medium amount (white) into another, and cells with the lowest concentration (red) into a third type, see Fig. 1. With the arrival of quantitative data, aspects of the

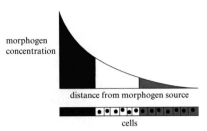

Fig. 1 Lewis Wolpert's French Flag model

French Flag model had to be modified, but the essence of the model has stood the test of time.

In the original publication, the source was included as a fixed boundary condition. No reactions were included in the domain, but morphogen removal was included implicitly by including an absorbing (zero concentration) boundary condition on the other side. The resulting steady-state gradient is linear and scales with the size of the domain, i.e., the relative pattern remains the same independent of the size of the domain. This is an important aspect as the patterning processes are typically robust to (small) differences in embryo size. Quantitative measurements have since shown that morphogen gradients are of exponential rather than linear shape, see [25]. The emergence of exponentially shaped gradients can be explained with morphogen turn-over in the tissue, cp. [19]. However, such steady-state exponential gradients have a fixed length scale and thus do not scale with a changing length of the patterning domain. Scaled steady-state patterns would require the diffusion coefficient, the reaction parameters, or the flux, to change with the domain size, cf. [26, 27]. At least, an appropriate change in the diffusion coefficient can be ruled out [25]. Intriguingly, pattern scaling is also observed on growing domains [28]. The observed dynamic scaling of the Dpp gradient can be explained with the pre-steady-state kinetics of a diffusion-based transport mechanism (rather than the steady-state gradient shape) and thus does not require any changes in the parameter values [29]. Finally, the quantitative measurements showed that the Dpp gradient amplitude increases continuously [28]. A threshold-based read-out as postulated by the French Flag model is nonetheless possible because the amplitude increase and the imperfect scaling of the pre-steady-state gradient compensate such that the Dpp concentration remains constant in the region of the domain where the Dpp-dependent pattern is defined, see [30]. In summary, current experimental evidence supports a French Flag-like mechanism where tissue is patterned by the threshold-based read-out of morphogen gradients. However, these gradients are not necessarily in steady state. Accordingly, dynamic models of morphogen gradients must be considered on growing domains. To do so, a mathematical formalism and simulations are required.

2.2 Mathematical Description of Diffusing Morphogens

Morphogen behavior can be modeled mathematically using the reaction diffusion equation, which we derive here. We assume, for the moment, that there is no tissue growth and the movement of the morphogen is a consequence of random motion. We denote the concentration of a morphogen in the domain $\Omega \subset \mathbb{R}^d$ as $c(\mathbf{x}, t)$, as it is dependent on time and its spatial position in the domain. Then the total concentration in Ω is $\int_\Omega c(\mathbf{x}, t)\mathrm{d}\mathbf{x}$ and the rate of change of the total concentration is

$$\frac{\mathrm{d}}{\mathrm{d}t} \int_\Omega c(\mathbf{x}, t)\mathrm{d}\mathbf{x}. \tag{1}$$

The rate of change of the total concentration in Ω is a result of interactions between the morphogens that impact their concentration and random movement of the morphogens. The driving force of diffusion is a decrease in Gibbs free energy or chemical potential difference. This means that a substance will generally move from an area of high concentration to an area of low concentration. The movement of $c(\mathbf{x}, t)$ is called the flux, i.e., the amount of substance that will flow through a unit area in a unit time interval. As the movement of the morphogen is assumed to be random, Fick's first law holds. The latter states that the magnitude of the morphogens movement from an area of high concentration to one of low concentration is proportional to that of the difference between the concentrations, or concentration gradient, i.e.,

$$\mathbf{j} = -D\nabla c(\mathbf{x}, t), \tag{2}$$

where D is the diffusion coefficient or diffusivity of the morphogen. This is a measure of how quickly the morphogen moves from a region of high concentration to a region of low concentration. The total flux out of Ω is then

$$\int_{\partial\Omega} \mathbf{j} \cdot \mathbf{n} \, \mathrm{d}S,$$

where $\mathrm{d}S$ is the surface element and $\mathbf{n}$ is the normal vector to the boundary. Reactions between the morphogens also affect the rate of change of $c(\mathbf{x}, t)$. We denote the reaction rate $R(c)$. The rate of change of the concentration in the domain Ω due to morphogen interactions is

$$\int_\Omega R(c)\mathrm{d}\mathbf{x}.$$

As the rate of change of the total concentration in Ω is the sum of the rate of change caused by morphogen interactions and the rate of change caused by random movement, we have

$$\frac{d}{dt}\int_{\Omega} c(\mathbf{x}, t)d\mathbf{x} = -\int_{\partial\Omega} \mathbf{j} \cdot \mathbf{n} \ dS + \int_{\Omega} R(c)d\mathbf{x}. \qquad (3)$$

Now, the Divergence Theorem yields

$$\int_{\partial\Omega} \mathbf{j} \cdot \mathbf{n} \, dS = \int_{\Omega} \nabla \cdot \mathbf{j} \, d\mathbf{x}. \qquad (4)$$

Substituting Eqs. 4 and 2 into Eq. 3 and exchanging the order of integration and differentiation using Leibniz's theorem gives

$$\int_{\Omega} \frac{\partial c}{\partial t} - D\Delta c - R(c)d\mathbf{x} = 0.$$

Taking into account that this equilibrium holds for any control volume $V \subset \Omega$, we obtain the classical reaction-diffusion equation

$$\frac{\partial c}{\partial t} = D\Delta c + R(c). \qquad (5)$$

This partial differential equation (PDE) can be solved on a continuous domain to study the behavior of morphogens in a fixed domain over time. If there is more than one morphogen, then their respective concentrations can be labelled $c_i(\mathbf{x}, t)$ for $i = 1, \ldots, N$ where N is the number of morphogens. The reaction term then describes the morphogen interactions i.e., $R = R(c_1, \ldots, c_N)$. This results in a coupled system of PDEs, which, depending on the reaction terms, can be nonlinear. The accurate solution of these equations can be difficult and computationally costly.

It is important to keep in mind that reaction-diffusion equations only describe the average behavior of a diffusing substance. This approach is therefore not suitable if the number of molecules is small. In that case stochastic effects dominate, and stochastic, rather than deterministic, techniques should be applied, see [31–33].

2.3 Morphogen Dynamics on Growing Domains

In the previous paragraph we introduced morphogen dynamics on a fixed domain. However, tissue growth plays a key role in morphogenesis and can play a crucial part in the patterning process of the organism [34, 35]. Growth can affect the distribution of the morphogens, transporting them via advection and impacting the concentration via dilution, see Fig. 2. In turn, morphogens can influence tissue shape change and growth, for example, by initiating cell death and cell proliferation, respectively. This results in a mutual feedback between tissue growth and morphogen concentration.

It is then necessary to modify Eq. 5 to account for growth. Applying Reynolds transport theorem to the left-hand side of Eq. 3 we get

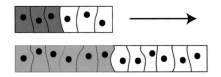

Fig. 2 Advection and dilution induced by a growing domain. The direction of growth is indicated by the arrow. Movement of the cells or tissue resulting from growth causes the morphogens to move (advection). Simultaneously cell division, or growth, dilutes morphogen molecules

$$\frac{\mathrm{d}}{\mathrm{d}t} \int_{\Omega_t} c(\mathbf{x}, t)\mathrm{d}\mathbf{x} = \int_{\Omega_t} \left(\frac{\partial c}{\partial t} + \mathrm{div}(c\mathbf{v})\right)\mathrm{d}\mathbf{x}.$$

For a more detailed derivation, we refer to [36]. This results in the reaction-diffusion equation on a growing domain:

$$\frac{\partial c}{\partial t} + \mathrm{div}(c\mathbf{v}) = D\Delta c + R(c). \tag{6}$$

By the Leibniz rule, there holds $\mathrm{div}(c\mathbf{v}) = c\mathrm{div}(\mathbf{v}) + \mathbf{v} \cdot \nabla c$. These terms can be interpreted as the *dilution*, i.e., the reduction in concentration of a solute in a solution, usually by adding more solvent, and *advection*, i.e., movement of a substance in a fluid caused by the movement of the fluid, respectively.

2.4 Modelling Tissue Growth

The details of the process of tissue growth still remain to be elucidated. It is therefore an open question of how best to incorporate it into a model. One approach considers the velocity field to be dependent on morphogen concentration, i.e., $\mathbf{v}(c, \mathbf{x}, t)$, where c is again the morphogen concentration present in the tissue [37, 38]. Another is "prescribed growth", in which the velocity field $\mathbf{v}(\mathbf{x}, t)$ of the tissue is specified and the initial domain is moved according to this velocity field. A detailed measurement of the velocity field can be obtained from experimental data. To this end, the tissue of interest can be stained and imaged at sequential developmental time points. These images can then be segmented to determine the shape of the domain. Displacement fields can be calculated by computing the distance between the domain boundary of one stage and that of the next. A velocity field in the domain can then be interpolated, for example, by assuming uniform growth between the center of mass and the nearest boundary point. To this extent, high quality image data is required to enable detailed measurements of the boundary to be extracted. For a detailed review on this process see [2, 3].

There are also other techniques to model tissue growth. If the local growth rate of the tissue is known, the Navier-Stokes equation can be used. Tissue is assumed to be an incompressible fluid and tissue growth can then be described with the Navier-Stokes equation for incompressible flow of Newtonian fluids, which reads

$$\rho(\partial_t \mathbf{v} + (\mathbf{v} \cdot \nabla)\mathbf{v}) = -\nabla p + \mu\left(\Delta \mathbf{v} + \frac{1}{3}\nabla(\operatorname{div} \mathbf{v})\right) + \mathbf{f}, \qquad (7)$$
$$\rho\operatorname{div} \mathbf{v} = \omega S,$$

where ρ is fluid density, μ dynamic viscosity, p internal pressure, $\mathbf{f}$ external force density, and $\mathbf{v}$ the fluid velocity field. The term ωS is the local mass production rate. The parameter ω is the molecular mass of the cells. The impact of cell signaling on growth can be modeled by having the source term S dependent on the morphogen concentration, i.e., $S = S(c)$. Note that the source term results in isotropic growth. External forces as implemented in the $\mathbf{f}$ term can induce anisotropic growth. Based on measurements, the Reynolds number for tissues is typically small, e.g., in embryonic tissue it is assumed to be of the order 10^{-14}, see [36]. Accordingly, the terms on the left-hand side of Eq. 7 can be neglected, as in Stokes' Flow.

The Navier-Stokes description, with a source term dependent on signaling, has been used in simulations of early vertebrate limb development, see [39]. An extended anisotropic formulation has been applied to *Drosophila* imaginal disc development in [40]. It has also been used to model bone development, cf. [41], and coupled with a travelling wave to simulate the developing *Drosophila* eye disc, see [42]. In the case of the developing limb, the proliferation rates were later determined, see [43]. They were then used as source terms in the isotropic Navier-Stokes tissue model. There was, however, a significant discrepancy between the predicted and actual growth. The shapes of the experimental and simulated developing limb were qualitatively different, and the actual expansion of the limb was much larger than expected. This shows that limb expansion must result from anisotropic processes, and suggested that the growth of the limb could in part be due to cell migration rather than solely local proliferation of cells.

2.5 Tissue Mechanics

Tissue expands and deforms during growth. Given its elastic properties, stresses must emerge in an expanding and deforming tissue. Cell rearrangements are able to dissipate these stresses and numerous experiments confirm the viscoelastic properties of tissues [13, 44–46]. Over long time scales, as characteristic for many developmental processes, tissue is therefore typically represented as a liquid, viscous material and is then described by the Stokes equation [36, 42, 47]. Over short time scales, however, tissues have mainly elastic properties. Continuum mechanical models are widely used to simulate the mechanical properties of tissues, see, e.g., [48, 49] and the references therein. Continuum mechanical descriptions usually consist of three parts: the *kinematics*, which describes the motions of objects, their displacements, velocity, and acceleration, the *constitutive equations*, which model the material laws and describe the response of the material to induced strains in terms of stress, and the underlying *balance principle*, which

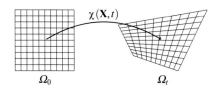

Fig. 3 Deformation of the domain Ω_0 via the deformation field χ

describes the governing physical equations. For a comprehensive introduction into continuum mechanics, we refer to [50, 51].

The mathematical representation of kinematics, i.e., the description of motion of points and bodies, is usually performed with respect to two different frameworks. They are called *Lagrangian* (or *material*) and *Eulerian* (or *spatial*) coordinates. The Lagrangian framework adopts a particle point of view, for example, the perspective of a single cell, and tracks its movement over time. In contrast to this, the Eulerian framework adopts the perspective of an entire body, for example, a tissue, and describes its position over time with respect to a given coordinate frame. More precisely, let $\Omega_0 \subset \mathbb{R}^d$ denote a body and

$$\chi : \Omega_0 \times [0, \infty) \to \mathbb{R}^d, \quad (\mathbf{X}, t) \mapsto \chi(\mathbf{X}, t)$$

a *deformation field*. The deformed body at a given time $t \in [0, \infty)$ is then denoted by $\Omega_t := \chi(\Omega_0, t)$, see Fig. 3 for a visualization. The position of the particle $\mathbf{X} \in \Omega_0$ at time $t \geq 0$ is therefore given by $\mathbf{x} = \chi(\mathbf{X}, t)$, which is the description in Eulerian coordinates. On the other hand, we can also consider $\mathbf{X} = \chi^{-1}(\mathbf{x}, t)$, which is the description in Lagrangian coordinates. More important than the deformation is the *displacement*

$$\mathbf{U}(\mathbf{X}, t) := \mathbf{x}(\mathbf{X}, t) - \mathbf{X} \quad \text{or} \quad \mathbf{u}(\mathbf{x}, t) := \mathbf{x} - \mathbf{X}(\mathbf{x}, t),$$

respectively. Based on the displacement, one can consider balance principles of the form

$$\operatorname{div}\sigma + \mathbf{f} = \rho\ddot{\mathbf{u}},$$

which is Newton's second law and is also known as Cauchy's first equation of motion, see, e.g., [51]. Herein, the tensor field σ characterizes the stresses inside the body, the vector field $\mathbf{f}$ summarizes internal forces, ρ denotes the mass density, and $\ddot{\mathbf{u}}$ is the acceleration. Thus, at steady state, the equation simplifies to

$$-\operatorname{div}\sigma = \mathbf{f}.$$

Note that the steady state in morphogen concentrations is reached very fast compared to the time scale on which growth happens.

Several models exist to describe material behavior. A material is called *elastic* if there exists a *response function* with

$$\sigma = \mathbf{g}(\mathbf{F}),$$

where

$$\mathbf{F}(\mathbf{X}, t) := \nabla\chi = [\partial_{X_j}\chi_i]_{i,j}$$

is the *deformation gradient. Linearly elastic* materials are described by *Hooke's law*. In this case, the function **g** is linear. For the description of tissues, nonlinear material responses are better suited. To that end, *hyperelastic* material models are used. They are characterized by the response function

$$\mathbf{g}(\mathbf{F}) = J^{-1}\frac{\partial W(\mathbf{F})}{\partial \mathbf{F}}\mathbf{F}^{\mathrm{T}},$$

where $J := \det\mathbf{F}$ and W is a scalar *strain energy density function*. For the modeling of soft tissues, *Fung-elastic* materials might be employed, see, e.g., [48, 49]. Here the strain energy density function is, for example, given by

$$W = \frac{C}{\alpha}\left[e^{\alpha(I_1-3)} - 1\right], \quad \text{with } C, \alpha > 0, \quad I_1 := \text{trace}(\mathbf{F}\mathbf{F}^{\mathrm{T}}).$$

In [49], this model is suggested to simulate the blastula stage of the sea urchin. Figure 4a shows the corresponding computational model. By considering only a cross section, the model can be reduced to two spatial dimensions. In Fig. 4b, pressure versus radius curves for different values of the parameter α are shown, where we assume that an interior pressure p acts on the interior wall. As can be seen, the material stiffens for increasing values of α. For the numerical simulations, the thickness of the blastula is set to $b_0 - a_0 = 75 - 50 = 25$ and we chose $C = 0.2$ [kPa], see also [49, 52].

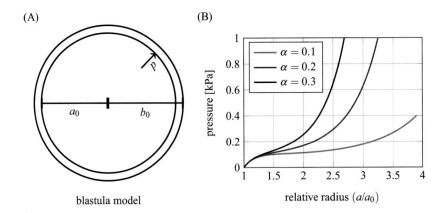

(A)   blastula model

(B)   relative radius (a/a_0)

Fig. 4 Numerical simulation of the blastula. (**a**) Model of the sea urchin blastula. (**b**) Pressure versus radius curves for different values of the parameter α/α_0

3   Cell-Based Simulation Frameworks

Cell-based simulations complement continuum models, and are important when cellular processes need to be included explicitly, i.e., cell adhesion, cell migration, cell polarity, cell division, and cell differentiation. Cell-based simulations are particularly valuable when local rules at the single-cell level give rise to global patterns. Such emergence phenomena have been studied with agent-based models in a range of fields, long before their introduction to biology. Early agent-based models of tissue dynamics were simulated on a lattice, and each cell was represented by a single, autonomous agent that moved and interacted with other agents according to a set of local rules. The effects of secreted morphogens or cytokines can be included by coupling the agent-based model with reaction-diffusion-based continuum models, as done in a model of the germinal center reaction during an immune response [53]. In this way, both direct cell-cell communication and long-range interactions can be realized.

A wide range of cell-based models has meanwhile been developed. The approaches differ greatly in their resolution of the underlying physical processes and of the cell geometries, and have been realized as both lattice-based and lattice-free models. In lattice-based models the spatial domain is represented by a one, two, or three dimensional lattice and a cell occupies a certain number of lattice sites. Cell growth can be included by increasing the number of lattice sites per cell and cell proliferation by adding new cells to the lattice. In off-lattice approaches, cells can occupy an unconstrained area in the domain. Similarly to on-lattice models, tissue growth can be implemented by modeling cell growth and proliferation. However, cell growth and proliferation is not restricted to discrete lattice sites. Among the off-lattice models it can be distinguished between center-based models, which represent cell dynamics via forces acting on the cell centers, and deformable cell models that resolve cell shapes. In case of a higher resolution of the cell geometry, the cell-based models can be coupled to signaling models, where components are restricted to the cells or the cell surface. A higher resolution and the independence from a lattice permit a more realistic description of the biological processes, but the resulting higher computational costs limit the tissue size and time frame that can be simulated.

In the following, we will provide a brief overview of the most widely used cell-based models for morphogenetic simulations, i.e., the *Cellular Potts model*, the *spheroid model*, the *Subcellular Element model*, the *vertex model*, and the *Immersed Boundary Cell model*, see Fig. 5 for their arrangement with respect to physical detail and spatial resolution. In the following, we will focus on the main ideas behind each model and name common software frameworks

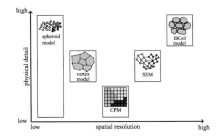

Fig. 5 Cell-based simulation frameworks and their arrangement with respect to physical detail and spatial resolution

that implement the aforementioned methods. A more detailed description of the models and their applications in biology can be found in [54–56].

The *Cellular Potts Model* (CPM) is a typical on-lattice approach as it originates from the Ising model, see [57]. It represents the tissue as a lattice where each lattice site carries a spin value representing the cell identity. The update algorithm of the CPM is the Metropolis algorithm, cf. [58], which aims to minimize the Hamiltonian energy function which is defined over the entire lattice. In the original CPM introduced in [59], the Hamiltonian includes a volume constriction term and a cell-cell adhesion term:

$$H = \sum_{\sigma} \lambda_v \left(V_\sigma - V_\sigma^T \right)^2 + \sum_{(\mathbf{x}, \mathbf{x}')} J\left(\tau(\sigma(\mathbf{x})), \tau(\sigma(\mathbf{x}')) \right) \cdot \left(1 - \delta(\sigma(\mathbf{x}), \sigma(\mathbf{x}')) \right).$$

The first term describes the volume constriction with λ_v being the coefficient controlling the energy penalization, V_σ being the actual volume, and V_σ^T the target volume of cell σ. The second term represents the cell-cell adhesion, where $J(\tau(\sigma(\mathbf{x})), \tau(\sigma(\mathbf{x}')))$ denotes the surface energy term between two cell types and $\delta(\cdot, \cdot)$ the Dirac δ-function. The CPM has been used to simulate various processes in morphogenesis, including kidney branching morphogenesis [60], somitogenesis [61], and chicken limb development [62]. As for many lattice-based algorithms, an advantage of the CPM is the efficient application of high performance computing and parallelization techniques, see, e.g., [55]. The main limitations of the CPM concern its high level of abstraction that limits the extent to which the simulations can be validated with experimental data: the interpretation of the temperature in the Metropolis algorithm is not straightforward and there is no direct translation between iteration steps and time. Moreover, the representation of cells and cell growth is coarse, and biophysical properties are difficult to directly relate to measurements. The open-source software framework CompuCell3D, see [63], is based on the CPM.

The *spheroid model* is an example for off-lattice *agent-based models* and represents cells as particle-like objects being a typical off-lattice approach. The cells are assumed to have a spherical shape being represented by a soft sphere interaction potential like the Johnson-Kendall-Roberts potential or the Hertz potential. The evolution of cells in time in the spheroid model can be performed in two different ways: deterministically by solving the equation of motion

$$\eta \frac{dx_i}{dt} = \sum_i \mathbf{F}_i,$$

where η is the mobility coefficient and $\mathbf{F}_i$ represent the forces acting on each particle, or by solving the stochastic Langevin equation. The simple representation of cells enables the simulation of a large number of cells with the spheroid model and further, to simulate tissues in 3D. However, as all cells are represented equally, there are no cellular details represented and thus also the coupling of tissue dynamics to morphogen dynamics is restricted. The 3D software framework CellSys, cf. [64], is built based on the spheroid model.

The *Subcellular Element Model* (SEM) represents a cell by many subcellular elements assuming that the inner of a cell, i.e., the cytoskeleton, can be subdivided. The elements are represented by point particles which interact via forces that are derived from interaction potentials such as the Morse potential. A typical equation of motion for the position $\mathbf{y}_{\alpha_i}$ of a subcellular element α_i of cell i reads, cf. [55],

$$\eta \frac{\partial \mathbf{y}_{\alpha_i}}{\partial t} = \zeta_{\alpha_i} - \nabla_{\alpha_i} \sum_{\beta_i \neq \alpha_i} V_{\text{intra}}(|\mathbf{y}_{\alpha_i} - \mathbf{y}_{\beta_i}|) - \nabla_{\alpha_i} \sum_{j \neq i} \sum_{\beta_j} V_{\text{inter}}(|\mathbf{y}_{\alpha_i} - \mathbf{y}_{\beta_j}|)$$

with η being the viscous damping coefficient and ζ_{α_i} Gaussian noise. The first term describes intra-cellular interactions between the subcellular element α_i and all other subcellular elements β_i of cell i. The second term represents inter-cellular interaction that takes into account all pair-interactions between subcellular elements β_j of neighboring cells j of cell i. For intra- and inter-cellular interaction potentials V_{intra}, V_{inter}, for example, the Morse potential can be used:

$$V(r) = U_0 \exp\left(\frac{-r}{\varepsilon_1}\right) - V_0 \exp\left(\frac{-r}{\varepsilon_2}\right),$$

where r is the distance between two subcellular elements, U_0, V_0 are the energy scale parameters, and ε_1, ε_2 are the length scale parameters defining the shape of the potential. Therefore, the SEM is very similar to agent-based models with the difference that each point particle represents parts of and not an entire cell. The SEM offers an explicit, detailed resolution of the cell shapes, further, a 3D implementation is straightforward. A disadvantage of the SEM is its high computational cost.

In the *vertex model*, cells are represented by polygons, where neighboring cells share edges and an intersection point of edges is a vertex. It was first used in 1980 to study epithelial sheet deformations [65]. The movement of the vertices is determined by forces acting on them, which can either be defined explicitly, see, e.g., [66], or are derived from energy potentials, cf. [67]. A typical energy function has the following form, see [55],

$$E(\mathbf{R}_i) = \sum_{\alpha} \frac{K_\alpha}{2} (A_\alpha - A_0)^2 + \sum_{\langle i,j \rangle} \Lambda_{ij} l_{ij} + \sum_{\alpha} \frac{\Gamma_\alpha}{2} L_\alpha^2$$

with $\mathbf{R}_i$ being the junctions direction of vertex i. The first term describes elastic deformations of a cell with K_α being the area elasticity coefficient, A_α the current area of a cell, and A_0 the resting area. The second term represents cell movements due to cell-cell adhesion via the line tension between neighboring vertices i and j with Λ_{ij} being the line tension coefficient and l_{ij} the edge length. The third term describes volumetric changes of a cell via the perimeter contractility, with Γ_α being the contractility coefficient, L_α the perimeter of a cell, see [68, 69] and the references therein for further details. Different approaches have been developed to move the vertices over time. The explicit cell shapes in the vertex model allow for a relatively high level of detail, however, cell-cell junction dynamics, cell rearrangements, etc. require a high level of abstraction. The vertex model is well suited to represent densely packed epithelial tissues, but there is no representation of the extracellular matrix. Computationally, the vertex model is still relatively efficient. The software framework Chaste, cp. [70, 71], is a collection of cell-based tissue model implementations that includes the vertex model among others.

In the *Immersed Boundary Cell Model* (IBCell model) the cell boundaries are discretized resulting in a representation of cells as finely resolved polygons. These polygons are immersed in a fluid and, in contrast to the vertex model, each cell has its own edge, cf. [72]. Therefore, there are two different fluids: fluid inside the cells representing the cytoplasm and fluid between the cells representing the inter-cellular space. The fluid-structure interaction is achieved as follows: iteratively, the fluid equations for intra- and extracellular fluids are solved, the velocity field to the cell geometries is interpolated, the cells are moved accordingly, the forces acting on the cell geometries are recomputed and distributed to the surrounding fluid, and the process restarts. In other words, the moving fluids exert forces on the cell membranes, and the cell membranes in turn exert forces on the fluids. Further, different force generating processes can be modeled on the cell boundaries, such as cell-cell junctions or membrane tensions, for example, by inserting Hookean spring forces between pairs of polygon vertices. The IBCell model offers a high level of detail in representing the cells, down to individual cell-cell junctions. Furthermore, the representation of tissue mechanics such as cell division or cell growth can be easily implemented. A disadvantage of the IBCell model is its inherent computational cost. The open-source software framework LBIBCell, cp. [73], is built on the combination of the IBCell model and the Lattice Boltzmann (LB) method, see Subheading 4.4. LBIBCell realizes the fluid-structure interaction by an iterative algorithm. Moreover, it allows for the coupled simulation of cell dynamics and biomolecular signaling.

4 Overview of Numerical Approaches

As we have seen so far, the mathematical description of biological processes leads to complex systems of reaction diffusion equations, which might even be defined with respect to growing domains. In this section, we give an overview of methods to solve these equations numerically. For the discretization of partial differential equations, we consider the finite element method, which is more flexible when it comes to complex geometries than, the also well known, finite volume and finite difference methods, see, e.g., [74, 75] and the references therein. The numerical treatment of growing domains can be incorporated by either the arbitrary Lagrangian-Eulerian method or the Diffuse-Domain method. Finally, we consider the Lattice Boltzmann method, which is feasible for the simulation of fluid dynamics and the simulation of reaction diffusion equations.

4.1 Finite Element Method

The finite element method is a versatile tool to treat partial differential equations in one to three spatial dimensions numerically. The method is heavily used in practice to solve engineering, physical, and biological problems. Finite elements were invented in the 1940s, see the pioneering work [76], and are textbook knowledge in the meantime, see, e.g., [75, 77–79]. Particularly, there exists a wide range of commercial and open-source software frameworks that implement the finite element method, e.g., *COMSOL, see* **Note 1**. *dune-fem, see* **Note 2**. and *FEniCS., see* **Note 3**.

The pivotal idea, the Ritz-Galerkin method, dates back to the beginning of the twentieth century, see [80] for a historical overview. The underlying principle is the fundamental lemma of calculus of variations: Let $g : (0, 1) \to \mathbb{R}$ be a continuous function. If

$$\int_0^1 gv\mathrm{d}x = 0 \quad \text{for all } v \in C_0^\infty (0, 1),$$

i.e., for all compactly supported and smooth functions v on $(0,1)$, then there holds $g \equiv 0$, cf. [81]. We can apply this principle to solve the second-order boundary value problem

$$-\frac{\mathrm{d}^2 u}{\mathrm{d}x^2} = f \ \text{ in } (0,1) \ \text{ and } \ u(0) = u(1) = 0 \qquad (8)$$

for a continuous function $f : (0,1) \to \mathbb{R}$, numerically. The fundamental lemma of calculus of variations yields

$$-\frac{\mathrm{d}^2 u}{\mathrm{d}x^2} = f \ \Leftrightarrow \ \int_0^1 \left(-\frac{\mathrm{d}^2 u}{\mathrm{d}x^2} - f\right) v \, \mathrm{d}x = 0 \quad \text{for all } v \in C_0^\infty(0,1).$$

Rearranging the second equation and integrating by parts then leads to

$$-\frac{\mathrm{d}^2 u}{\mathrm{d}x^2} = f \ \Leftrightarrow \ \int_0^1 \frac{\mathrm{d}u}{\mathrm{d}x}\frac{\mathrm{d}v}{\mathrm{d}x} \, \mathrm{d}x = \int_0^1 f v \, \mathrm{d}x \quad \text{for all } v \in C_0^\infty(0,1).$$

Depending on the right-hand side, the above equation does not necessarily have a solution $u \in C^2(0,1)$. However, solvability is guaranteed in the more general function space $H_0^1(0,1)$, which consists of all weakly differentiable functions with square integrable derivatives. Then, introducing the bilinear form

$$a : H_0^1(0,1) \times H_0^1(0,1) \to \mathbb{R}, \quad a(u,v) := \int_0^1 \frac{\mathrm{d}u}{\mathrm{d}x}\frac{\mathrm{d}v}{\mathrm{d}x} \, \mathrm{d}x,$$

and the linear form

$$\ell : H_0^1(0,1) \to \mathbb{R}, \quad \ell(v) := \int_0^1 f v \, \mathrm{d}x,$$

the *variational formulation* of Eq. 8 reads

$$a(u,v) = \ell(v) \quad \text{for all } v \in H_0^1(0,1).$$

The idea of the *Ritz-Galerkin method* is now to look for the solution to the boundary value problem only in a finite dimensional subspace $V_N \subset H_0^1(0,1)$:

$$\text{Find } u_N \in V_N \text{ such that}$$

$$a(u_N, v_N) = \ell(v_N) \quad \text{for all } v_N \in V_N.$$

Let $\varphi_1, \ldots, \varphi_N$ be a basis of V_N. Then, there holds $u(x) \approx \sum_{j=1}^{N} u_j \varphi_j(x)$. Moreover, due to linearity, it is sufficient to

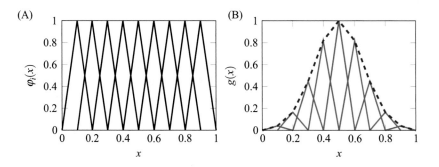

Fig. 6 Approximation by linear finite elements. (**a**) Linear basis functions φ_i on the unit interval. (**b**) Finite element approximation of the Gaussian $g(x)$

consider only the basis functions φ_i as *test functions* $v_N \in V_N$. Thus, we obtain

$$a\left(\sum_{j=1}^{N} u_j \varphi_j, \varphi_i\right) = \sum_{j=1}^{N} a(\varphi_j, \varphi_i) u_j = \ell(\varphi_i) \quad \text{for } i = 1, \ldots, N$$

and consequently, by setting $\mathbf{A} := [a(\varphi_j, \varphi_i)]_{i,j=1}^{N} \in \mathbb{R}^{N \times N}$, $\mathbf{u} := [u_i]_{i=1}^{N} \in \mathbb{R}^{N}$ and $\mathbf{f} := [\ell(\varphi_i)]_{i=1}^{N} \in \mathbb{R}^{N}$, we end up with the linear system of equations

$$\mathbf{Au} = \mathbf{f}.$$

The latter can now be solved by standard techniques from linear algebra.

A suitable basis in one spatial dimension is, for instance, given by the linear hat functions, see Fig. 6a for a visualization. Figure 6b shows how the function $g(x) := \exp(-20(x - 0.5)^2)$ is represented in this basis on the grid $x_i = i/10$ for $i = 1, \ldots, 9$. For the particular choice of the hat functions, the coefficients are given by the evaluations of g at the nodes x_i, i.e.,

$$g(x) \approx \sum_{i=1}^{9} g_i \varphi_i(x) \quad \text{with} \quad g_i := \exp(-20(x_i - 0.5)^2).$$

Note that using the hat functions as a basis results in a tridiagonal matrix $\mathbf{A}$.

The presented approach can be transferred one-to-one to two and three spatial dimensions and also to more complex (partial) differential equations. In practice, the *ansatz space* V_N is obtained by introducing a triangular mesh for the domain in two spatial dimensions or a tetrahedral mesh in three spatial dimensions, and then considering piecewise polynomial functions with respect to this mesh. To reduce the computational effort, the basis functions are usually locally supported, since this results in a sparse pattern for the matrix $\mathbf{A}$.

4.2 Arbitrary Lagrangian-Eulerian (ALE)

The underlying idea of the *Arbitrary Lagrangian-Eulerian* (ALE) description of motion is to decouple the movement of a given body $\Omega_0 \subset \mathbb{R}^d$ from the motion of the underlying mesh that is used for the numerical discretization. We refer to [82] and the references therein for a comprehensive introduction into ALE. As motivated by the paragraph on continuum mechanics, we start from a deformation field

$$\chi : \Omega_0 \times [0, \infty) \to \mathbb{R}^d,$$

which describes how the body $\Omega_t = \chi(\Omega_0, t)$ evolves and moves over time. Remember that the position of a particle $\mathbf{X} \in \Omega_0$ at time $t \geq 0$ in Eulerian coordinates is given by $\mathbf{x} = \chi(\mathbf{X}, t)$, whereas its Lagrangian coordinates read $\mathbf{X} = \chi^{-1}(\mathbf{x}, t)$. Analogously, the corresponding *velocity fields* for Ω_t are then given by

$$\mathbf{V}(\mathbf{X}, t) = \frac{\partial}{\partial t} \chi(\mathbf{X}, t)$$

in Lagrangian coordinates and by

$$\mathbf{v}(\mathbf{x}, t) := \mathbf{V}\big(\chi^{-1}(\mathbf{X}, t), t\big)$$

in spatial coordinates, respectively.

Usually, the representation of quantities of interest changes with their description in either spatial or material coordinates. Consider the scalar field $u : \Omega_t \to \mathbb{R}$. We set

$$\dot{u}(\mathbf{x}, t) := \frac{\partial}{\partial t} u(\chi(\mathbf{X}, t), t)\Big|_{\mathbf{X} = \chi^{-1}(\mathbf{x}, t)},$$

i.e., $\dot{u}(\mathbf{x}, t)$ is the time derivative of u where we keep the material point $\mathbf{X}$ fixed. Therefore, $\dot{u}(\mathbf{x}, t)$ is referred to as *material derivative* of u. The chain rule of differentiation now yields

$$\dot{u}(\mathbf{x}, t) = \frac{\partial}{\partial t} u(\mathbf{x}, t) + \mathbf{v}(\mathbf{x}, t) \cdot \nabla u(\mathbf{x}, t). \tag{9}$$

Thus, the material derivative $\dot{u}(\mathbf{x}, t)$ is comprised of the *spatial derivative* $\frac{\partial}{\partial t} u(\mathbf{x}, t)$ and the advection term $\mathbf{v}(\mathbf{x}, t) \cdot \nabla u(\mathbf{x}, t)$, see, e.g., [51] for further details.

As a consequence, given the fixed computational mesh in the Eulerian framework, the domain Ω_t moves over time. The Eulerian description is well suited to capture large distortions. However, the resolution of interfaces and details becomes rather costly. On the other hand, in the Lagrangian framework, it is easy to track free surfaces and interfaces, whereas it is difficult to handle large distortions, which usually require frequent remeshing of the domain Ω_0. In order to bypass the drawbacks of both frameworks, the ALE method has been introduced, see [82, 83]. Here, the movement of the mesh is decoupled from the movement of the particles $\mathbf{X} \in \Omega_0$. The mesh might, for example, be kept fixed as in the Eulerian

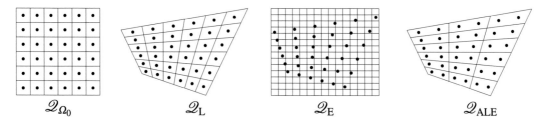

Fig. 7 Comparison of the different frameworks: Reference mesh with particles (Q_{Ω_0}), deformed mesh with particles in the Lagrangian framework (Q_L), fixed mesh with particles in the Eulerian framework (Q_E), modified mesh with particles in the ALE framework (Q_{ALE})

framework or be moved as in the Lagrangian framework or even be handled in a completely different manner, see Fig. 7 for a visualization.

Within the ALE framework, the reaction diffusion equation on growing domains can be written as

$$\frac{\partial}{\partial t}c + \mathbf{w}\nabla c + c\,\mathrm{div}\mathbf{v} = D\Delta c + R(c).$$

Herein, $\mathbf{w} = \mathbf{v} - \mathbf{u}$ the relative velocity between the material velocity $\mathbf{v}$ and the mesh velocity $\mathbf{u}$. If the velocity of the mesh and the material coincide, i.e., $\mathbf{u} = \mathbf{v}$, the Lagrangian formulation is recovered. On the other hand, setting $\mathbf{u} = \mathbf{0}$, we retrieve the Eulerian formulation, i.e., Eq. 6, cf. [36, 84]. In practice, the mesh velocity is chosen such that one obtains a Lagrangian description in the vicinity of moving boundaries and an Eulerian description in static region, where a smooth transition between the corresponding velocities is desirable, cf. [82]. We remark that, in this view, the treatment of composite domains demands for additional care, particularly, when the subdomains move with different velocities, see [85, 86].

4.3 Diffuse-Domain Method

The ALE method facilitates the modeling of moving and growing domains. Due to the underlying discretization of the simulation domain, the possible deformation is still limited and topological changes cannot be handled. The *Diffuse-Domain* method, introduced in [87], decouples the simulation domain from the underlying discretization. A diffuse implicit interface-capturing method is used to represent the boundary instead of implicitly representing the domain boundaries by a mesh.

The general idea is appealingly simple. We extend the integration domain to a larger computational bounding box and introduce an auxiliary field variable ϕ to represent the simulation domain. A level set of ϕ describes the implicit surface of the domain, see Fig. 8 for a visualization. To restrict the partial differential equations to the bulk and/or surface, we multiply those equations in the weak form by the characteristic functions of the corresponding domain.

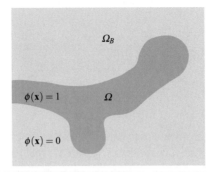

Fig. 8 Visualization of the computational domain Ω, the bounding box Ω_B, and the phase field ϕ

To deform and grow the geometry, an additional equation with an advective term is solved to update the auxiliary field, cf. [88]. Following the example given in [88], let us consider the classical Poisson equation with Neumann boundary conditions in the domain Ω.

$$\begin{aligned} \Delta u &= f \quad \text{in} \quad \Omega, \\ \nabla u \cdot \mathbf{n} &= g \quad \text{on} \quad \Gamma_\Omega. \end{aligned} \tag{10}$$

First, we extend the Poisson equation into the bounding box $\Omega_B \supset \Omega$. Thus, Eq. 10 becomes

$$\text{div}(\phi \nabla u) + \text{B.C.} = \phi f \quad \text{in} \quad \Omega. \tag{11}$$

Now, the Neumann boundary conditions can be enforced by replacing the B.C.-term by $\text{B.C.} = g|\nabla\phi|$ or $\text{B.C.} = \varepsilon g|\nabla\phi|^2$, where $|\nabla\phi|$ and $\varepsilon|\nabla\phi|^2$ approximate the Dirac δ-function. Dirichlet and Robin boundary conditions can be dealt with similarly, see [88]. It is possible to show that in the limit $\varepsilon \to 0$ the explicit formulation Eq. 10 is recovered, cf. [88, 89]. Depending on the approximation of the δ-function and the type of boundary conditions used, the Diffuse-Domain method is first- or second-order accurate. The treatment of dynamics on the surface is more intricate as we do not have an explicit boundary anymore. The general idea is to extend the dependent variables constantly in surface-normal direction over the interface, cf. [90, 91].

Several methods exist to track the diffuse interface of the computation domain. In the level set method, the interface Γ is represented by the zero isosurface of the signed-distance function to the surface Γ_Ω. In contrast to this artificial level set formulation, phase fields are constructed by a physical description of the free energy of the underlying system. Several such physical descriptions exist, but probably the most famous derivations are known under the Allen-Cahn and Cahn-Hilliard equations. The former formulation does not conserve the phase areas (or volumes in 3D) whereas the latter conserves the total concentration of the phases. Starting with a two phase system without mixing, the free energy can be described by the Ginzburg-Landau energy according to

$$E_\sigma(\phi) := \int_\Omega \frac{1}{\varepsilon} W(\phi) + \frac{\varepsilon}{2} |\nabla\phi|^2 \mathrm{dx},$$

where Ω is the domain, ε controls the thickness of the interface, and ϕ is the phase field, cf. [92]. Moreover, the function $W(\phi)$ is a double-well potential having its two minima in the values representing the bulk surfaces, e.g.,

$$W(\phi) = \frac{1}{4}\phi^2(1-\phi)^2,$$

having its two minima at $\phi = 0$ and $\phi = 1$. The second term of the Ginzburg-Landau energy penalizes gradients in the concentration field and thus can be interpreted as the free energy of the phase transition, see [93]. With $W(\phi)$ defined as above, E_σ is also referred to as surface energy or Cahn-Hilliard energy, cp. [92]. Minimizing the energy E_σ with respect to ϕ results in solving the equation

$$\frac{\mathrm{d}E_\sigma}{\mathrm{d}\phi} = \frac{1}{\varepsilon} W'(\phi) - \varepsilon\Delta\phi \overset{!}{=} 0.$$

The solution of this equation has two homogenous bulks describing the two phases and a tanh -profile in between, see, e.g., [92]. The Allen-Cahn equation then reads as

$$\frac{\partial\phi}{\partial t} + \mathbf{v}\cdot\nabla\phi = -M\mu(\phi),$$

$$\mu(\phi) = \frac{1}{\varepsilon} W'(\phi) - \varepsilon\Delta\phi,$$

where M is a mobility parameter describing the stiffness of the interface dynamic and $\mathbf{u}$ the velocity field moving and deforming the phases again. An interface Γ can be represented by the $\phi(\mathbf{x}) = 0.5$ contour line (isosurface in 3D). Based on the interface profile, the surface normals can be approximated by

$$\mathbf{n} = \frac{\nabla\phi}{|\nabla\phi|}$$

and the mean curvature by

$$\kappa = \nabla\cdot\mathbf{n},$$

cf. [92].

The above description of the physical two-phase system minimizes the area of the interface and thus exhibits unwanted self-dynamics for an interface tracking method. In [94] it is proposed to add the correction term $-\varepsilon^2\kappa(\phi)|\nabla\phi|$ to the right-hand side, since $\kappa(\phi)$ depends on the local curvature, canceling out this effect, which is known as the Allen-Cahn law. Hence, the right-hand side becomes

$$\mu(\phi) = \frac{1}{\varepsilon} W'(\phi) - \varepsilon \text{div}(\mathbf{nn} \cdot \nabla\phi).$$

The Diffuse-Domain method has been successfully applied to several biological problems, for example, for modeling mechanically induced deformation of bones [95] or simulating endocytosis [90]. It is even possible to couple surface and bulk reactions for modeling transport, diffusion, and adsorption of any material quantity in [96]. Two-phase flows with soluble nanoparticles or soluble surfactants have been modeled with the Diffuse-Domain method in [97] and [98], respectively. A reaction-advection-diffusion problem combining volume and surface diffusion and surface reaction has been solved with the Diffuse-Domain method in [91, 99] to address a patterning problem in murine lung development.

4.4   Lattice Boltzmann Method

The Lattice Boltzmann method (LBM) is a numerical scheme to simulate fluid dynamics. It evolved from the field of cellular automata, more precisely lattice gas cellular automata (LGCA), in the 1980s. The first LGCA that could simulate fluid flow was proposed in 1986, cp. [100]. However, LGCA were facing several problems such as a relatively high fixed viscosity and an intrinsic stochastic noise, see [101]. In 1988, the LBM was introduced as an independent numerical method for fluid flow simulations, when tackling the noise problem of the LGCA method, cf. [102]. A detailed description of LGCA and the development of the LBM from it can be found in [103].

The LBM is applied in many different areas that study various different types of fluid dynamics, for example, incompressible, iso-thermal, non-isothermal, single- and multi-phase flows, etc., as well as biological flows, see [101]. As we have seen in the previous sections, biological fluid dynamics can be the dynamics of biomolecules, i.e., morphogens, being solved in a fluid or the dynamics of a biological tissue which can be approximated by a viscous fluid, cp. [13]. Further, the LBM can be used to simulate signaling dynamics of biomolecules by solving reaction-diffusion equations on the lattice.

In contrast to many conventional approaches modeling fluid flow on the macroscopic scale, for example, by solving the Navier-Stokes equations, the LBM is a mesoscopic approach. It is also advantageous for parallel computations because of its local dynamics. LBM models the fluid as fictive particles that propagate and collide on a discrete lattice, where the incompressible Navier-Stokes equations can be captured in the nearly incompressible limit of the LBM, see [104].

The Boltzmann equation originates from statistical physics and describes the temporal evolution of a probability density distribution function $f(\mathbf{x},\mathbf{v},t)$ defining the probability of finding a particle with velocity $\mathbf{v}$ at location $\mathbf{x}$ at time t. In the presence of an external force $\mathbf{F}$ acting on the particles and considering two processes, i.e., propagation of particles and their collision, the temporal evolution of $f(\mathbf{x},\mathbf{v},t)$ is defined by

$$\frac{\partial f}{\partial t} + \mathbf{v} \cdot \nabla_{\mathbf{x}} f + \mathbf{F} \cdot \nabla_{\mathbf{v}} f = Q(f).$$

The most commonly used collision term is the single-relaxation-time Bhatnagar-Gross-Krook (BGK) collision operator

$$Q(f) = Q_{\text{BGK}} := -\frac{1}{\tau}(f - f^{\text{eq}}),$$

with f^{eq} being the Maxwell-Boltzmann equilibrium distribution function with a characteristic time scale τ. The discretized Boltzmann equation is then

$$\frac{\partial f_i}{\partial t} + \mathbf{v}_{i,\alpha} \frac{\partial f_i}{\partial x_\alpha} = -\frac{1}{\tau}(f_i - f_i^{\text{eq}}) + \mathbf{F}_i,$$

where i determines the number of discrete velocities and α the spatial dimensionality of the system. The discrete LB equation is given by

$$\begin{aligned} f_i(\mathbf{x} + \mathbf{v}_i \Delta t, t + \Delta t) &- f_i(\mathbf{x}, t) \\ &= -\frac{1}{\tau}\left(f_i(\mathbf{x}, t) - f_i^{\text{eq}}(\mathbf{x}, t)\right) + \mathbf{F}_i. \end{aligned} \tag{12}$$

The equation implicates a two step algorithm for the LBM. In the first step referring to the left-hand side of Eq. 12, the probability density distribution functions perform a free flight to the next lattice point. In the second step referring to the right-hand side of Eq. 12, the collision of the incoming probability density distribution functions on each lattice point is computed, followed by the relaxation towards a local equilibrium distribution function. As the collision step is calculated on the lattice points only, the computations of the LBM are local, rendering it well suited for parallelization. For further notes on the computational cost of the LBM, see [101].

The equilibrium function is defined by a second-order expansion of the Maxwell equation in terms of low fluid velocity

$$f^{\text{eq}} = \rho w_i \left[1 + \frac{\mathbf{v}_i \cdot \mathbf{u}}{c_s^2} + \frac{(\mathbf{v}_i \cdot \mathbf{u})^2}{2c_s^4} - \frac{\mathbf{u} \cdot \mathbf{u}}{c_s^2} \right]$$

with the fluid velocity $\mathbf{u}$, the fluid density ρ, the speed of sound c_s, and the weights w_i that are given according to the chosen lattice, cf. [105]. The lattices used in LBM are regular and characterized as $DdQq$, where d indicates the spatial dimension and q the number of discrete velocities. For simulations, commonly used lattices are $D1Q3$, $D2Q9$, and $D3Q19$, see [101]. The macroscopic quantities, i.e., density ρ and momentum density $\rho\mathbf{u}$, are defined by the first few moments of the probability density distribution function f_i, i.e.,

$$\rho = \sum_{i=0}^{q} f_i, \qquad \rho\mathbf{u} = \sum_{i=0}^{q} f_i \mathbf{v}_i.$$

The fluid pressure p is related to the mass density ρ via the equation for an ideal gas $p = \rho c_s^2$. To simulate fluid-structure interactions, the LBM can be combined with the Immersed Boundary method (IBM), see [106], that represents elastic structures being immersed in a fluid. A detailed description of the IBM can be found in [55]. The LBM and IBM were first combined in [107], and later used, for instance, to simulate red blood cells in flow, cf. [108], and to simulate the coupled tissue and signaling dynamics in morphogenetic processes with cellular resolution, cf. [73].

5   Conclusion

In this chapter, we have given an overview of the mathematical modeling of tissue dynamics, growth, and mechanics in the context of morphogenesis. These different aspects of morphogenesis can be modeled either on a microscopic scale, for example, by agent-based models, or on a macroscopic scale by continuum approaches. In addition, we have discussed several numerical approaches to solve these models. Due to the complex nature and coupling of the different aspects that are required to obtain realistic models, the numerical solution of these models is computationally expensive. When it comes to incorporating measurement data, these large computational efforts can place a severe limitation. Algorithms for parameter estimation are usually based on gradient descent or sampling and therefore require frequent solutions of the model. As a consequence, incorporating measurement data may quickly become infeasible and efficient algorithms have to be devised.

References

1. Iber D (2011) Inferring Biological Mechanisms by Data-Based Mathematical Modelling: Compartment-Specific Gene Activation during Sporulation in Bacillus subtilis as a Test Case. Adv Bioinformatics 2011:1–12

2. Iber D, Karimaddini Z, Ünal E (2015) Image-based modelling of organogenesis. Brief Bioinform.

3. Gómez HF, Georgieva L, Michos O, Iber D (2017) Image-based in silico models of organogenesis. In: Systems Biology, vol 6.

4. Mogilner A, Odde D (2011) Modeling cellular processes in 3d. Trends Cell Biol 21 (12):692–700

5. Sbalzarini IF (2013) Modeling and simulation of biological systems from image data: Prospects & Overviews. BioEssays 35 (5):482–490

6. Alberts B, Johnson A, Lewis J, Morgan D, Raff M, Roberts K, Walter P (2014) Molecular biology of the cell, 6th edn. Taylor & Francis, London

7. van den Hurk R, Zhao J (2005) Formation of mammalian oocytes and their growth, differentiation and maturation within ovarian follicles. Theriogenology 63(6):1717–1751

8. Worley MI, Setiawan L, Hariharan IK (2013) Tie-dye: a combinatorial marking system to

visualize and genetically manipulate clones during development in drosophila melanogaster. Development 140(15):3275–3284

9. Ricklefs RE (2010) Embryo growth rates in birds and mammals. Funct Ecol 24 (3):588–596

10. Liang X, Michael M, Gomez GA (2016) Measurement of mechanical tension at cell-cell junctions using two-photon laser ablation. Bio Protoc 6(24):e2068

11. Eden E, Geva-Zatorsky N, Issaeva I, Cohen A, Dekel E, Danon T, Cohen L, Mayo A, Alon U (2011) Proteome half-life dynamics in living human cells. Science 331 (6018):764–768

12. Müller P, Rogers KW, Shuizi RY, Brand M, Schier AF (2013) Morphogen transport. Development 140(8):1621–1638

13. Forgacs G, Foty RA, Shafrir Y, Steinberg MS (1998) Viscoelastic properties of living embryonic tissues: a quantitative study. Biophys J 74(5):2227–2234

14. Turing AM (1952) The chemical basis of morphogenesis. Philos Trans R Soc B 237 (641):37–72

15. Dierick HA, Bejsovec A (1998) Functional analysis of wingless reveals a link between intercellular ligand transport and dorsal-cell-specific signaling. Development 125 (23):4729–4738

16. Ramírez-Weber FA, Kornberg TB (1999) Cytonemes: cellular processes that project to the principal signaling center in drosophila imaginal discs. Cell 97(5):599–607

17. Rodman J, Mercer R, Stahl P (1990) Endocytosis and transcytosis. Curr Opin Cell Biol 2 (4):664–672

18. Entchev EV, Schwabedissen A, González-Gaitán M (2000) Gradient formation of the TGF-β homolog dpp. Cell 103(6):981–992

19. Lander AD, Nie Q, Wan FY (2002) Do morphogen gradients arise by diffusion? Dev Cell 2(6):785–796

20. Schwank G, Dalessi S, Yang SF, Yagi R, de Lachapelle AM, Affolter M, Bergmann S, Basler K (2011) Formation of the long range dpp morphogen gradient. PLoS Biol 9(7):1–13

21. Kornberg TB, Roy S (2014) Cytonemes as specialized signaling filopodia. Development 141(4):729–736

22. Bischoff M, Gradilla AC, Seijo I, Andrés G, Rodríguez-Navas C, González-Méndez L, Guerrero I (2013) Cytonemes are required for the establishment of a normal hedgehog morphogen gradient in drosophila epithelia. Nat Cell Biol 15(11):1269–1281

23. Sanders TA, Llagostera E, Barna M (2013) Specialized filopodia direct long-range transport of shh during vertebrate tissue patterning. Nature 497(7451):628–632

24. Wolpert L (1969) Positional information and the spatial pattern of cellular differentiation. J Theor Biol 25(1):1–47

25. Gregor T, Bialek W, de Ruyter van Steveninck RR, Tank DW, Wieschaus EF (2005) Diffusion and scaling during early embryonic pattern formation. Proc Natl Acad Sci USA 102 (51):18403–18407

26. Umulis DM, Othmer HG (2013) Mechanisms of scaling in pattern formation. Development 140(24):4830–4843

27. Umulis DM (2009) Analysis of dynamic morphogen scale invariance. J R Soc Interface

28. Wartlick O, Mumcu P, Kicheva A, Bittig T, Seum C, Jülicher F, Gonzalez-Gaitan M (2011) Dynamics of dpp signaling and proliferation control. Science 331 (6021):1154–1159

29. Fried P, Iber D (2014) Dynamic scaling of morphogen gradients on growing domains. Nat Commun 5:5077

30. Fried P, Iber D (2015) Read-out of dynamic morphogen gradients on growing domains. PloS ONE 10(11):e0143226

31. Gillespie DT (2007) Stochastic simulation of chemical kinetics. Annu Rev Phys Chem 58:35–55

32. Berg HC (1993) Random walks in biology. Princeton University Press, Princeton

33. Gillespie DT (1977) Exact stochastic simulation of coupled chemical reactions. J Phys Chem 81(25):2340–2361

34. Kondo S, Asai R (1995) A reaction-diffusion wave on the skin of the marine angelfish pomacanthus. Nature 376(6543):765–768

35. Henderson J, Carter D (2002) Mechanical induction in limb morphogenesis: the role of growth-generated strains and pressures

36. Iber D, Tanaka S, Fried P, Germann P, Menshykau D (2014) Simulating tissue morphogenesis and signaling. In: Nelson CM (ed) Tissue morphogenesis: methods and protocols. Springer, New York, pp 323–338

37. Iber D, Menshykau D (2013) The control of branching morphogenesis. Open Biol 3 (9):130088

38. Menshykau D, Iber D (2013) Kidney branching morphogenesis under the control of a ligan-receptor-based Turing mechanism. Phys Biol 10(4):046003

39. Dillon R, Gadgil C, Othmer HG (2003) Short-and long-range effects of sonic

hedgehog in limb development. Proc Natl Acad Sci 100(18):10152–10157

40. Bittig T, Wartlick O, Kicheva A, González-Gaitán M, Jülicher F (2008) Dynamics of anisotropic tissue growth. New J Phys 10 (6):063001

41. Tanaka S, Iber D (2013) Inter-dependent tissue growth and Turing patterning in a model for long bone development. Phys Biol 10 (5):056009

42. Fried P, Sánchez-Aragón M, Aguilar-Hidalgo D, Lehtinen B, Casares F, Iber D (2016) A model of the spatio-temporal dynamics of drosophila eye disc development. PLoS Comput Biol 12(9)

43. Boehm B, Westerberg H, Lesnicar-Pucko G, Raja S, Rautschka M, Cotterell J, Swoger J, Sharpe J (2010) The role of spatially controlled cell proliferation in limb bud morphogenesis. PLoS Biol 8(7):e1000420

44. Forgacs G (1998) Surface tension and viscoelastic properties of embryonic tissues depend on the cytoskeleton. Biol Bull 194:328–329

45. Foty RA, Forgacs G, Pfleger CM, Steinberg MS (1994) Liquid properties of embryonic tissues: measurement of interfacial tensions. Phys Rev Lett 72(14):2298–2301

46. Marmottant P, Mgharbel A, Käfer J, Audren B, Rieu JP, Vial JC, van der Sanden B, Mareé AFM, Graner F, Delano-ë-Ayari H (2001) The role of fluctuations and stress on the effective viscosity of cell aggregates. Proc Natl Acad Sci USA 106 (41):17271–17275

47. Dillon RH, Othmer HG (1999) A mathematical model for outgrowth and spatial patterning of the vertebrate limb bud. J Theor Biol 197:295–330

48. Fung YC (1993) Biomechanics: mechanical properties of living tissues. Springer, New York

49. Taber LA (2004) Nonlinear theory of elasticity: applications in biomechanics. World Scientific, Singapore

50. Ciarlet PG (1988) Mathematical elasticity volume 1: three-dimensional elasticity. Elsevier, Amsterdam

51. Holzapfel G (2000) Nonlinear solid mechanics: a continuum approach for engineering. Wiley, Chichester

52. Peters MD, Iber D (2017) Simulating organogenesis in COMSOL: tissue mechanics. arXiv:1710.00553v2

53. Meyer-Hermann ME, Maini PK, Iber D (2006) An analysis of b cell selection mechanisms in germinal centers. Math Med Biol 23 (3):255–277

54. Merks RMH, Koolwijk P (2009) Modeling morphogenesis in silico and in vitro : towards quantitative, predictive, cell-based modeling. Math Model Nat Phenom 4(4):149–171

55. Tanaka S (2015) Simulation frameworks for morphogenetic problems. Computation 3 (2):197–221

56. Van Liedekerke P, Palm MM, Jagiella N, Drasdo D (2015) Simulating tissue mechanics with agent-based models: concepts, perspectives and some novel results. Comput Part Mech 2(4):401–444

57. Ising E (1925) Beitrag zur Theorie des Ferromagnetismus. Z Phys 31(1):253–258

58. Metropolis N, Rosenbluth AW, Rosenbluth MN, Teller AH, Teller E (1953) Equation of state calculations by fast computing machines. J Chem Phys 21(6):1087–1092

59. Graner F, Glazier JA (1992) Simulation of biological cell sorting using a two-dimensional extended Potts model. Phys Rev Lett 69(13):2013–2016

60. Hirashima T, Iwasa Y, Morishita Y (2009) Dynamic modeling of branching morphogenesis of ureteric bud in early kidney development. J Theor Biol 259(1):58–66

61. Hester SD, Belmonte JM, Gens JS, Clendenon SG, Glazier JA (2011) A multi-cell, multi-scale model of vertebrate segmentation and somite formation. PLoS Comput Biol 7 (10):e1002155

62. Poplawski NJ, Swat M, Gens JS, Glazier JA (2007) Adhesion between cells, diffusion of growth factors, and elasticity of the AER produce the paddle shape of the chick limb. Physica A 373:521–532

63. Swat MH, Thomas GL, Belmonte JM, Shirinifard A, Hmeljak D, Glazier JA (2012) Multi-scale modeling of tissues using CompuCell3D. Methods Cell Biol 110:325–366

64. Hoehme S, Drasdo D (2010) A cell-based simulation software for multi-cellular systems. Bioinformatics 26(20):2641–2642

65. Honda H, Eguchi G (1980) How much does the cell boundary contract in a monolayered cell sheet? J Theor Biol 84(3):575–588

66. Weliky M, Oster G (1990) The mechanical basis of cell rearrangement. I. Epithelial morphogenesis during Fundulus epiboly. Development 109(2):373–386

67. Nagai T, Honda H (2001) A dynamic cell model for the formation of epithelial tissues. Philos Mag Part B 81(7):699–719

68. Fletcher AG, Osborne JM, Maini PK, Gavaghan DJ (2013) Implementing vertex dynamics models of cell populations in biology

within a consistent computational framework. Prog Biophys Mol Biol 113(2):299–326

69. Fletcher AG, Osterfield M, Baker RE, Shvartsman SY (2014) Vertex models of epithelial morphogenesis. Biophys J 106 (11):2291–2304

70. Pitt-Francis J, Pathmanathan P, Bernabeu MO, Bordas R, Cooper J, Fletcher AG, Mirams GR, Murray P, Osborne JM, Walter A, Chapman, S.J., Garny A, van Leeuwen IMM., Maini PK, Rodríguez B, Waters SL, Whiteley JP, Byrne HM, Gavaghan DJ (2009) Chaste: a test-driven approach to software development for biological modelling. Comput Phys Commun 180(12):2452–2471

71. Mirams GR, Arthurs CJ, Bernabeu MO, Bordas R, Cooper J, Corrias, A., Davit Y, Dunn SJ, Fletcher AG, Harvey DG, Marsh ME, Osborne JM, Pathmanathan P, Pitt-Francis J, Southern J, Zemzemi N, Gavaghan DJ (2013) Chaste: an open source C++ library for computational physiology and biology. PLoS Comput Biol 9(3):e1002970

72. Rejniak KA (2007) An immersed boundary framework for modelling the growth of individual cells: an application to the early tumour development. J Theor Biol 247(1):186–204

73. Tanaka S, Sichau D, Iber D (2015) LBIBCell: a cell-based simulation environment for morphogenetic problems. Bioinformatics 31 (14):2340–2347

74. Eymard R, Gallouët T, Herbin R (2000) Finite volume methods. In: Solution of equation in $\mathbb{R}^n$ (Part 3), techniques of scientific computing (Part 3). Volume 7 of handbook of numerical analysis. Elsevier, New York, pp 713–1018

75. Braess D (2007) Finite elements: theory, fast solvers, and applications in solid mechanics, 3rd edn. Cambridge University Press, Cambridge

76. Courant R (1943) Variational methods for the solution of problems of equilibrium and vibrations. Bull Am Math Soc 49(1):1–23

77. Brenner S, Scott LR (2008) The mathematical theory of finite element methods. Texts in applied mathematics. Springer, New York

78. Szabo BA, Babuška I (1991) Finite element analysis. Wiley, Chichester

79. Zienkiewicz O, Taylor R, Zhu JZ (2013) The finite element method: its basis and fundamentals, 7th edn. Butterworth-Heinemann, Oxford

80. Gander MJ, Wanner G (2012) From Euler, Ritz, and Galerkin to modern computing. SIAM Rev 54(4):627–666

81. Gelfand IM, Fomin SV (1963) Calculus of variations. Prentice-Hall, Upper Saddle River

82. Donea J, Huerta A, Ponthot J, Rodríguez-Ferran A (2004) Arbitrary Lagrangian–Eulerian methods. In: Stein E, de Borst R, Hughes TJR (eds) Encyclopedia of computational mechanics. Wiley, Hoboken

83. Hirt CW, Amsden AA, Cook JL (1974) An arbitrary Lagrangian-Eulerian computing method for all flow speeds. J Comput Phys 14(3):227–253

84. MacDonald G, Mackenzie J, Nolan M, Insall R (2016) A computational method for the coupled solution of reaction-diffusion equations on evolving domains and manifolds: application to a model of cell migration and chemotaxis. J Comput Phys 309:207–226

85. Karimaddini Z, Unal E, Menshykau D, Iber D (2014) Simulating organogenesis in COMSOL: image-based modeling. arXiv:1610.09189v1

86. Menshykau D, Iber D (2012) Simulation organogenesis in COMSOL: deforming and interacting domains. arXiv:1210.0810

87. Kockelkoren J, Levine H, Rappel WJ (2003) Computational approach for modeling intra- and extracellular dynamics. Phys Rev E Stat Nonlin Soft Matter Phys 68(3–2)

88. Li X, Lowengrub J, Rätz A, Voigt A (2009) Solving PDEs in complex geometries: a diffuse domain approach. Commun Math Sci 7 (1):81–107

89. Lervåg KY, Lowengrub J (2014) Analysis of the diffuse-domain method for solving PDEs in complex geometries. arXiv:1407.7480v3

90. Lowengrub J, Allard J, Aland S (2016) Numerical simulation of endocytosis: viscous flow driven by membranes with non-uniformly distributed curvature-inducing molecules. J Comput Phys 309:112–128

91. Wittwer LD, Croce R, Aland S, Iber D (2016) Simulating organogenesis in COMSOL: phase-field based simulations of embryonic lung branching morphogenesis. arXiv:1610.09189v1

92. Aland S (2012) Modelling of two-phase flow with surface active particles. PhD thesis, Technische Universität Dresden

93. Eck C, Garcke H, Knabner P (2011) Mathematische Modellierung. Springer, Berlin

94. Folch R, Casademunt J, Hernández-Machado A, Ramírez-Piscina L (1999) Phase-field model for Hele-Shaw flows with arbitrary viscosity contrast. I. Theoretical approach. Phys Rev E Stat Phys Plasmas

Fluids Relat Interdiscip Topics 60 (2-B):1724–1733

95. Aland S, Landsberg C, Müller R, Stenger F, Bobeth M, Langheinrich AC, Voigt A (2014) Adaptive diffuse domain approach for calculating mechanically induced deformation of trabecular bone. Comput Methods Biomech Biomed Engin 17(1):31–38

96. Teigen KE, Li X, Lowengrub J, Wang F, Voigt A (2009) A diffuse-interface approach for modeling transport, diffusion and adsorption/desorption of material quantities on a deformable interface. Commun Math Sci 4 (7):1009–1037

97. Aland S, Lowengrub J, Voigt A (2011) A continuum model of colloid-stabilized interfaces. Phys Fluids 23:062103

98. Teigen KE, Song P, Lowengrub J, Voigt A (2011) A diffuse-interface method for two-phase flows with soluble surfactants. J Comput Phys 230(2):375–393

99. Wittwer LD, Peters M, Aland S, Iber D (2017) Simulating organogenesis in COMSOL: comparison of methods for simulating branching morphogenesis. arXiv:1710.02876v1

100. Frisch U, Hasslacher B, Pomeau Y (1986) Lattice-gas automata for the Navier-Stokes equation. Phys Rev Lett 56(14):1505–1508

101. Frouzakis CE (2011) Lattice Boltzmann methods for reactive and other flows. In: Echekki T, Mastorakos E (eds) Turbulent combustion modeling, fluid mechanics and its applications. Springer Science+Business Media, Berlin

102. McNamara GR, Zanetti G (1988) Use of the Boltzmann equation to simulate lattice-gas automata. Phys Rev Lett 61(20):2332–2335

103. Wolf-Gladrow DA (2000) Lattice-gas cellular automata and lattice boltzmann models - an introduction. Springer, Berlin

104. Chen S, Doolen GD (1998) Lattice Boltzmann method for fluid flows. Annu Rev Fluid Mech 30:329–364

105. He X, Luo L (1997) A priori derivation of the lattice Boltzmann equation. Phys Rev E 55(6)

106. Peskin C (2002) The immersed boundary method. Acta Numer 11:479–517

107. Feng ZG, Michaelides EE (2004) The immersed boundary-lattice Boltzmann method for solving fluid–particles interaction problems. J Comput Phys 195(2):602–628

108. Zhang J, Johnson PC, Popel AS (2007) An immersed boundary lattice Boltzmann approach to simulate deformable liquid capsules and its application to microscopic blood flows. Phys Biol 4(4):285–295

Chapter 14

Mechanisms and Measurements of Scale Invariance of Morphogen Gradients

Yan Huang and David Umulis

Abstract

Morphogen gradients provide positional information to underlying cells that translate the information into differential gene expression and eventually different cell fates. Scale invariance is the property where the gradients of the morphogen adjust proportionately to the size of the domain. Scale invariance of morphogen gradients or patterns of differentiation is a common phenomenon observed between individuals within the same species and between homologous tissues or structures in different species. To determine whether or not a pattern is scale invariant, others and we have developed definitions and measurements of gradient scaling. These include point-wise and global scaling errors as well as global scaling power. Furthermore, there are a number of mathematical conditions for scale invariance of advection–diffusion–reaction models that inform mechanisms of scaling. Herein we provide a deeper perspective on modeling and measurement of scale invariance of morphogen gradients.

Key words Scale invariance, Morphogen gradient, Reaction–diffusion, Mathematical model, Morphogenesis

1 Introduction

1.1 Definition of Scale Invariance

A morphogen is a chemical component that forms a nonuniform spatial distribution over a field of cells that drives differential gene expression in relation to its concentration and ultimately patterns the underlying tissue or organ. In nature, many organisms and their tissues and organs vary substantially in size but differ little in morphology, appearing to be scaled versions of a common template or pattern [1]. This preservation of proportion is called scaling, or scale invariance. Pattern scale invariance is achieved by the scaling of morphogen gradients with system sizes.

1.2 Significance of Studying Scaling

It is only a slight overestimate to say that the most important attribute of an animal, both physiologically and ecologically, is its size. Size constrains virtually every aspect of structure and function and strongly influences the nature of most inter- and intraspecific interactions. Body mass, which in any given taxon

Julien Dubrulle (ed.), *Morphogen Gradients: Methods and Protocols*, Methods in Molecular Biology, vol. 1863, https://doi.org/10.1007/978-1-4939-8772-6_14, © Springer Science+Business Media, LLC, part of Springer Nature 2018

is a close correlate of size, is the most widely useful predictor of physiological rates.—G. A. Bartholomew 1981, Insect thermoregulation, p. 46.

Scaling achieves precise and robust patterning in proportion to perturbations of system size and is a very important feature for animals to survive [2–4]. Failure of scaling leads to developmental defects or even death. In Xenopus dorsal–ventral (DV) patterning, for example, the ventral half fails to scale with size perturbation after bisection at blastula stage and results in a mass of ventral tissues without DV axis structure, whereas the dorsal grows into proportionally patterned DV axis [5]. An intriguing aspect of scaling is that the reaction, diffusion, and advection equations that capture the biophysics of morphogen gradient pattern formation DO NOT automatically provide scaling and in fact the equations do not naturally have any information about the size of the system that is being patterned. Each process or balance of processes—diffusion, advection, and morphogen capture/decay—contains a biophysical scale that depends on the parameters of the process but not on the size of the system. Processes have evolved to tune the biophysical parameters to adjust them in some relation to the system's size. Studies of scale invariance often focus on how biology solved the problem by feedback that modifies the rates of diffusion or chemical reaction to adjust to the size of the system. Knowledge of scaling mechanisms provides information on how cells communicate over distances, sense their positions, and how morphogen gradients work. This has impact in our understanding of the regeneration of tissues and organs, and patterning at multiple levels of organization from the cellular to the organism to the ecosystem scale [6].

1.3 Types and Examples of Scaling

Pattern scale invariance among individuals of different species is called *interspecies scaling*. For example, as shown in Fig. 1a, the interspecies scaling of patterns of gap and pair-rule gene expression can be traced back to the interspecies scaling of Bcd gradients with blastoderm embryo size of *Lucilia sericata*, *Drosophila melanogaster*, and *Drosophila busckii*, reproduced from Gregor et al. with permission from PNAS [7]. The scaling of Bcd gradients most likely depends on the species-specific effective lifetime of Bcd. Pattern scale invariance among different individuals within a species is called *intraspecies scaling*. For example, intraspecies scaling of A-P axis development in *Drosophila melanogaster* is suggested as being achieved by scaling of anterior Bcd production rate with embryo volume [8].

As a tissue or organ develops over time, the morphogen concentration changes until it reaches quasi-steady state. There are two types of scale invariance depending on the dynamics of this process: *steady-state scaling* and *dynamic scaling*. Steady-state scaling is the scaling of morphogen gradients of different individuals when morphogen concentration no longer appreciably changes with time.

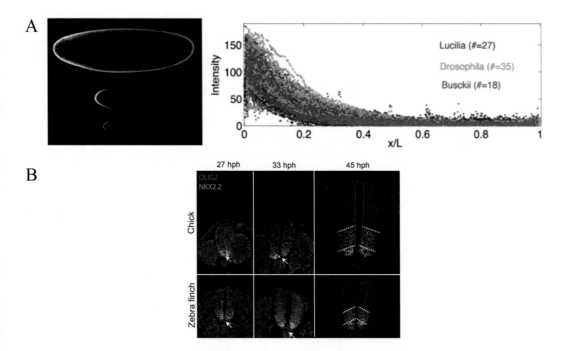

Fig. 1 Examples of three types of scale invariance. (**a**) Interspecies scaling of the Bcd protein distribution in blastoderm embryos of *Lucilia sericata*, *Drosophila melanogaster*, and *Drosophila busckii*, reproduced from Gregor et al. [7] with permission from PNAS, Copyright (2005) National Academy of Sciences, USA [7]; (**b**) interspecies scaling of neural tube development patterned by SHH in the zebra finch and the chick, reproduced from Uygur et al. [15] with permission from Developmental Cell [15]

Dynamic scaling is the scaling of a morphogen gradient of the same individual with respect to tissue/organ size as the organ grows. Dynamic scaling also includes the scaling of morphogen gradients of different individuals developing in a population such as sibling embryos in a single clutch at different time points before steady state. One example of dynamic scaling reports that the Decapentaplegic (Dpp) gradient of Drosophila during the growth of the wing imaginal disc depends on the regulation of Dpp capture and decay to ensure dynamic scaling [9]. That is, the gradient of Dpp appears to change in direct proportion to the imaginal disc size.

Numerous examples of interspecies and intraspecies scaling have contributed to our understanding of how scaling is achieved and how to measure systems to test for scaling. The first experimental evidence of scaling in development was observed by Hans Spemann back in 1923. Spemann bisected amphibian embryos into dorsal and ventral halves and found that the dorsal half of the embryo developed normally up through the tadpole stage [10]. This simple experiment demonstrated the intrinsic robustness to size perturbation. Scaling is also observed in Drosophila DV patterning, skin pattern formation in fish pigmentation, and under bisection in goldfish embryo development [5, 11–14]. Interspecies scaling of neural tube patterning between the zebra finch and the chick has also been reported to be regulated by activating and repressive transcription factors of the morphogen Sonic Hedgehog (SHH) (Fig. 1b), reproduced from Uygur et al. with permission from Developmental Cell [15]. Measuring scale invariance and determining the mechanisms that lead to scaling have provided a number of useful metrics to quantify scaling. In this paper, we take a theoretical view of the advection–diffusion–reaction model for morphogen patterning and conditions that inform mechanisms of scaling. Then we offer perspectives on point-wise and global measurement approaches of morphogen gradient scaling that others and we have developed.

2   Methods

2.1   Mechanisms of Scale Invariance for Morphogen-Mediated Patterning of Advection–Diffusion–Reaction Model

Theoretical studies on morphogen-mediated patterning provide important inspiration on possible mechanisms of scale invariance and measures of scaling. In order to identify mechanisms of scaling, mathematical equations are formulated for how morphogen concentration changes throughout space as a function of time. In so doing, we can understand how different processes interact, how system size is involved in these processes and find mathematical conditions on how scaling is achieved. A brief introduction for advection, diffusion, reaction equations is provided below.

The dispersal of a morphogen involves reaction and physical transport by diffusion and advection. Suppose the molecular diffusion rate is independent of space; then in a one-dimensional system of length L the reaction transport equation for a morphogen system in Cartesian coordinates is as follows [1]:

$$\frac{\partial m}{\partial t} = D_m \frac{\partial^2 m}{\partial x^2} - v_x \frac{\partial m}{\partial x} - k_m R(m) \text{ where } m = m(x,t) \quad (1)$$

$$q_{\text{in}} = -D_m \frac{\partial m(0,t)}{\partial x} + v_x m(0,t) \quad \text{or} \quad m(0,t) = m_0 \quad (2)$$

$$q_{\text{out}} = -D_m \frac{\partial m(L,t)}{\partial x} + v_x m(L,t) \quad \text{or} \quad m(L,t) = m_1 \quad (3)$$

$$m(x,0) = f(x) \quad (4)$$

Here m is the concentration of the morphogen, t is time, x is the vector of the Cartesian coordinate, D_m is the diffusion coefficient of the morphogen, v_x is the velocity of cytoplasm or growing tissue that contributes to advection in the x direction, k_m is the first-order decay rate of the morphogen, $R(m)$ contains all the reaction steps that affect m, q_{in} is the input molecular flux of the morphogen, and q_{out} is the output molecular flux of the morphogen.

In many cases, the velocity of cytoplasm and the growing tissue is negligible, and the transport of morphogen is diffusion-dominated. In these situations, the advection term can be neglected and Eq. 1 can be simplified to:

$$\frac{\partial m}{\partial t} = \left(D_m \frac{\partial^2 m}{\partial x^2} \right) - k_m R(m) \tag{5}$$

Setting $x = 0$ as the source and $x = L$ as a no flux boundary leads to Eqs. 6 and 7

$$-D_m \frac{\partial m(x,\ t)}{\partial x} = q, \quad x = 0 \tag{6}$$

$$\frac{\partial m(x,t)}{\partial x} = 0, \quad x = L \tag{7}$$

Assume an initial spatial distribution of zero morphogen, Eq. 4 is simplified to:

$$m(x,0) = 0 \tag{8}$$

In order to identify general mechanisms that lead to scale invariance, we introduce system size into the equations by defining a dimensionless space variable $\xi = \frac{x}{L}$. A dimensionless time variable $\tau = \frac{t}{T}$ is also defined for time scale. Eqs. 5–7 are then rewritten in terms of the two dimensionless variables:

$$\left(\frac{1}{k_m T} \right) \frac{\partial m}{\partial \tau} = \left(\frac{D_m}{k_m L^2} \right) \frac{\partial^2 m}{\partial \xi^2} - R(m) \tag{9}$$

$$-\frac{\partial m(\xi, \tau)}{\partial \xi} = \frac{q_{in} L}{D_m}, \quad \xi = 0 \tag{10}$$

$$\frac{\partial m(\xi, \tau)}{\partial \xi} = 0, \quad \xi = 1 \tag{11}$$

Scaling will be achieved if and only if there is no explicit dependence on L in Eqs. 9–11, i.e., the system length terms in the diffusion, morphogen reaction as well as morphogen flux processes can be canceled out. Here, scale invariance is achieved if these three terms $k_m T, \frac{D_m}{k_m L^2}$ and $\frac{q_{in} L}{D_m}$ are independent of L.

In the scenario of steady-state or quasi-steady state $\left(\frac{\partial m}{\partial \tau} = 0 \right)$, assume linear decay, Eqs. 9–11 can be written as:

$$0 = \left(\frac{D_m}{L^2} \right) \frac{d^2 m}{d\xi^2} - k_m m \tag{12}$$

$$-\frac{dm(\xi)}{d\xi} = \frac{q_{in} L}{D_m}, \quad \xi = 0 \tag{13}$$

$$\frac{dm(\xi)}{d\xi} = 0, \quad \xi = 1 \tag{14}$$

For large k_m, small D_m, the solution is approximately:

$$m(\xi) \approx \frac{q_{\text{in}}}{\sqrt{k_m D_m}} e^{\left(-\sqrt{\frac{k_m L^2}{D_m}}\, \xi\right)} \tag{15}$$

The system scales if $\frac{k_m}{D_m} \propto L^{-2}$, and $q_{\text{in}} \propto \sqrt{k_m D_m}$. Thus scaling is achieved by changes in the ratio of the effective morphogen removal rate k_m or the effective morphogen diffusion rate D_m.

2.1.2 Advection-Dominated Transport

Advection is dominated when the morphogen is not diffusive and mainly transported by the growing tissue. This is the case if the morphogen or morphogen–ligand complex is bound to the surface on cells or resides inside cells [16–18]. For example, advection in a cell-bound ligand has been reported to contribute to ensure dynamic scaling of the Dpp gradient in the Drosophila wing imaginal disc [19], and if the dispersal of morphogen is dominated by advection, and the decay is linear to morphogen concentration, Eqs. 1–4 can be rewritten as follows, introducing variable $\xi = \frac{x}{L}$ and variable $\tau = \frac{t}{T}$ yields:

$$\frac{1}{T}\frac{\partial m}{\partial \tau} = -\frac{v_x}{L}\frac{\partial m}{\partial \xi} - k_m m \tag{16}$$

$$q_{\text{in}} = v_x m(0, \tau) \tag{17}$$

$$q_{\text{out}} \approx 0 \tag{18}$$

Consider steady state $\left(\frac{\partial m}{\partial \tau} = 0\right)$:

$$\frac{dm}{d\xi} = -\frac{k_m L}{v_x} m \tag{19}$$

$$q_{\text{in}} = v_x m(0) \tag{20}$$

$$q_{\text{out}} \approx 0 \tag{21}$$

This has the solution:

$$m(\xi) = \frac{q_{\text{in}}}{v_x} e^{\left(-\frac{k_m L}{v_x}\, \xi\right)} \tag{22}$$

The system scales if $k_m \propto L^{-1}$ or $(v_x \propto L) \wedge (q_{\text{in}} \propto L)$. This means that for scaling to occur, biology must have targeted k_m to decrease as the domain gets larger. In both of these derivations the requirements for scaling are captured by parameter groupings that dictate the amplitude and the shape of the gradient and this is useful in global measures of scale invariance discussed later herein.

2.2 Measures of Scaling

The most intuitive way to quickly estimate scaling is by plotting the morphogen gradients on absolute x and normalized x/L graphs and test whether or not the normalization of position of x to

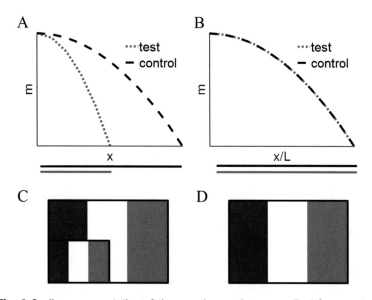

Fig. 2 Scaling representation of the sample morphogen gradient from control (black) and test (magenta). The morphogen gradient profiles as a function of x (**a**) or x/L (**b**). m is the morphogen concentration. The system length of the test (magenta) is half of the control (black), shown in (**a**) and scaled to equal dimensionless length (**b**). The analogous French flags for scaling in absolute scale (**c**) and relative scale (**d**)

relative position on x/L plot improves the overlay of morphogen gradients around a mean morphogen profile [8, 20, 21]. Figure 2 shows an example of morphogen scaling in test and control systems. The French flag is often used as an illustration of pattern scale invariance in accordance with the scaled morphogen gradients between a test and control. The smaller French Flag is half the size of the larger French Flags in Fig. 2c, in correspondence to the relative length of test and control systems in Fig. 2a. The lengths of the control and the test after normalizing the position scale to an arbitrary unit of 1 are the same (*see* **Note 1**), as shown by the magenta and black lines in Fig. 2b. Compared to significant difference between morphogen gradients when plotted against absolute position x in Fig. 2a, the x/L graph in Fig. 2b shows that the two morphogen gradients converge. Consistent in Fig. 2d, the patterns of the two flags overlay one another after normalizing x to relative positions. Thus the two systems appear to scale.

2.2.1 Point-Wise Scaling Error

A few metrics have been developed to measure the scaling error in order to quantify scaling and compare the degree of scaling among different systems. One of the methods, point-wise scaling error, measures position shifts of the morphogen gradients of two systems at certain points in x. Suppose there are two one-dimensional systems, one is the wild-type system S of size L, and the other is the test system S' of size $L'(L' \neq L)$. Figure 3a, b display two

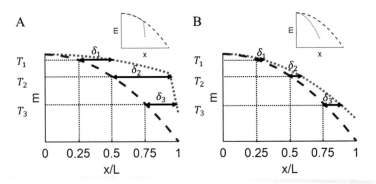

Fig. 3 An example of measuring point-wise scaling error. (**a**) Morphogen profile of the control system S (blue) and a test system S' (magenta) system on x/L plot, with the profile on x plot on the right-top corner. (**b**). Morphogen profile of the control (blue) and another test (red) system on x/L plot, with the profile on x plot on the right-top corner. The black horizontal arrow lines labeled δ_1, δ_2, δ_3 are the scaling error of the test morphogen gradient at the relative position 0.25, 0.5, and 0.75, respectively, in reference of the control. Taking the mean of these three scaling errors gives us the average scaling error. The average scaling error of **b** is smaller than that of **a**

examples. In Fig. 3a, b, the figure in the center shows the morphogen gradients of S (in dashed line in blue) and S' (in dashed line in magenta) on x/L plot, whereas the figure at the upper right corner provides the morphogen gradients on unnormalized graph. Point-wise scaling error is measured by the position shift of the same morphogen level between S and S' at selected relative positions in the x/L plot. Here, we present the measurement of point-wise scaling error at the relative positions $x_1 = 0.25$, $x_2 = 0.5$, and $x_3 = 0.75$:

1. Use the x/L plot of morphogen gradient of S to measure the morphogen concentrations at three relative positions, x_1, x_2 and x_3 and denote them as T_1, T_2, and T_3 accordingly. T_1, T_2, and T_3 are labeled on the left of Fig. 3a, b.

2. In the x/L plot of morphogen gradient of S', find the relative positions corresponding to morphogen levels T_1, T_2, and T_3, denoted as x'_1, x'_2 and x'_3. The distance between the relative positions of the two systems is called the point-wise scaling error, denoted as $\delta_i = x_i - x'_i, i \in \{1, 2, 3\}$. δ_1, δ_2, and δ_3 are highlighted with double arrowheaded lines in black in both Fig. 3a and b. The point-wise scaling errors of the example in Fig. 3b are overall smaller than those of the example in Fig. 3a, and thus S and S' of Fig. 3b scale better than S and S' of Fig. 3a.

3. The average scaling error is measured by $\delta = \frac{1}{3} \sum_{i=1}^{3} |\delta_i|$. In the case of perfect scaling, point-wise scaling errors and the average scaling error are all zero [22]. The average scaling error of Fig. 3b is smaller than that of Fig. 3a.

The advantage of point-wise scaling error is that it is easy to understand and to calculate. It is very useful for evaluating scaling for smooth simulation results. However, point-wise scaling error cannot be measured if the morphogen level of S at a selected position does not exist on the morphogen gradient of S'. In addition, one condition required to make sure the point-wise scaling error method is accurate is that the points chosen to calculate scaling error can represent the general shape of the morphogen gradients. The value of scaling error highly depends on the relative positions chosen. The scaling error can appear to be small if the local positions are selected in such a way that the morphogen gradients are close to each other, i.e., δ is small.

2.2.2 Sensitivity Factor

Instead of focusing on several local positions like point-wise scaling error, global methods utilize the overall shape of the morphogen gradient. One of the global measurement methods calculates a parametric sensitivity with respect to the length L and is called the sensitivity factor. The sensitivity factor measures the position shifts continuously using the global function of morphogen gradients. The sensitivity error is a measure of the local sensitivity of pattern deformation corresponding to the changes of system size on continuous morphogen profiles. Assume that the relative plot of a morphogen profile can be depicted by functions $u\,(\xi, L_1)$ and $u\,(\xi, L_2)$. For example, in Fig. 4a, the control (black) and test (magenta) correspond to $u\,(\xi, L_1)$ and $u\,(\xi, L_2)$ respectively. Here $\xi = \frac{x}{L}$ is the dimensionless variable, L_1 is the length of system number 1 and L_2 is the length of system number 2. Given any $\xi_1 (0 \leq \xi_1 \leq 1)$, find the ξ_2 where $u\,(\xi_1, L_1) = u\,(\xi_2, L_2) = T$. As shown in Fig. 4a, find T using the red dashed line according to position ξ_1 and then find ξ_2 using the red perpendicular line starting from the intersection of the red dashed line and the magenta line.

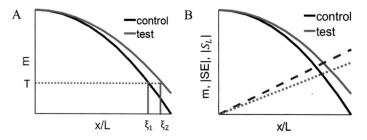

Fig. 4 An example of the sensitivity factor and scaling coefficient. The morphogen gradient profile on the *x*/*L* plot of the control is plotted in black line and that of the test is plotted in magenta. (**a**). Given $\xi_1 (0 \leq \xi_1 \leq 1)$, the morphogen concentration of the control at ξ_1 is T (red dashed line). Find ξ_2 where $u\,(\xi_1, L_1) = u\,(\xi_2, L_2) = T$. (**b**) The absolute values of the test morphogen profiles' sensitivity factor $|SE|$ is plotted in blue dashed line and the scaling coefficient $|S_L|$ is plotted in red dashed line. The system length of the test is 80% of the control

The relative position of the same morphogen level for perfect scaling is the same, i.e., $\xi_1 = \xi_2$. We measure the sensitivity of position shift in response to the changes of system size $\frac{\xi_2 - \xi_1}{L_2 - L_1}$ as the "sensitivity error," SE. If we assume a small variation in L and in ξ, then we can apply a linear approximation: $u(\xi_2, L_2) = u(\xi_1, L_1) + u'_\xi(\xi_2 - \xi_1) + u'_L(L_2 - L_1)$. Thus, at positions where $u_0(\xi_1, L_1) = u_0(\xi_2, L_2)$, $\text{SE} = \frac{\xi_2 - \xi_1}{L_2 - L_1} = -\frac{u'_L}{u'_\xi}$. $\text{SE} \equiv 0$ for perfect scaling [23].

This sensitivity factor can be regarded as a correction of the scaling coefficient, which is given as $S_L = -\left(\frac{\partial u}{\partial L}\right)\left(\frac{\partial u}{\partial x}\right)^{-1}\frac{L}{x}$ [24, 25]. Figure 4b shows the corresponding sensitivity factor and scaling coefficient between the control and test morphogen gradients. As the relative position increases, the difference between the morphogen concentration of the control and the test go up, both the sensitivity factor and the scaling coefficient increase. This method extracts the information of the whole morphogen gradient profiles and is quantitative. However, this method only works when the three following conditions are satisfied:

1. The variations of system lengths are small.

2. The position shifts of the morphogen gradient profiles are small.

3. The morphogen gradient profiles can be modeled as a function $m = u(\xi, L)$.

2.2.3 Scaling Power

Another global method to determine scaling calculates a quantity called the scaling power. According to He et al. [26], when an exponential function is used to fit the Bcd morphogen gradient $m(\xi, L) = Ae^{-\Gamma\xi}$, where m is Bcd morphogen concentration, then scaling can be determined by examination of A (the amplitude parameter) and Γ (the slope parameter). The scaling power is defined for the amplitude parameter as $n_A = \frac{\partial A/A}{\partial L/L}$ and the scaling power of the slope parameter is defined as $n_\Gamma = \frac{\partial \Gamma/\Gamma}{\partial L/L}$. Then the analytical solution of Eq. 23 is explored:

$$\frac{\partial m(L, \xi, t)}{\partial L} = 0 \qquad (23)$$

Two cases are discussed. When scaling powers are 0, all positions satisfy Eq. 23. In this case scale invariance is achieved. When scaling powers are not 0, only one position satisfies Eq. 23 and this position is called the critical position.

In practice, we can derive scale invariance by showing that the scaling power of the amplitude and the slope parameter are zero:

1. Find a function that fits best the morphogen gradients in study.

2. Generate (A, L) and (Γ, L) for each sample, where A is the amplitude parameter and Γ is the slope parameter.

3. Derive the equations for $n_A=0$ by correlation between A and L. If A does not significantly correlates with L, then $n_A=0$.

4. Similarly, derive $n_\Gamma=0$.

5. Explore solutions for Eq. 23 and prove that (Eq. 23) is always satisfied for any position of ξ within $(0, 1)$ when $n_A=0$ and $n_\Gamma = 0$. This result means the system is scale invariant.

The scaling power method is useful in evaluating scale invariance because it takes the overall gradient shape into account and is straightforward by mathematical derivation. However, the method is harder to extend to general gradient shapes that do not achieve the same separation of amplitude and exponent that occurs for the classical exponential gradient shape used to develop the scaling power.

3   Note

1. Normalization of the morphogen gradient impacts the calculation of scaling. Using different normalization methods can lead to different conclusions. One of the normalization methods called the "anchor point normalization" pins the minimum and maximum of individual profiles within a population and this may lead to erroneous attributions of "scaling" as discussed by Brooks et al. [27]. Methods that integrate the area under the concentration curve for a gradient and use that integral value for scaling systematically alter scale invariance conclusions that are a byproduct of data normalization [28]. An alternative method for gradient data normalization, "control-normalization" can be applied to normalize samples stained and imaged on different days with different settings by multiplying each set of samples with a scalar value [28]. This scalar value is calculated using controls. For each set of samples, a group of controls should be imaged in conjunction with corresponding experimental condition. The scalar value for sample sets imaged on different days is determined by minimizing the sum of the error among the controls imaged on corresponding days. After control-normalization, the gradient range of the population average of each control group is rescaled to be within [0, 1].

References

1. Umulis DM, Othmer HG (2013) Mechanisms of scaling in pattern formation. Development 140(24):4830–4843

2. Waddington CH (1942) Canalization of development and the inheritance of acquired characters. Nature 150:563–565

3. Patel NH, Lall S (2002) Precision patterning. Nature 415:748–749

4. Lander AD (2011) Pattern, growth, and control. Cell 144:955–969

5. Inomata H, Shibata T, Haraguchi T, Sasai Y (2013) Scaling of dorsal-ventral patterning by

embryo size-dependent degradation of spemann's organizer signals. Cell 153 (6):1296–1311

6. James H Brown, Geoffrey B West, Brian J Enquist (2000) Scaling in biology: patterns and processes, causes and consequences. Scaling in Biology 1–24

7. Gregor T, Bialek W, Steveninck RRDRV, Tank DW, Wieschaus EF (2005) Diffusion and scaling during early embryonic pattern formation. Proc Natl Acad Sci U S A 102 (51):18403–18407

8. Cheung D, Miles C, Kreitman M, Ma J (2011) Scaling of the Bicoid morphogen gradient by a volume dependent production rate. Development 138:2741–2749

9. Wartlick O, Mumcu P, Kicheva A, Bitting T, Seum C, Julicher F, Gonzalez-Gaitan M (2011) Dynamics of Dpp signaling and proliferation control. Science 331 (6021):1154–1159

10. Spemann H, Mangold H (2001) Induction of embryonic primordia by implantation of organizers from a different species. 1923. Int J Dev Biol 45(1):13–38

11. Umulis D, O'Connor MB, Othmer HG (2008) Robustness of embryonic spatial patterning in Drosophila melanogaster. Curr Top Dev Biol 81(81):65

12. Benzvi D, Shilo BZ, Fainsod A, Barkai N (2008) Scaling of the bmp activation gradient in xenopus embryos. Nature 453 (7199):1205–1211

13. Mizuno T, Yamaha E, Yamazaki F (1997) Localized axis determinant in the early cleavage embryo of the goldfish, Carassius auratus. Dev Genes Evol 206(6):389–396

14. Tung TC, Tung YFY (1944) The development of egg-fragments, isolated blastomeres and fused eggs in the goldfish. Proc Zool Soc London 114:46–64

15. Uygur A, Young J, Huycke TR, Koska M, Briscoe J, Tabin CJ (2016) Scaling pattern to variations in size during development of the vertebrate neural tube. Dev Cell 37(2):127

16. Kicheva A et al (2007) Kinetics of morphogen gradients formation. Science 315:512–525

17. Zhou S et al (2012) Free extracellular diffusion creates the Dpp morphogen gradient of the Drosophila wing disc. Curr Biol 22:668–675

18. Teleman AA, Cohen SM (2000) Dpp gradient formation in the Drosophila wing imaginal disc. Cell 103:971–980

19. Fried P, Iber D (2014) Dynamic scaling of morphogen gradients on growing domains. Nat Commun 5:5077

20. Deng J, Wang W, Lu LJ, Ma J (2010) A two-dimensional simulation model of the Bicoid gradient in Drosophila. PLoS One 5 (4):e10275

21. He F, Wen Y, Deng J, Lin X, Lu J, Jiao R, Ma J (2008) Probing intrinsic properties of a robust morphogen gradient in Drosophila. Dev Cell 15:558–567

22. Ben-Zvi D, Barkai N (2010) Scaling of morphogen gradients by an expansion-repression integral feedback control. PNAS 107 (15):6924–6929

23. Rasolonjananhary M, Vasiev B (2016) Scaling of morphogenetic patterns in reaction-diffusion systems. J Theor Biol 404:109–119

24. He F, Saunders TE, Wen Y et al (2010) Shaping a morphogen gradient for positional precision. Biophys J 99(3):697–707

25. de Lachapelle AM, Bergmann S (2010) Precision and scaling in morphogen gradient readout. Mol Sys Biol 6(1):351

26. He F, Wei C, Wu H, Cheung D, Jiao R, Ma J (2015) Fundamental origins and limits for scaling a maternal morphogen gradient. Nat Commun 6:6679

27. Brooks A, Dou W, Yang X, Brosnan T, Pargett M, Raftery LA, Umulis DM (2012) BMP signaling in wing development: a critical perspective on quantitative image analysis. FEBS Lett 586:1942–1952

28. Zinski J, Bu Y, Wang X, Dou W, Umulis D, Mullins M (2017) Systems biology derived source-sink mechanism of BMP gradient formation. eLife 6:e22199

Chapter 15

Scaling of Morphogenetic Patterns

Manan'Iarivo Rasolonjanahary and Bakhtier Vasiev

Abstract

Mathematical studies of morphogenetic pattern formation are commonly performed by using reaction–diffusion equations that describe the dynamics of morphogen concentration. Various features of the modeled patterns, including their ability to scale, are analyzed to justify constructed models and to understand the processes responsible for these features in nature. In this chapter, we introduce a method for evaluation of scaling for patterns arising in mathematical models and demonstrate its use by applying it to a set of different models. We introduce a quantity representing the sensitivity of a pattern to changes in the size of the domain, where it forms, and we show how to use it to perform a formal analysis of scaling for chemical patterns forming in continuous systems.

Key words Mathematical modeling, Pattern formation, Robustness and scaling

1 Introduction

An important property of developmental processes in biology is their robustness with respect to the developmental conditions and particularly the scaling of patterns as they form with the size of the developing object. A simple illustration of a pattern forming in response to changes in morphogen concentration and exhibiting scaling is given by the "French flag" model [1]. This model demonstrates how a simple linear concentration profile defines domains of cellular differentiation in a homogenous tissue. The linear concentration profiles can form naturally in various settings. The simplest case is when the production and degradation of morphogen take place outside the tissue on its opposing sides and the morphogen passively diffuses along the tissue from the side where it is produced to the side where it is degraded. Mathematically, the concentration of the morphogen in this system should obey the so-called Laplace's equation under certain boundary conditions. For a tissue represented by a one-dimensional domain of length L under Dirichlet boundary conditions (i.e., the concentration of morphogen is buffered on the boundaries of the tissue), it is given as:

Julien Dubrulle (ed.), *Morphogen Gradients: Methods and Protocols*, Methods in Molecular Biology, vol. 1863,
https://doi.org/10.1007/978-1-4939-8772-6_15, © Springer Science+Business Media, LLC, part of Springer Nature 2018

$$D\frac{d^2u}{dx^2} = 0; \quad u(0) = u_0; \quad u(L) = u_L, \tag{1}$$

where u is the morphogen concentration, D—its diffusion coefficient, x defines the location inside the tissue, u_0 and u_L are the morphogen concentrations on the opposite sides of the tissue. The solution of (1) is given by a linear profile

$$u = u_0 + (u_L - u_0)\frac{x}{L}, \tag{2}$$

The profile (2) scales since a point with any preset concentration divides domains of different sizes in the same proportion. This can be seen by replacing the absolute coordinate x by the relative coordinate $\xi = x/L$ so that the coordinate within the tissue varies between 0 and 1 and the profile defined in terms of ξ does not depend on the size, L:

$$u = u_0 + (u_L - u_0)\xi. \tag{3}$$

In the French-flag model, the domains of cellular determination are defined by the threshold concentrations of morphogen and if, for example, the size of the tissue is doubled then the size of all of the domains will also be doubled. Thus, the patterns formed in the French-flag model scale.

According to experimental observations, the morphogen concentration profiles are commonly exponential rather than linear. A typical example of an exponential profile is presented by the transcriptional factor Bicoid in the fly embryo [2, 3]. It can be shown mathematically that an exponential profile forms if it is assumed that the morphogen not only diffuses but also decays inside the domain; that is, it forms in the so-called diffusion–decay systems [4]:

$$\frac{\partial u}{\partial t} = D_u\frac{\partial^2 u}{\partial x^2} - k_u u. \tag{4}$$

where D_u and k_u are the morphogen's diffusion and decay coefficients respectively. Various boundary conditions can be imposed on this system. For example, one can assume (as for system (1)) that the concentration of morphogen is buffered on the boundaries of the tissue. Alternatively, we can assume that the tissue is isolated (no flows on the boundaries) and the production of the morphogen takes place in a restricted area inside the domain. This assumption is perfectly reasonable in many cases. For example, the maternal Bicoid mRNA in the fly embryo is localized in a small region on its apical side and the Bicoid protein produced in this region diffusively spreads and decays along the entire embryo, in which case equation (4) should be solved under the Neumann boundary conditions:

$$\frac{\partial u}{\partial x}\bigg|_{x=0} = -q; \quad \frac{\partial u}{\partial x}\bigg|_{x=L} = 0 \tag{5}$$

The solution of (4) is given by a superposition of two exponents, $u(x) = Ae^{-\frac{x}{\lambda}} + Be^{\frac{x}{\lambda}}$, where $\lambda=(D_u/k_u)^{1/2}$ is the so-called characteristic length of the exponential profile and the constants A and B are found from the boundary conditions. In the case of the boundary conditions (5) and for sufficiently large tissue ($L \gg \lambda$), the solution can be approximated as

$$u(x) = q\lambda e^{-\frac{x}{\lambda}}, \tag{6}$$

where $q\lambda$ is the concentration of the morphogen at the source boundary ($x=0$).

The exponential profile (6) does not scale: if the domains of cellular determination are defined by the threshold concentration, $u=T$, then they are separated at $x(u=T)=\lambda\ln(q\lambda/T))$ and the location of this border does not depend on the size of the tissue. Note that for the scaled pattern, the sizes of the domains (and correspondingly the coordinate of the border) would be proportional to the size of the tissue. A possible scaling mechanism can be provided by another chemical agent (a so-called modulator [5]) which may affect the exponential u-profile such that it scales. Thus, the scaled exponential profile can be obtained in a two-variable system represented by two interlinked equations, one for rate of change in the concentration of morphogen, u, and the second—for the concentration of modulator, M. The modulator should affect the dynamics (the diffusion and/or the kinetics) of the morphogen:

$$\frac{\partial u}{\partial t} = \frac{\partial}{\partial x}\left(D_u(M)\frac{\partial u}{\partial x}\right) + k_u(M)u. \tag{7}$$

Examples in the literature include cases in which both the diffusion and reaction terms of the morphogen are functions of the modulator. On the other hand, the morphogen may or may not affect the dynamics of the modulator, scenarios which are called active and passive moderation correspondingly [5]. Several mechanisms of scaling have been proposed for systems with active moderation [6–10]. Passive moderation is considerably simpler [11, 12] especially in the case when the characteristic length of the modulator profile is sufficiently large $\left(\lambda_M = \sqrt{D_M/k_M} \gg L\right)$ and its concentration does not change very much within the domain being considered. In this case, scaling can be achieved if the modulator's concentration depends on the medium size L, for example, if it is directly or inversely proportional to this size. An illustration of the model resulting from the inverse proportionality is presented in the Considered Models section.

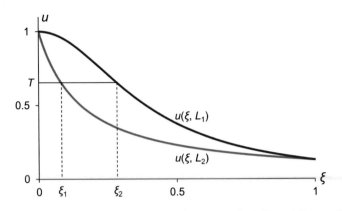

Fig. 1 Plot of the morphogen concentration, u, against the relative position, $\xi = x/L$. The blue, $u(\xi, L_1)$, and red, $u(\xi, L_2)$, plots illustrate different concentration profiles corresponding to the domains of different sizes. The relative position ξ varies between 0 and 1 for both profiles. The relative positions ξ_1 and ξ_2 indicate the points with the same value of u ($u = T$) for the two profiles (i.e., T corresponds to a concentration threshold defining cellular differentiation)

For analysis of scaling properties of morphogen gradients, a quantity defining scaling precision is introduced in the following manner [11]. Note that the morphogenetic profile can be represented as a function of two variables: the location within the object, which can be given by the relative coordinate, $\xi=x/L$, and the object size, L. Assume that the morphogenetic profiles occurring in two objects of different sizes L_1 and L_2 are described by the differentiable functions $u(\xi, L_1)$ and $u(\xi, L_2)$ (*see* Fig. 1). Note that if the profile is scaled across the two objects, then $u(\xi, L_1)=u(\xi, L_2)$ at any point ξ. This is not true in general and from $u(\xi_1, L_1)=u(\xi_2, L_2)$, it does not follow that $\xi_1=\xi_2$ (*see* Fig. 1). If the difference between L_1 and L_2 is small then the difference between ξ_1 and ξ_2 is also small and we can use the linear approximation: $u(\xi_2, L_2) = u(\xi_1, L_1) + u'_\xi(\xi_2 - \xi_1) + u'_L(L_2 - L_1)$. From $u(\xi_1, L_1)=u(\xi_2, L_2)$ it follows that $\xi_2 - \xi_1 = -\frac{u'_L}{u'_\xi}(L_2 - L_1)$, i.e., the deformation of the profile is proportional to the change in the size of the system with the coefficient of proportionality representing the local sensitivity which is called the "sensitivity factor":

$$S(\xi, L) = -\left(\frac{\partial u}{\partial L}\right)\left(\frac{\partial u}{\partial \xi}\right)^{-1} \tag{8}$$

The sensitivity factor defines the rate at which a point with a given level of morphogen concentration, u, shifts when the medium size is varied. In the case of perfect scaling, there is no shift and therefore $S(\xi, L)\equiv0$. Furthermore, a positive sign of the sensitivity factor implies that the profile is stretched (to the right)—a characteristic known as hyperscaling. A negative sign of the sensitivity factor implies that the profile is contracted (to the left), corresponding to hyposcaling. The smaller the absolute value of the sensitivity factor the better the scaling of the profile.

Other quantities to describe the scaling properties of concentration profiles have also been introduced. For example, a quantity named the "scaling coefficient" was introduced in [13, 14] as:

$$S_L = -\frac{dx}{dL}\frac{L}{x} \tag{9}$$

Assuming that the threshold concentration is fixed (implying $du = (\partial u/\partial x)dx + (\partial u/\partial L)dL = 0$) where u is the morphogen concentration and L is the size of the embryo, we obtain $\frac{dx}{dL} = -\frac{\partial u}{\partial L}\left(\frac{\partial u}{\partial x}\right)^{-1}$ and therefore, (9) can be rewritten as

$$S_L = -\frac{\partial u}{\partial L}\left(\frac{\partial u}{\partial x}\right)^{-1}\frac{L}{x}. \tag{10}$$

Perfect scaling, according to formula (10), occurs when $S_L=1$; $S_L<1$ corresponds to the case when the stretch/shrink of the profile does not compensate for the stretch/shrink of the medium (hyposcaling) while $S_L>1$ corresponds to the case of overcompensation (hyperscaling) [14]. The definition of the scaling coefficient in (10) is given in terms of the absolute value of coordinate, x, and this comes with certain drawbacks—at $x=0$, the scaling coefficient is undefined. Also, definition (10) introduces a certain asymmetry: the scaling factor is different depending on which edge of the domain is chosen to correspond to $x = 0$ (i.e., it depends on the choice of the domain edge from which the coordinate is measured). The sensitivity factor given by (8) is free from these drawbacks.

Another way to quantify scaling of the morphogenetic profile is associated with the so-called "sensitivity coefficient" introduced for engineering systems in [15] and defined as follows:

$$S_R = \lim_{\Delta p \to 0}\frac{\Delta f/f}{\Delta p/p} = \frac{d\ln f}{d\ln p},$$

where p is an input parameter and f is the output function. In our case, the input parameter would be the medium size L and the output function would be the morphogen concentration $u(\xi)$. Thus, the above formula can be translated as

$$S_R(\xi, L) = \lim_{\Delta L \to 0}\frac{\Delta u\ (\xi, L)/u\ (\xi, L)}{\Delta L/L}$$

$$= \frac{L}{u\ (\xi, L)}\frac{du\ (\xi, L)}{dL} = \frac{d\ln u\ (\xi, L)}{d\ln L} \tag{11}$$

According to this definition, $S_R=1$ means that the variation in the input is fully compensated by the change of the output. Furthermore, $S_R<1$ would correspond to an undercompensation (hyposcaling) while $S_R>1$ corresponds to an overcompensation (hyperscaling). The sensitivity coefficient, S_R, as defined by equation (11) has certain advantages compared with the sensitivity factor, S_L, defined by equation (10), i.e., unlike S_L in (10), the sensitivity coefficient S_R for a symmetrical profile is also symmetric. The drawback of definition (11) is that the calculation of S_R is far more difficult and time consuming than the calculations associated with definitions (8) and (10). In what follows, we analyze scaling properties of concentration profiles using the definition of scaling factor given by (8).

2   Considered Models

In this section, we introduce models describing diffusion–decay and activator–inhibitor systems involving passive modulation. We will consider the simplest scaling scenario in which the concentration of the modulator, M, is roughly constant over the modeled domain and is proportional to the size of the medium. In what follows, the modulator's dynamics are described by the diffusion–decay equation. The concentration of the morphogen is leveled up over the entire domain if the characteristic length, λ_M, is considerably larger than the domain size, L, $\left(\text{i.e.} \lambda_M = \sqrt{D_M/k_M} \gg L\right)$. Furthermore, to ensure proportionality between the modulator concentration and the domain size, we postulate that although the modulator is produced everywhere inside the domain, it decays only in a small area (located, for example, in the middle of the domain). Thus, the dynamics of the modulator are described by the following equation:

$$\frac{\partial M}{\partial t} = D_M \frac{\partial^2 M}{\partial x^2} - k(x)M + p, \quad k(x)$$

$$= \begin{cases} k_M & \text{for} \quad \dfrac{L-a}{2} < x < \dfrac{L+a}{2} \\ 0 & \text{otherwise} \end{cases} \tag{12}$$

where D_M is the modulator's diffusion coefficient, p—its production rate, a—the size of the degradation area, and k_M is the degradation rate. This model is illustrated by Fig. 2. To model the isolated tissue, we should solve equation (12) under the zero flux (Neumann) boundary conditions at both ends: $\frac{\partial M}{\partial x}\big|_{x=0} = \frac{\partial M}{\partial x}\big|_{x=L} = 0$.

Furthermore, we consider another morphogen whose concentration is denoted by u and which is responsible for scaling in the system. We will illustrate our method on two kinds of morphogen

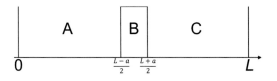

Fig. 2 Illustration of the model described by equation (12). The modulator (with concentration M) is produced and diffuses all over the domain (regions **A–C**) and decays only in region **B**

profile. In the simplest model, the morphogen profile is exponential and its dynamics are also described by the diffusion–decay equation but with the decay term affected by the modulator. For example, this dynamic can be described by the equation:

$$D_u \frac{d^2 u}{dx^2} - k_u M^n u = 0; \quad u(x=0) = 1; \quad \left. \frac{du}{dx} \right|_{x=L} = 0, \quad (13)$$

where D_u is the diffusion coefficient and the factor $k_u M^n$ defines the decay rate of the morphogen (if n was a positive integer, this would mean that the morphogen degradation takes place when one molecule of the morphogen collides with n molecules of the modulator). Boundary conditions in (13) imply that the morphogen is entering the domain at $x = 0$, either by means of external flow or due to the local production in a small area near the left edge of the domain, while there is no flux on the right edge of the domain. In the system described by equations (12) and (13), the morphogen profile is exponential and the parameter n defines the strength of the modulator's influence to this profile. If in (13) $n = 0$, then the decay is not affected by the modulator (no modulation) and no scaling mechanism is in place. In the Methods section, we will show that the morphogenetic profiles forming in this system scale when n is negative: $n = -1$ corresponds to the case of hyposcaling, $n = -2$ corresponds to perfect scaling and $n = -3$ to hyperscaling.

We will also analyze the scaling properties of Turing patterns forming in activator–inhibitor systems affected by the modulator, whose concentration (described by equation (12)) is leveled off. The effect of passive modulation on Turing patterns has been widely studied and in particular, it has been shown that it can result in a conservation of the number of spikes in a medium of variable size and thus to scaling of the pattern [11, 12]. Turing patterns form in systems of interacting morphogens described by reaction-diffusion equations. In the simplest case, the system involves only two morphogens: an activator, concentration u, and an inhibitor, concentration v, whose kinetics are described by cubic and linear terms:

$$\frac{\partial u}{\partial t} = D_u \frac{\partial^2 u}{\partial x^2} + \left(a(u - u_0) + b(v - v_0) - \alpha_u(u - u_0)^3 \right) M^n, \quad \left. \frac{\partial u}{\partial x} \right|_{x=0} = \left. \frac{\partial u}{\partial x} \right|_{x=L} = 0;$$

$$\frac{\partial v}{\partial t} = D_v \frac{\partial^2 v}{\partial x^2} + \left(c(u - u_0) + d(v - v_0) \right) M^n, \quad \left. \frac{\partial v}{\partial x} \right|_{x=0} = \left. \frac{\partial v}{\partial x} \right|_{x=L} = 0.$$

$$(14)$$

Here, D_u and D_v are the diffusion coefficients for morphogen and inhibitor respectively; a, b, c and d are constants whose particular values allow the formation of Turing patterns and determine the distance between the forming stripes [16, 17] while the constant, α_u, associated with the cubic term, defines the amplitude of the concertation's variations in the pattern. Parameters u_0 and v_0 define average values of morphogen concentrations in the system; to ensure that the solution of (14) is represented by positive functions (associated with biologically meaningful positive concentrations) the point (u_0, v_0) should be safely placed in the first quadrant so that the variation range for both concentrations is positive. Finally, the factor M^n links system (14) with equation (12) and defines the effect of the modulator, M, on the Turing pattern. In similarity to the previous discussion of the system (12) and (13), $n=0$ corresponds to the case when the modulator has no effect on system (14) and therefore no scaling mechanism is in place (number of stripes is proportional to the size of the medium). $n = -1$ corresponds to the case when the number of stripes increases more slowly than the size of the medium (hyposcaling); for $n = -2$ the number of strips does not depend on the size of the medium (perfect scaling) and for $n = -3$; the number of stripes increases faster than the size of the medium (hyperscaling).

The model (14) (with $n = 0$ and therefore detached from equation (12)) is generic and derived to satisfy minimal requirements for Turing pattern formation. The biologically meaningful model describing the regeneration of hydra proposed by Gierer and Meinhardt [18], has nonlinear terms in both equations and its common representation [19] extended by passive modulation by M is given as:

$$\frac{\partial u}{\partial t} = D_u \frac{\partial^2 u}{\partial x^2} + \left(\frac{u^2}{v} - u + \sigma \right) M^n \quad \left. \frac{\partial u}{\partial x} \right|_{x=0} = \left. \frac{\partial u}{\partial x} \right|_{x=L} = 0;$$

$$\frac{\partial v}{\partial t} = D_v \frac{\partial^2 v}{\partial x^2} + \rho(u^2 - v) M^n \quad \left. \frac{\partial v}{\partial x} \right|_{x=0} = \left. \frac{\partial v}{\partial x} \right|_{x=L} = 0.$$

$$(15)$$

Here, parameter ρ defines the kinetics rate of the second morphogen (whose concentration is denoted by v) and is called the cross-reaction coefficient while parameter σ stands for the basic production term in the kinetics of the first morphogen (whose concentration is denoted by u). Another model allowing formation of Turing patterns is given by the modified FitzHugh–Nagumo system [20] and the equations (including passive modulation given by the factors, M^n) representing this model are the following:

$$\frac{\partial u}{\partial t} = D_u \frac{\partial^2 u}{\partial x^2} - \left(ku(u-1)(u-a)+v \right) M^n \qquad \left.\frac{\partial u}{\partial x}\right|_{x=0} = \left.\frac{\partial u}{\partial x}\right|_{x=L} = 0;$$

$$\frac{\partial v}{\partial t} = D_v \frac{\partial^2 v}{\partial x^2} + \varepsilon(u-v)M^n \qquad \left.\frac{\partial v}{\partial x}\right|_{x=0} = \left.\frac{\partial v}{\partial x}\right|_{x=L} = 0.$$

$$(16)$$

FitzHugh–Nagumo model was originally designed as a prototype model describing propagation of electrical activity along nerve fibers and its parameters can be described in the terms associated with this process: k represents the conductance of cell membrane, a is the threshold potential for opening ion channels and ε defines the rate of channel closure.

3   Analytical and Numerical Studies of Scaling

We can compute analytically the steady state solution of equation (4) by equating it to zero.

$$D_u \frac{d^2 u}{dx^2} - k_u u = 0 \qquad (17)$$

This is a linear second order differential equation whose solution is represented by the superposition of two exponents [21]:

$$u(x, L) = A(L)e^{-\frac{x}{\lambda}} + B(L)e^{\frac{x}{\lambda}} \text{ or } u(\xi, L) = A(L)e^{-\frac{L\xi}{\lambda}} + B(L)e^{\frac{L\xi}{\lambda}}$$

$$(18)$$

where $\lambda = (D_u/k_u)^{1/2}$, $\xi = x/L$, and the functions $A(L)$ and $B(L)$ are found from the boundary conditions imposed to equation (17). This equation can be subjected to Dirichlet, mixed, or Neumann boundary conditions. Under the Dirichlet boundary conditions, $u(x=0) = u_0$ and $u(x=L) = 0$, the full solution of (17) is given as:

$$u(\xi, L) = u_0 \frac{\exp\left(\frac{1-\xi}{\lambda}L\right) - \exp\left(\frac{\xi-1}{\lambda}L\right)}{\exp\left(\frac{L}{\lambda}\right) - \exp\left(-\frac{L}{\lambda}\right)}. \qquad (19)$$

The solution of (17) under the mixed boundary condition, $\partial u/\partial x(x=0) = 0$ and $u(x=L) = u_L$, is:

$$u(\xi, L) = u_L \frac{\exp\left(\frac{\xi}{\lambda}L\right) + \exp\left(-\frac{\xi}{\lambda}L\right)}{\exp\left(\frac{L}{\lambda}\right) + \exp\left(-\frac{L}{\lambda}\right)} \qquad (20)$$

And finally, the solution of (17) under the Neumann boundary conditions (5) is given as:

$$u(\xi, L) = q\lambda \frac{\exp\left(\frac{\xi-1}{\lambda}L\right) + \exp\left(\frac{1-\xi}{\lambda}L\right)}{\exp\left(\frac{L}{\lambda}\right) - \exp\left(-\frac{L}{\lambda}\right)} \tag{21}$$

In order to compute the sensitivity factor S defined by equation (8), one would need to compute the partial derivatives $\partial u/\partial L$ and $\partial u/\partial \xi$ and then find their ratio. This can be done manually or using an appropriate software package, such as Maple or Matlab. For example, in Maple, the profile (21) can be written as

$u := q * lambda * (\exp(L*(xi-1)/lambda) + \exp(L*(1-xi)/lambda))/$
$\left(\exp(L/lambda) - \exp(-L/lambda)\right)$

where xi and $lambda$ stand for ξ and λ respectively. The partial derivatives $\partial u/\partial L$ and $\partial u/\partial \xi$ then can be found using the standard Maple subroutine diff($*$,$*$) as **diff(u, L)** and **diff(u, xi)** respectively. Thus, the sensitivity factor S is given by the Maple expression:

$$S := -diff(u, L)/diff(u, xi).$$

The sensitivity factors for profiles (19)–(21) as computed by Maple are respectively:

$$S = -\frac{(\xi-2)\left(e^{\frac{\xi L}{\lambda}} - e^{-\frac{\xi L}{\lambda}}\right) + \xi\left(e^{-\frac{L(\xi-2)}{\lambda}} - e^{\frac{L(\xi-2)}{\lambda}}\right)}{L\left(e^{\frac{L}{\lambda}} - e^{-\frac{L}{\lambda}}\right)\left(e^{\frac{L(\xi-1)}{\lambda}} + e^{-\frac{L(\xi-1)}{\lambda}}\right)}, \tag{22}$$

$$S = -\frac{(\xi-1)\left(e^{\frac{L(1+\xi)}{\lambda}} - e^{-\frac{L(1+\xi)}{\lambda}}\right) + (\xi+1)\left(e^{\frac{L(\xi-1)}{\lambda}} - e^{\frac{L(1-\xi)}{\lambda}}\right)}{L\left(e^{\frac{L}{\lambda}} + e^{-\frac{L}{\lambda}}\right)\left(e^{\frac{L\xi}{\lambda}} - e^{-\frac{L\xi}{\lambda}}\right)} \tag{23}$$

and

$$S = -\frac{(\xi-2)\left(e^{\frac{\xi L}{\lambda}} + e^{-\frac{\xi L}{\lambda}}\right) - \xi\left(e^{\frac{L(\xi-2)}{\lambda}} + e^{-\frac{L(\xi-2)}{\lambda}}\right)}{L\left(e^{\frac{L}{\lambda}} - e^{-\frac{L}{\lambda}}\right)\left(e^{\frac{L(\xi-1)}{\lambda}} - e^{-\frac{L(\xi-1)}{\lambda}}\right)} \tag{24}$$

The plots of solutions (19)–(21) and their sensitivity factors (22)–(24) (*see* **Note 1**): are shown in Fig. 3. Note that the profiles for media of different sizes under Dirichlet boundary conditions (solid and dotted blue in panel **A**) do not differ very much and the corresponding scaling factor (shown in blue in panel **B**) is close to zero. Another important observation is that the profiles forming under Neumann boundary conditions (red profiles in panel **A**) are shallow and fit in the smallest range. Furthermore, we note that under certain conditions, namely when the morphogen diffuses quickly and/or decays slowly, that is, when λ is large ($\lambda \gg L$), this profile levels off. Indeed, under the condition $\lambda \gg L$, we can take the linear approximations of the exponential terms in formula (21):

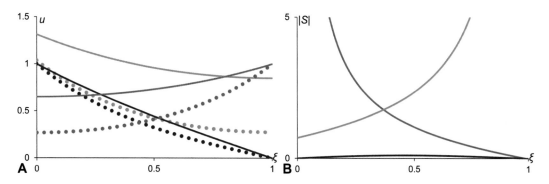

Fig. 3 Scaling in a diffusion–decay system with no modulation. Morphogen profiles and their sensitivity factors in the diffusion–decay system (4) are shown. (**a**) The blue, red, and green profiles correspond to the profiles forming under the Dirichlet, mixed, and Neumann boundary conditions and given by formulas (19)–(21) respectively. (**b**) The blue, red, and green profiles represent the absolute value of sensitivity factors for the profiles forming under the Dirichlet, mixed, and Neumann boundary conditions and given by formulas (22)–(24) respectively. The parameters are as follows: $D = k = 1$, $L = 1$ (solid lines) and $L = 2$ (dotted lines). Plots are produced using Excel (Microsoft Office)

$$u(\xi, L) = q\lambda \frac{\exp\left(L\frac{\xi-1}{\lambda}\right) + \exp\left(L\frac{1-\xi}{\lambda}\right)}{\exp\left(\frac{L}{\lambda}\right) - \exp\left(-\frac{L}{\lambda}\right)} \approx q\lambda \frac{1 + L\left(\frac{\xi-1}{\lambda}\right) + 1 + L\left(\frac{1-\xi}{\lambda}\right)}{1 + \frac{L}{\lambda} - 1 + \frac{L}{\lambda} = \frac{q\lambda^2}{L}}.$$

Thus, the first order approximation of (21) does not depend on the coordinate, ξ, that is, the concentration is roughly the same everywhere and, besides, it is inversely proportional to L and, therefore, can be used as a measure of the medium size.

Formulas (19–24) represent the morphogen profiles and corresponding scaling factors for equation (4). These results can be extended to the profiles forming in system (12). The stationary profile forming in system (12) is a solution of the following differential equations, each describing the profile on one of the three domains denoted as A, B, and C in Fig. 2:

$$D_M \frac{d^2 M_1}{dx^2} + p = 0, \quad \text{for } 0 \le x \le \frac{L - a}{2}; \tag{25}$$

$$D_M \frac{d^2 M_2}{dx^2} - k_M M_2 + p = 0, \quad \text{for } \frac{L - a}{2} < x < \frac{L + a}{2}; \tag{26}$$

$$D_M \frac{d^2 M_3}{dx^2} + p = 0, \quad \text{for } \frac{L + a}{2} \le x \le L. \tag{27}$$

Solutions of equations (25) and (27) are represented by the quadratic functions of the coordinate x, while the solution of (26) by the superposition of the exponential functions and a constant:

$$M_1 = -\frac{p}{2D_M}x^2 + A_1 x + A_2, \quad \text{for } 0 \le x \le \frac{L - a}{2};$$

$$M_2 = A_3 e^{\frac{x}{\lambda_M}} + A_4 e^{-\frac{x}{\lambda_M}} + \frac{p}{k_M}, \quad \text{for} \quad \frac{L-a}{2} < x < \frac{L+a}{2};$$

$$M_3 = -\frac{p}{2D_M} x^2 + A_5 x + A_6, \quad \text{for} \quad \frac{L+a}{2} \le x \le L.$$

where $\lambda_M = (D_M/k_M)^{1/2}$. In order to determine the six coefficients (A_1, A_2, A_3, A_4, A_5 and A_6), we need to apply six conditions. The first two conditions are the zero-flux Neumann boundary conditions at both ends of the entire domain. The next two conditions are the continuity of the profiles at the border between the regions A and B ($x = (L-a)/2$) and the border between B and C ($x = (L+a)/2$). The last two conditions are the continuity of the first derivative of the profile at these two borders. After long calculations, the three full solutions are computed as (*see* **Note 2**):

$$M_1 = p\left(-\frac{x^2}{2D_M} - \frac{\lambda_M(L-a)\left(2 + e^{-\frac{a}{\lambda_M}} + e^{\frac{a}{\lambda_M}}\right)}{2D_M\left(e^{-\frac{a}{\lambda_M}} - e^{\frac{a}{\lambda_M}}\right)} + \frac{1}{k_M} + \frac{(L-a)^2}{8D_M} \right);$$

$$M_2 = p\left(-\frac{\lambda_M(L-a)\left(e^{\frac{L+a}{2\lambda_M}} + e^{\frac{L-a}{2\lambda_M}}\right)\left(e^{\frac{x}{\lambda_M}} + e^{\frac{L-x}{\lambda_M}}\right)}{2D_M\left(e^{\frac{L-a}{\lambda_M}} - e^{\frac{L+a}{\lambda_M}}\right)} + \frac{1}{k_M} \right);$$

$$M_3 = p\left(-\frac{x^2}{2D_M} + \frac{Lx}{D_M} - \frac{\lambda_M(L-a)\left(2 + e^{-\frac{a}{\lambda_M}} + e^{\frac{a}{\lambda_M}}\right)}{2D_M\left(e^{-\frac{a}{\lambda_M}} - e^{\frac{a}{\lambda_M}}\right)} + \frac{1}{k_M} \right.$$
$$\left. + \frac{(L+a)^2}{8D_M} - \frac{L(L+a)}{2D_M} \right).$$

$$(28)$$

Solution (28) can be simplified by assuming that the diffusion coefficient D_M is large (*i.e.* $\sqrt{D_M/p} \gg L$):

$$M_1 \approx p\left(-\frac{\lambda_M(L-a)\left(2 + e^{-\frac{a}{\lambda_M}} + e^{\frac{a}{\lambda_M}}\right)}{2D_M\left(e^{-\frac{a}{\lambda_M}} - e^{\frac{a}{\lambda_M}}\right)} + \frac{1}{k_M} \right),$$

$$M_2 = p\left(-\frac{\lambda_M(L-a)\left(e^{\frac{L+a}{2\lambda_M}} + e^{\frac{L-a}{2\lambda_M}}\right)\left(e^{\frac{x}{\lambda_M}} + e^{\frac{L-x}{\lambda_M}}\right)}{2D_M\left(e^{\frac{L-a}{\lambda_M}} - e^{\frac{L+a}{\lambda_M}}\right)} + \frac{1}{k_M} \right),$$

$$M_3 \approx p\left(-\frac{\lambda_M(L-a)\left(2 + e^{-\frac{a}{\lambda_M}} + e^{\frac{a}{\lambda_M}}\right)}{2D_M\left(e^{-\frac{a}{\lambda_M}} - e^{\frac{a}{\lambda_M}}\right)} + \frac{1}{k_M} \right).$$

We note that the approximations of M_1 and M_3 are identical. After performing the linear approximation of the exponential terms (under the condition $\lambda_M = \sqrt{D_M/k_M} \gg L$ as done in (21)), all the three solutions can be simplified to

$$M \approx \frac{pL}{ak_M} \tag{29}$$

We can see that the concentration of modulator levels off and is directly proportional to the medium size.

We can estimate the solution of (13): (under the conditions used to derive (29)) by substituting (29) in (13). The stationary solution of equation (13) is given (as in the case of equation (17)) by a superposition of two exponents. However, in the case of a sufficiently large medium (large as compared with the characteristic length of the morphogen: $L \gg \lambda_u = \sqrt{D_u/k_u}$) this solution can be approximated to by a single exponent:

$$u \approx u_0 e^{-x\sqrt{\frac{k_u}{D_u}M^n}} = u_0 e^{-\xi L \sqrt{\frac{k_u}{D_u}\left(\frac{pL}{ak_M}\right)^n}}. \tag{30}$$

The above equation can be rewritten as $u \approx u_0 e^{-\xi\frac{L}{\lambda_u^*}}$ where:

$$\lambda_u^* = \sqrt{\frac{D_u}{k_u}\left(\frac{pL}{ak_M}\right)^{-n}} \tag{31}$$

is the characteristic length of the morphogen profile in system (12 and 13). In the case of $n = 0$, the characteristic length, λ_u^*, does not depend on the medium size indicating the absence of any scaling mechanism in the system:

$$u \approx u_0 e^{-x\sqrt{\frac{k_u}{D_u}}} = u_0 e^{-\xi L\sqrt{\frac{k_u}{D_u}}} \tag{32}$$

In the case of $n = -1$, the characteristic length, λ_u^*, is proportional to the square root of the medium size. The morphogen profile is stretched when the medium size is increased, but the stretch of the profile does not fully compensate for the increase of the medium size (hyposcaling):

$$u \approx u_0 e^{-\xi\sqrt{L}\sqrt{\frac{ak_M}{p}\frac{k_u}{D_u}}}. \tag{33}$$

In the case of $n = -2$, the characteristic length is proportional to the medium size and, consequently, the morphogen profile does not depend on the length of the medium, L:

$$u \approx u_0 e^{-\xi L\sqrt{\frac{k_u}{D_u}\left(\frac{pL}{ak_M}\right)^{-2}}} = u_0 e^{-\xi\frac{ak_M}{p}\sqrt{\frac{k_u}{D_u}}}. \tag{34}$$

Thus, the morphogen profile scales perfectly well if $n = -2$. Finally, in the case of $n = -3$ the dependence of the characteristic length, λ_u^*, on the medium size is stronger (than linear) and the changes in the morphogen profile overcompensate (hyperscaling) for the changes in the medium size:

$$u \approx u_0 e^{-\frac{\xi}{\sqrt{L}}\sqrt{\left(\frac{ak_M}{p}\right)^3 \frac{k_u}{D_u}}}. \qquad (35)$$

Sensitivity factors for profiles, given by formula (30) as defined by (8) (and found analytically) are $S = -\frac{\xi}{L}, -\frac{\xi}{2L}, 0$ and $\frac{\xi}{2L}$ for the cases $n = 0, -1, -2$, and -3 respectively. The extension of this result for the case of any n is given by the following formula:

$$S_n = -\frac{(2+n)\xi}{2L}. \qquad (36)$$

The profiles obtained by numerical integration (using C-codes as described below) of the system (12 and 13) for the cases $n = 0$, $-1, -2$ and -3 and corresponding sensitivity factors (as defined by (8) and found by calculation of the partial derivatives $\partial u/\partial L$ and $\partial u/\partial \xi$ and their ratio $\partial u/\partial L$ by $\partial u/\partial \xi$ using standard Maple subroutines) are plotted in Fig. 4. Profiles are shown for media of two different sizes and we see that the profile corresponding to $n = -2$ scales perfectly well (profiles for media of different sizes coincide) and its sensitivity factor is close to zero. The highest discrepancy between profiles is observed for $n = 0$ (shown in green) and the absolute value of corresponding scaling factor is the largest of the four. Thus, the scaling property of the profile forming in the diffusion–decay system without modulation is the worst of the cases shown.

Finding scaling factors for Turing patterns is far more challenging. Unlike equations (12) and (13), equations (14–16) cannot be solved analytically. However, they can be solved numerically using Matlab or Visual C++. Here, we briefly describe a source code written in C++ which we used for numerical integration of system (15) (Meinhardt model). A one-dimensional medium (segment of a straight line) where a Turing's pattern forms is discretized into collection of NI grid points uniformly distributed (with a distance $HI=L/NI$ between neighboring points) over the modeled segment and each is characterized by the concentrations of activator, inhibitor, and modulator. These concentrations can be initiated using the following loop:

```
for(i=0;i<NI;i++)
{
        u[i]=0.5;    // initial concentration of activator is 0.5
everywhere
        v[i]=0.5;    // initial concentration of inhibitor is 0.5
everywhere
        M[i]=1.0;    // initial concentration of modulator is 1.0
everywhere
}
```

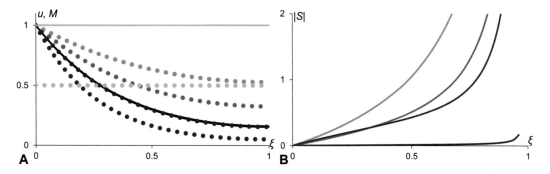

Fig. 4 Scaling in a diffusion–decay system with modulation. Morphogen profiles (panel **a**) and the absolute value of their sensitivity factors (panel **b**) are shown for the system (12) and (13) with $n = 0$, -1, -2 and -3 in equation (13). The solid black line corresponds to u-profile when $L = 1$ for all four cases of n while dotted lines – to u-profiles when $L = 0.5$. Sensitivity factors (panel **b**) are shown for $L = 1$. The green, red, blue, and purple colors correspond to $n = 0$, -1, -2 and -3 respectively. Solid and dotted orange lines represent profiles of the modulator, M, (given by equation (12)) when $L = 1$ and 0.5 respectively. The values of model parameters: $D_M = 10^6$, $D_u = 1$, $k_M = 1$, $k_u = 6.4$ and $p = a = 6$. Plots are produced using Excel (Microsoft Office)

The dynamics of these concentrations are calculated (in all points $i = 1, \ldots, NI{-}2$, except for the boundary points $i = 0$, $NI{-}1$) over a discrete time steps of duration HT using the explicit Euler's method with central differencing scheme for diffusion as shown by the following piece of C-code:

```
for(i=1;i<NI-1;i++)
{
if(i>NI/2-3 && i<NI/2+3) // area of M-degradation
hM[i]=M[i]+HT*(DM*(M[i+1]-2*M[i]+M[i-1])-kM*M[i]+p);
else // area where M does not degrade, and only diffuses and is
produced
hM[i]=M[i]+HT*(DM*(M[i+1]-2*M[i]+M[i-1])+p);hu[i]=u[i]+HT*
(Du*(u[i+1]-2*u[i]+u[i-1])+((u[i]*u[i]/v[i])-u[i]+sigma)/M
[i]/M[i]/coeff);hv[i]=v[i]+HT*(Dv*(v[i+1]-2*v[i]+v[i-1])+rho*
(u[i]*u[i]-v[i]))/M[i]/M[i]/coeff);if(time<0.1) // adding
noice for a short time to expose instablity
hu[i]+=N*(float)rand()/RAND_MAX);
}
```

where u, v, and M denote concentrations at time step n, while hu, hv, and hM—at time step $n + 1$. The terms Du, Dv, and DM are diffusion coefficients D_u, D_v, and D_M divided by HI^2. The coefficient N (found in the last line) defines the strength of noise applied for a short initial time ($t < 0.1$) on u-dynamics, which is needed to switch Turing instability on. After these computations are done, we need to update all three variables:

```
for(i=1;i<NI-1;i++)
{
    u[i]=hu[i];
    v[i]=hv[i];
    M[i]=hM[i];
}
```

and apply boundary conditions. Neumann boundary conditions at $x = 0$ and $x = L$ are implemented in the following way:

```
u[0]=u[1];      v[0]=v[1];     M[0]=M[1];     // at x=0
u[NI-1]=u[NI-2]; v[NI-1]=v[NI-2]; M[NI-1]=M[NI-2]; // at x=L
```

At the end of the simulation, we save the stationary morphogen profiles obtained (set of u- and v-concentrations in all grid points) on a file:

```
CStdioFile fFile("xxxxx.txt", CFile::modeCreate | CFile::mode-
Write);
// The above code shows the creation of a file containing its
name(first argument) and its mode (second argument).
        for(i=0;i<NI;i+=1)
        {
                str.Format("%4.2f    %17.15f       %17.15f
\n", float(i)/float(NI-1), u[i], v[i]);
                    fFile.Write(str, str.GetLength());
        }
```

The C++ code to run the numerical simulation of pattern formation in Turing model is provided in the supplementary file Turing.rar. In order to run the code, one would need to have Microsoft Visual Studio 2005 (or later) installed on the computer. Turing.rar contains a Microsoft Visual C++ project which can be extracted into the folder called "Turing." In order to open the project in Microsoft Visual Studio, click on the file Turing.sln. In the Studio environment, you can build and run the solution. When you run the solution, a screen called Turing will open and display the graphical output of simulation. Upon the end of simulation, the data representing the final pattern will be stored in a text file called "turing_patterns.txt". This file contains a few columns each giving values for different variables. The first column represents the relative position ξ, the second and third columns—the morphogen concentrations u and v respectively and the fourth column-the modulator concentration M. The file "turing_patterns.txt" can be opened in Excel to plot simulated pattern outside the Microsoft Studio environment.

Since we do not have an analytical solution for (15), we will not be able to derive analytically the sensitivity factor for the obtained

Turing pattern. Finding the sensitivity factor, S, defined by (8) requires calculations of the partial derivatives $\partial u/\partial L$ and $\partial u/\partial \xi$ (or alternatively $\partial v/\partial L$ and $\partial v/\partial \xi$). $\partial u/\partial \xi$ can be found from what was stored in the Excel file as a ratio of the difference between the two consecutive numbers in column B (or C) and the distance between grid points (in terms of the relative positions ξ: $\Delta \xi = NI^{-1}$). To find the derivative $\partial u/\partial L$, we need to find the numerical solution (i.e., perform the second simulation) for the medium size $L + \Delta L$. This solution (concentrations of u and v) can be stored in the same Excel file, say, in columns D and E respectively. To calculate $\partial u/\partial L$ for each point, ξ, we take the concentrations stored in the same row of columns B and D, find the difference between them and then divide it by the length difference, ΔL.

Results of simulations using the C++ code and Excel procedure described above are shown in Fig. 5. There are four stripes in the Turing pattern forming in the medium of size $L = 1$ while when $L = 2$, there are eight, five and a half, four, and three and a half stripes for $n = 0, -1, -2$, and -3 respectively. For $n = 0$, we have a system without a modulator where the characteristic length of the Turing pattern does not depend on the size of the medium and therefore the number of spikes in the pattern is proportional to the medium size. For $n = -1$, the characteristic length of Turing pattern is increasing with the medium size but not quick enough to conserve the number of spikes. For $n = -2$, we have a modulation resulting in perfect scaling when the characteristic length is proportional to the medium size and the number of spikes in the periodic pattern forming due to Turing instability does not depend on the medium size, L. Finally, for $n = -3$, the modulation results to overstretching of the Turing pattern so that the number of spikes gets smaller in the larger medium.

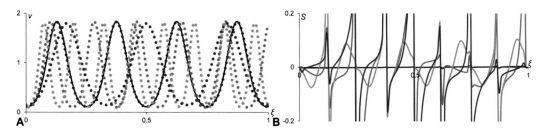

Fig. 5 Scaling of Turing pattern forming in Meinhardt model with passive modulation. Morphogen profiles (panel **a**) and their sensitivity factors (panel **b**) are shown for $n = 0, -1, -2$ and -3 (in green, red, blue, and purple respectively) in system (15). (**a**) The black line represents the profile (coinciding for all four values of n) when $L = 1$. Dotted lines represent profiles for $L = 2$. The values of model parameters: $D_u = 1$, $D_v = 50$, $\rho = 5$ and $\sigma = 0$. Plots are produced using Excel (Microsoft Office)

4   Notes

1. Derivation of the sensitivity factors (22)–(24) was done using Maple.

2. The analytical solution (28) was derived analytically. Finding six coefficients determined by six conditions is a long procedure.

Acknowledgments

This work has been supported by the EPSRC scholarship to M.R. and BBSRC grant BB/K002430/1 to B.V.

References

1. Wolpert L (1969) Positional information and spatial pattern of cellular differentiation. J Theor Biol 25(1):1–47

2. Driever W, Nussleinvolhard C (1988) A gradient of bicoid protein in drosophila embryos. Cell 54(1):83–93

3. Gregor T et al (2007) Stability and nuclear dynamics of the bicoid morphogen gradient. Cell 130(1):141–152

4. Lander AD et al (2005) Diverse paths to morphogen gradient robustness.

5. Umulis DM, Othmer HG (2013) Mechanisms of scaling in pattern formation. Development 140(24):4830–4843

6. Ben-Zvi D, Barkai N (2010) Scaling of morphogen gradients by an expansion-repression integral feedback control. Proc Natl Acad Sci U S A 107(15):6924–6929

7. Umulis DM, Othmer HG (2012) Scale invariance of morphogen-mediated patterning by flux optimization. In Biomedical Engineering and Informatics (BMEI), 2012 5th International Conference on 2012.

8. Gregor T et al (2005) Diffusion and scaling during early embryonic pattern formation. Proc Natl Acad Sci U S A 102 (51):18403–18407

9. Averbukh I et al (2014) Scaling morphogen gradients during tissue growth by a cell division rule. Development 141(10):2150–2156

10. Ben-Zvi D et al (2008) Scaling of the BMP activation gradient in Xenopus embryos. Nature 453(7199):1205–1211

11. Rasolonjanahary MI, Vasiev B (2016) Scaling of morphogenetic patterns in reaction-diffusion systems. J Theor Biol 404:109–119

12. Ishihara S, Kaneko K (2006) Turing pattern with proportion preservation. J Theor Biol 238(3):683–693

13. de Lachapelle AM, Bergmann S (2010) Precision and scaling in morphogen gradient readout. Mol Syst Biol 6:351

14. De Lachapelle AM, Bergmann B (2011) A new scaling measure quantifies the conservation of proportions of gene expression profiles. Developing ORGANIC SHapes.

15. Reeves GT, Fraser SE (2009) Biological systems from an engineer's point of view. PLoS Biol 7(1):e21

16. Murray JD (2003) Mathematical biology II: spatial models and biomedical applications, 3rd edn. Springer, New York

17. Murray JD (1982) Parameter space for turing instability in reaction diffusion mechanisms: a comparison of models. J Theor Biol 98 (1):143–163

18. Meinhardt H (1982) Models of biological pattern formation. Academic Press, London

19. Koch AJ, Meinhardt H (1994) Biological pattern formation: from basic mechanisms to complex structures. Rev Mod Phys 66:1481–1507

20. Vasiev BN (2004) Classification of patterns in excitable systems with lateral inhibition. Physics Letters A 323(3–4):194–203

21. Weiglhofer WS, Lindsay KA (1999) Ordinary differential equations and applications: mathematical methods for applied mathematicians, physicists, engineers, bioscientists. Vol. Horwood series in mathematics & applications. Chichester: Albion.

Chapter 16

Modelling Time-Dependent Acquisition of Positional Information

Laurent Jutras-Dubé, Adrien Henry, and Paul François

Abstract

Theoretical and computational modelling are crucial to understand dynamics of embryonic development. In this tutorial chapter, we describe two models of gene networks performing time-dependent acquisition of positional information under control of a dynamic morphogen: a toy-model of a bistable gene under control of a morphogen, allowing for the numerical computation of a simple Waddington's epigenetic landscape, and a recently published model of gap genes in Tribolium under control of multiple enhancers. We present detailed commented implementations of the models using `python` and `jupyter` notebooks.

Key words Modelling, Ordinary differential equations, Positional information, Multistability, Seg-mentation, Python

1 Introduction

In 1952, the mathematician Alan Turing coined the term "morphogen" to propose the existence of a chemical substance "reacting together and diffusing through a tissue," and "adequate to account for the main phenomena of morphogenesis" [1]. His original focus was explicitly chemical (in opposition to a mechanical view of morphogenesis), which explains that his original formalism relied on coupled ordinary differential equations (ODE) aiming to model biochemical interactions within the cells. Turing's proposal was purely theoretical, and inspired many subsequent theoretical and experimental works that explored how those morphogens could influence development. An example is Wolpert's French Flag Model [2] introducing the idea of a morphogen gradient defining different thresholds of activation and associated fates. The observed dynamic aspect of morphogenesis was also more explicitly taken into account. For instance, in the mid-1970s Cooke and Zeeman

Electronic supplementary material: The online version of this chapter (https://doi.org/10.1007/978-1-4939-8772-6_16) contains supplementary material, which is available to authorized users.

proposed that a "wavefront" coupled to a clock could explain segment formation [3], while Meinhardt [4] showed how a reaction–diffusion model could oscillate and stabilize to define metameric units.

As we entered the age of molecular biology, many predicted aspects of those models were confirmed experimentally. The segmentation cascade controlled by *bicoid* first appeared as an almost canonical example of the French Flag Model [2]. The existence of a clock controlling segmentation in vertebrates has been confirmed in the late 1990s [5], visualized in real time [6], and its many puzzling properties are under current scrutiny [7, 8]. Limb bud formation and patterning have been recently shown to be controlled by a Turing patterning mechanism [9].

In all these works, the ODE-based formalism first proposed by Turing has proved crucial to reconcile theory with experimental observations. Mathematical modelling of gene interaction networks allows not only to describe the most complex aspects of biological dynamics [10–12], but also to assess crucial questions such as system drift [13] or predictions of mutants (sometimes even without explicit gene networks, see, e.g., [14]). Another important aspect is evolution itself: by changing parameters of models, one can infer and simulate ancestral forms [15] and transitions between modes of patterning [16, 17], or even propose scenarios for macroevolutionary transitions leading to the emergence of new dynamic features [18].

In this chapter, we aim to describe simple basic principles for the simulation of gene networks acquiring time-dependent positional information under the control of dynamic morphogens. The principles and examples described are not exhaustive—in particular we do not attempt to model cell–cell communication—but constitute a good introduction to more complex formalisms and systems. We describe how to simulate simple gene networks with explicit implementation using `python` and `jupyter` notebooks. We first start with a toy-model network encoding bistability of fates depending on the dynamics of a morphogen gradient, then move to a more realistic model of Tribolium segmentation with wave-like dynamics of stabilizing gene expressions, similar to the model published in [17]. Both networks encode time-dependent positional information where the final state of a cell depends on the duration and level of exposure to a morphogen.

2   Materials

The code examples use the `python3` programming language. The choice of `python3` is motivated by its shallow learning curve, the numerous packages available and maintained by one of the largest communities of users, the user-friendly package manager `pip`, and its free license which makes it an ideal language for educational purposes. In addition, `python3` code can be executed within a `jupyter` notebook. This allows writing code and formatted comments—using the `Markdown` language—in a standalone document that can be easily shared. A notebook runs in a web browser and is composed of a succession of cells of type code or `Markdown`.

The decomposition of the code in different cells makes the interaction with the code very natural since after editing a code cell one triggers its execution by clicking a run button (or pressing Shift+Enter) and the output of the execution is displayed under the cell. `Markdown` cells can be inserted to add formatted explanations between code cells. For example, the `Markdown` language allows the use of different levels of titles, bold and italic, and LaTeX formulas. To install `python` one can choose to install the software alone but we recommend the use of the `anaconda` distribution (https://www.anaconda.com/download/) because it installs both `python` and several useful packages. Note that in this chapter we use `python` version > 3 since version 2.7 is no longer maintained. All the packages used in this chapter are included in `anaconda`:

- `jupyter`: web-based notebook.
- `scipy`: the `python` scientific library. We will use its integration tools.
- `numpy`: sub-library of `scipy` used for handling large arrays of numbers and random numbers.
- `matplotlib`: the most extensive `python` plotting library.

All notebooks detailed in this chapter are available on GitHub at the following address https://github.com/prfrancois/modelling_chapter. To run a notebook, from the `anaconda` navigator, simply launch `jupyter`, which opens a new tab in your web browser, from which you can browse your local files to find and run the desired notebook.

3   Methods

3.1   Modelling Dynamics in a Single Cell

We start with a toy-model implementing the dynamics of a simple bistable network in a single cell, under the control of an external morphogen, Fig. 1. The network is made of a single gene activating its own transcription, we call P the concentration of the corresponding protein in the cell. A time-varying morphogen can also control the transcription of this gene in an additive way. The equation for this system is

$$\frac{dP}{dt} = M(t) + r\frac{P^n}{P^n + P_0^n} - P \tag{1}$$

The left-hand side of this equation describes the rate of change for the concentration of protein P that we impose. $rP^n/(P^n + P_0^n)$ is the production rate of P, and $-P$ represents the degradation of the protein. We use here a Hill function for autoactivation, and assume

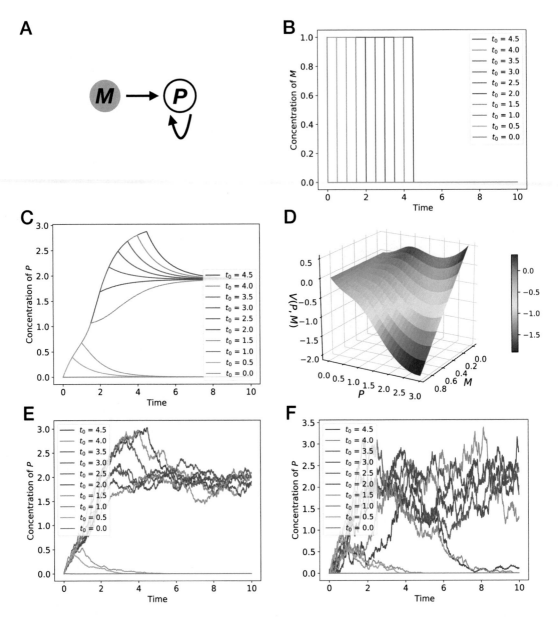

Fig. 1 Simulation of a toy-model of a bistable gene activated by a time-dependent morphogen. (**a**) Sketch of the model. (**b**) Time-dependent profiles of the morphogen used for the simulations. (**c**) Concentration of protein P as a function of time for the corresponding profiles of panel (**b**). (**d**) Epigenetic landscape for the model of panel (**a**). (**e**) Stochastic simulations of P with $\Omega = 100$. (**f**) Stochastic simulations of P with $\Omega = 10$

that proteins are degraded with a constant rate (rescaled to 1), *see* **Note 1**. A self-activating gene is well-known to produce bistable behavior, i.e., at steady state, the system can exhibit two states, one with $P = 0$ and another one with high P. Many biological systems indeed use this motif to define distinct cell fates (see, e.g., terminal selector genes [19]). The third term $M(t)$ describes the morphogen influence, and we assume that it acts in an additive way by increasing the production rate of P. We include a time-dependency with the goal of studying how the dynamics of the morphogen influences the final fate of the system.

1. In the first code cell of our notebook, we import the standard numerical methods numpy and define a function, taking as input arguments the concentrations of a protein P and of a morphogen M, and returning the value of the derivative defined in Eq. 1. Parameters were chosen to have bistability for P in the absence of M, i.e., the system can stabilize at two different concentrations depending on the past history of the system, *see* **Note 2**.

```
import numpy as np

# Definition of the parameters
dyn_parameters = {
    "r" : 2,    # Maximum production rate
    "n" : 5,    # Hill coefficients
    "p0": 1     # Level of p for half-activation
}

def dynamic(p, morphogen, param):

    r = param["r"]; p0 = param["p0"]; n = param["n"]
    dp = morphogen +r*p**n/(p**n+p0**n) -p

    return dp
```

2. We aim to study how the level of the protein P in a single cell depends on the time evolution of a morphogen M. We thus need to define a temporal evolution for the morphogen M. We will choose a very simple dynamics where the protein M is expressed at level M_{max}, and then suddenly disappears at time t_0 (*see* Fig. 1b for an illustration of the dynamics). In the second code cell, we define this dynamics, by taking as inputs the current time t, t_0 and M_{max}, and returning the corresponding value of the morphogen.

```
def compute_morphogen(t, t_0, m_max):

    m = m_max
    if (t > t_0):     m = 0.0

    return m
```

3. We now have everything we need to integrate the time course and do our simulations. We define a new function that will be returning an array of times, and corresponding morphogen and protein concentrations as a function of time corresponding to the simulated differential equation 1. The following function is an implementation of the Euler algorithm, the simplest numerical integrator. The idea of the Euler algorithm is to notice that the equality

$$P(t + d\tau) \simeq P(t) + d\tau \frac{dP(t)}{dt} \qquad (2)$$

becomes asymptotically exact as $d\tau$ gets smaller. So if we know $P(t)$ and $\frac{dP(t)}{dt}$, we can get a value for $P(t + d\tau)$, and from there

iterate and get a complete time course. The function below is
implementing this idea with $d\tau = 0.01$, *see* **Note 3**.

```
def compute_trajectory(t_0, m_max):

    init_time = 0.0
    total_time = 10.0
    time_step = 0.01  # Time-step of the Euler algorithm
    # Time points array: [0, 0.01, ..., 9.99]
    times = np.arange(init_time, init_time+total_time, time_step)

    p = 0.0  # Initial value of P
    # This list will be filled with the successive values of P(t)
    results = [p]
    # This list will be filled with the successive values of M(t)
    morphogen = [compute_morphogen(init_time, t_0, m_max)]

    # We iterate our process until the end of the 'times' array
    for t in times[1:]:

        # We compute the value of the morphogen concentration
        morpho = compute_morphogen(t, t_0, m_max)
        # We put this value in the corresponding list
        morphogen.append(morpho)

        dp = dynamic(p, morpho, dyn_parameters)  # We compute dP/dt
        p = p +time_step*dp  # We iterate P using the Euler algorithm
        results.append(p)     # We put this value in the corresponding list

    return times, morphogen, results
```

4. So far we only defined functions to perform the integration, now
   we actually integrate the equations, and plot the time course. We
   compute trajectories here for different values of t_0 corresponding
   to different times when the morphogen disappears. After import-
   ing the graphical library `matplotlib`, we define a list of times
   when we want to shut off the morphogen for comparison. Then,
   we compute the whole trajectories corresponding to those different times, and simply plot them all on the same graph, *see* **Note 4**.
   We also exemplify how to export the data to a csv file so that
   another program can use them.

```
import matplotlib.pyplot as plt
# Display plots in the notebook
%matplotlib inline

fig, ax = plt.subplots()  # We create a figure with 1 subplot

# We iterate over the different times at which M is shut off
times_off = np.arange(0.0, 5.0, 0.5)
for t_0 in times_off:

    # This is where we actually compute the trajectories of P
    times, morphogen, results = compute_trajectory(t_0, 1.0)
    # We plot the trajectories with different values of t_0
    ax.plot(times, results, label='$t_0$=%s'%t_0)

ax.legend(loc=2)
ax.set_xlabel('Time')
ax.set_ylabel('Concentration of $P$')

fig.savefig('Trajectories.png', dpi=300)

# Reshape the data: 1st column is t, 2nd column is P(t)
data_to_save = np.asarray([times, results]).transpose()
# Save the data in csv format, as a spreadsheet
# (elements of a row separated by ",")
np.savetxt("data.csv", data_to_save, delimiter=",")
# Load the data that we just saved
read_data = np.genfromtxt("data.csv", delimiter=",")
```

The output of this program is presented in Fig. 1c: we can see that if the morphogen disappears early, P starts increasing before dying when the morphogen disappears; however, if the morphogen is expressed for a longer time, we clearly see that P is high enough to actually turn on and sustain its own production, stabilizing in a high state. This is thus a simple model of differentiation relying on a time-dependent morphogen.

5. An interesting aspect of the simple model displayed here is that we can make a simple connection to classical notions such as Waddington's epigenetic landscape [20]. The reason is that the dynamics of P can be written as the derivative of a potential

$$\frac{dP}{dt} = -\frac{\partial V(P, M)}{\partial P} \tag{3}$$

Such dynamics can be interpreted as a damped system evolving in an energy landscape defined by V. Therefore, it is worth visualizing V as a function of M to see how this landscape is "tilted" with increasing M, leading to a change of fate (corresponding here to different values of P). This is done in the following code cell, which, for each value of the morphogen M and each value of P (corresponding to coordinates X and Y), plots the value of the landscape $V(M, P)$. Intuitively, the system is driven towards the values of P corresponding to lower values of the landscape (like energy minima in physics), colder colors in Fig. 1d, *see* **Note 5**.

```
from scipy.integrate import quad
from mpl_toolkits.mplot3d import axes3d
from matplotlib import cm

# X corresponds to morphogen concentrations
X_lin = np.arange(0.0, 1.0, 0.1)
nx = len(X_lin)
# Y corresponds to protein concentrations
Y_lin = np.arange(0.0, 3.0, 0.1)
ny = len(Y_lin)
X, Y = np.meshgrid(X_lin, Y_lin)
# Z corresponds to the potential V(M,P)
Z = np.zeros((ny, nx))

for i in range(nx):

    # Derivative of the potential with fixed morphogen
    g = lambda p: -dynamic(p, X_lin[i], dyn_parameters)

    for j in range(ny):

        # Integration of the potential
        value, error = quad(g, 0, Y_lin[j])
        Z[j,i] = value

# We specify the 3D plot by adding the '3d' keyword
fig, ax = plt.subplots(subplot_kw={"projection":"3d"})

surf = ax.plot_surface(X, Y, Z, cmap=cm.coolwarm, linewidth=0, antialiased
    =False)

fig.colorbar(surf, shrink=0.5, aspect=15)
ax.view_init(elev=20, azim=30)
ax.set_xlabel('$M$')
ax.set_ylabel('$P$')
ax.set_zlabel('$V(P,M)$')

fig.savefig('Landscape.png', dpi=300)
```

6. Lastly, this example is simple enough to study in a straightforward way what happens if we include biological noise. We will use the τ-leaping approximation [21] to integrate P as a function of time with the modified stochastic equation

$$P(t + d\tau) \simeq P(t) + d\tau(M(t) + r\frac{P^n}{P^n + P_0^n} - P)$$
$$+ \left(\sqrt{\frac{d\tau}{\Omega}}\sqrt{M(t) + r\frac{P^n}{P^n + P_0^n} + P}\right)\mathcal{N}_t(0, 1) \tag{4}$$

In the previous equation, $\mathcal{N}_t(0, 1)$ represents a Gaussian random number with average 0 and variance 1 (subscript t indicates that a new Gaussian number needs to be randomly drawn at each time step), and Ω is a concentration scale defining the typical maximal number of proteins in the system, *see* **Note 6**. We can make minimal modifications to functions defined previously to perform the stochastic integration, and vary the typical concentration scale Ω to increase or decrease the noise value, *see* Fig. 1e, f.

```
def compute_noise(p, morphogen, param):

    r = param["r"]; p0 = param["p0"]; n = param["n"]
    noise = np.random.normal()
    intensity = np.sqrt(morphogen +r*p**n/(p**n+p0**n) +p)

    return noise*intensity


def compute_trajectory_stat(t_0, m_max, typical_concentration):

    init_time = 0.0
    total_time = 10.0
    time_step = 0.01
    times = np.arange(init_time, init_time+total_time, time_step)

    p = 0.0
    results=[p]
    morphogen=[compute_morphogen(init_time, t_0, m_max)]

    for t in times[1:]:

        morpho = compute_morphogen(t, t_0, m_max)
        morphogen.append(morpho)

        dp = dynamic(p, morpho, dyn_parameters)
        noise = np.sqrt(time_step/typical_concentration)*compute_noise(p,
    morpho, dyn_parameters)
        p = p +time_step*dp +float(noise)
        if (p < 0.0):    p = 0.0
        results.append(p)

    return times, morphogen, results


fig, ax = plt.subplots()

for t_0 in times_off:
    times, morphogen, results = compute_trajectory_stat(t_0, 1.0, 10.0)
    ax.plot(times, results, label='$t_0$=%s'%t_0)

ax.legend(loc=2)
ax.set_xlabel('Time')
ax.set_ylabel('Concentration of $P$')

fig.savefig('Noise_10.png', dpi=300)
```

3.2 Modelling Dynamics in an Embryo

In this second part, we simulate a more realistic model of acquisition of time-dependent positional information. We show how to reproduce the results of [17] on Tribolium segmentation and generate corresponding kymographs (time-space maps of genetic expression). The network that we simulate is displayed on Fig. 2a. It encodes a "clock-like" behavior with a cascade of genes expressed sequentially, before stabilizing as the *caudal* morphogen gradient disappears.

1. For this problem, we need to make the distinction between different cells seeing different dynamics of a morphogen (here, *caudal*). We model an embryo as a line of 200 cells representing the tissue undergoing a patterning process. Each cell is an array with one variable corresponding to each protein in the network. We initialize all concentrations at 0, except for the first gene, *hb*, initialized at 0.2, *see* **Note 7**.

```
import numpy as np

n_cells = 200
ap_positions = np.linspace(0.0, 1.0, n_cells)

n_species = 5
species_names = ['hunchback', 'Kruppel', 'mille-pattes', 'giant', 'X']
species_colors = ['tab:blue', 'tab:red', 'tab:green', 'tab:orange', 'k:']

init_conc = np.zeros((n_cells, n_species))
init_conc[:,0] = 0.2
```

2. Similarly to the previous section, we define functions encoding the differential equations describing the interactions between proteins inside a single cell. Formally, the time derivative of the system can be written as

$$\frac{d\vec{P}}{dt} = D(\vec{P}, M(t)) + S(\vec{P}, M(t)) - \vec{P} \qquad (5)$$

where $\vec{P}$ is a vector of protein concentrations corresponding to genes *hb*, *Kr*, *mlpt*, *gt* and a "terminator" gene X shutting off *gt* eventually.

For this problem the derivative function is made of two parts: a dynamic network of interactions is encoded in $D(\vec{P}, M(t))$, and produces sequential gene activation. A static network of interactions $S(\vec{P}, M(t))$ amplifies and stabilizes an initial bias in the expression of one of the genes, while also attenuating the expression of the other genes. The static network is multistable, defining different stable cellular fates for $M(t) = 0$. Intuitively, the static and dynamic networks correspond to two families of transcriptional enhancers. The morphogen $M(t)$ controls the relative weight of each enhancer, so that the dynamics of the morphogen encodes the future fate.

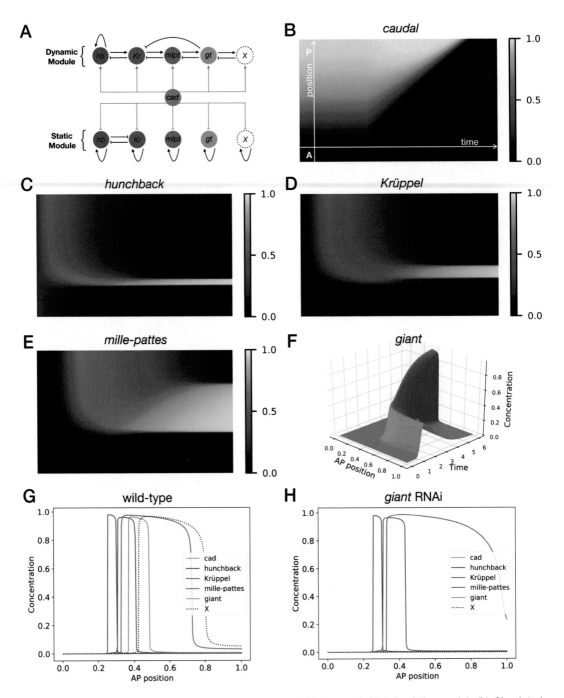

Fig. 2 Simulation of a model of gap-gene dynamics in Tribolium. (**a**) Sketch of the model. (**b**) Simulated kymograph for the imposed maternal gene *caudal*. (**c–e**) Simulated kymographs for the gap genes *hunchback*, *Krüppel*, and *mille-pattes*. (**f**) Simulated 3D surface for the gap gene *giant*. (**g**) Steady-state profile for the simulation of the wild-type network showing correct relative positioning of the gap genes. (**h**) Steady-state profile for a simulation of an RNAi perturbation

The functions require three input arguments. The first is the vector of concentrations (one component for each species), and the second the morphogen concentration. Notice here that we define the dynamics for one cell: the spatial difference only enters the equations through the morphogen dynamics that will be included below. We also include the possibility to model RNAi experiments by using a vector named `rna_i`, *see* **Note 8**.

```python
# Dynamic part of the derivative
def dynamic(conc_vec, morphogen, rna_i):

    c = morphogen/(1.0+morphogen)
    hb = conc_vec[0]
    kr = conc_vec[1]
    mlpt = conc_vec[2]
    gt = conc_vec[3]
    x = conc_vec[4]

    d_hb = 3.0*rna_i[0]*c*(hb/0.2)**5/(1.0 +(hb/0.2)**5)/(1.0 +(kr/0.12)
    **5)
    d_kr = 3.0*rna_i[1]*c*(hb/0.4)**5/(1.0 +(hb/0.4)**5)/(1.0 +(mlpt/0.25)
    **5)/(1.0 +(gt/0.01)**5)
    d_mlpt = 3.0*rna_i[2]*c*(kr/0.4)**5/(1.0 +(kr/0.4)**5)/(1.0 +(gt/0.3)
    **5)
    d_gt = 3.0*rna_i[3]*c*(mlpt/0.4)**5/(1.0 +(mlpt/0.4)**5)/(1.0 +(x
    /0.08)**5)
    d_x = 3.0*rna_i[4]*c*(gt/0.4)**5/(1.0 +(gt/0.4)**5)

    return np.array([d_hb, d_kr, d_mlpt, d_gt, d_x])


# Static part of the derivative
def static(conc_vec, morphogen, rna_i):

    c = 1.0/(1.0+morphogen)
    hb = conc_vec[0]
    kr = conc_vec[1]
    mlpt = conc_vec[2]
    gt = conc_vec[3]
    x = conc_vec[4]

    d_hb = rna_i[0]*c*(hb/0.4)**5/(1.0 +(hb/0.4)**5)/(1.0 +(kr/0.4)**5)
    d_kr = rna_i[1]*c*(kr/0.4)**5/(1.0 +(kr/0.4)**5)/(1.0 +(hb/0.4)**5)
    d_mlpt = rna_i[2]*c*(mlpt/0.4)**5/(1.0 +(mlpt/0.4)**5)
    d_gt = rna_i[3]*c*(gt/0.4)**5/(1.0 +(gt/0.4)**5)
    d_x = rna_i[4]*c*(x/0.4)**5/(1.0 +(x/0.4)**5)

    return np.array([d_hb, d_kr, d_mlpt, d_gt, d_x])


# Derivative function joining the two components
def derivative_single_cell(conc_vec, morphogen, rna_i):

    return dynamic(conc_vec, morphogen, rna_i) +static(conc_vec, morphogen
    , rna_i) -conc_vec
```

3. Again similarly to the previous section, we need to define the dynamics of a morphogen, and this time we need to take into account the spatial dependency of the morphogen, which is one extra parameter of the corresponding function. We model the dynamics of the morphogen as follows: the gradient remains static for a given amount of time (`time_static_morphogen`), *see* **Note 9**. Then it starts to retract spatially (via a linear increase of `flat_start`) while also becoming steeper (via an exponential increase of the exponent n), *see* **Note 10**.

```
def compute_morphogen(x, t, morphogen_parameters):

    peak = morphogen_parameters["peak"]
    shift = morphogen_parameters["shift"]
    time_static_morphogen = morphogen_parameters["time_static_morphogen"]
    retracting_speed = morphogen_parameters["retracting_speed"]

    if (t < time_static_morphogen):

        flat_start = morphogen_parameters["flat_start"]
        n = morphogen_parameters["n"]

    else:

        flat_start = morphogen_parameters["flat_start"]+retracting_speed*(
t-time_static_morphogen)
        n = morphogen_parameters["n"]*np.exp(t-time_static_morphogen)
        if (n > 100.0):     n = 100.0

    morphogen = peak*((x-shift)/flat_start)**n/(1.0 +((x-shift)/flat_start
)**n)
    morphogen[(x-shift) < 0.0] = 0.0

    return morphogen
```

4. We now define the function to integrate the dynamics over the entire embryo. So far we have defined a function that computes the derivative of a single cell that takes an array of five elements (the five different proteins) as an argument. However, the global derivative function must process a vector of size n_cells×n_species, corresponding to the n_species= 5 protein concentrations in each cell of the modelled embryo. The role of the global derivative function is to split the global concentration vector into n_cells smaller vectors that correspond to the individual cells and can be passed to derivative_single_cell, *see* **Note 11**.

   Furthermore, two objects are considered as parameters of the global derivative function: the vector for the morphogen profile across the whole tissue and the rna_i vector for RNAi simulations.

```
def derivative_whole_tissue(t, conc_vec, morphogen_parameters, rna_i):

    conc_vec = conc_vec.reshape(n_cells, n_species).transpose()
    morphogen = compute_morphogen(ap_positions, t, morphogen_parameters)

    return derivative_single_cell(conc_vec, morphogen, rna_i).transpose().
reshape(n_cells*n_species)
```

5. We now define a function taking various arguments and performing the actual numerical integration. The first argument is the matrix of initial concentrations of every species in every cell. The second argument is a dictionary specifying the initial time, the total duration, and the time step of integration. The third argument is another dictionary specifying the

different parameters required for modelling the morphogen gradient at each time point. The last argument is the array specifying if we perform RNAi experiments on one of the genes if needed.

The function first sets the integration options, and then performs the integration, using the `scipy.integrate` numerical `ode` integrator, *see* **Note 12**. It returns three outputs. The first output is a `numpy` array consisting of the results of the simulation: for every cell, we have a matrix for which each line represents the temporal evolution of the concentration of one protein inside that specific cell. The second output is the morphogen gradient across the tissue at each time point. The third output is the array of times used for the integration.

```python
from scipy.integrate import ode

def integration(init_conc, integration_time, morphogen_parameters, rna_i):

    # Set the integration options
    integrator = ode(derivative_whole_tissue)
    integrator.set_integrator('lsoda', with_jacobian=False, rtol=1e-5,
    nsteps=1000)
    integrator.set_f_params(morphogen_parameters, rna_i)

    total_time = integration_time["total_time"]
    time_step = integration_time["time_step"]
    init_time = integration_time["init_time"]
    times = np.arange(init_time, init_time+total_time, time_step)

    init_conc = init_conc.reshape(n_cells*n_species)
    integrator.set_initial_value(init_conc, init_time)

    results = [init_conc]
    morphogen_gradient = [compute_morphogen(ap_positions, init_time,
    morphogen_parameters)]


    # Perform the integration
    for t in times[1:]:

        integrator.integrate(t)

        if (integrator.successful):

            results.append(integrator.y)
            morphogen_gradient.append(compute_morphogen(ap_positions,
    integrator.t, morphogen_parameters))

        else:
            print("An error occurred during the integration.")
            break


    # Output the results
    results = np.array(results).transpose()
    results = results.reshape(n_cells, n_species, len(results[0]))
    morphogen_gradient = np.array(morphogen_gradient).transpose()

    return results, morphogen_gradient, times
```

6. Now that we have defined all integration tools, we build some routine allowing to visualize the spatiotemporal dynamics of the

system. Very simply, the spatiotemporal dynamics of each gene can be encoded into a matrix, where one dimension corresponds to time and the other to space. A kymograph is a visualization of such a matrix. We can rescale the concentrations of each protein between 0 and 1 and define a color map continuously associating colors to concentrations, which defines the kymograph. We can do this for each protein, producing the kymographs of Fig. 2b–e, *see* **Note 13**.

```python
import matplotlib.pyplot as plt
from matplotlib import colors
%matplotlib inline

def plot_kymographs(results, morphogen_gradient):

    fig, ax = plt.subplots(3,2)

    # Plot kymographs for the different species
    for species_index in range(n_species):

        i = int((species_index)/2)  # Row index in the plot grid
        j = species_index%2  # Column index in the plot grid

        ax[i,j].axis('off')

        normal = colors.Normalize(vmin=0.0, vmax=1.0)
        graph = ax[i,j].imshow(results[:,species_index,:], norm=normal,
origin='lower', interpolation='bilinear', aspect='auto')

        colorbar = fig.colorbar(graph, ticks=[1.0,0.5,0.0], ax=ax[i,j])
        colorbar.ax.tick_params(labelsize=8)
        ax[i,j].set_title(species_names[species_index], fontsize=10)


    # Plot a kymograph for the morphogen gradient
    ax[2,1].axis('off')

    normal = colors.Normalize(vmin=0.0, vmax=1.0)
    graph = ax[2,1].imshow(morphogen_gradient, norm=normal, origin='lower'
, interpolation='bilinear', aspect='auto')

    colorbar = fig.colorbar(graph, ticks=[1.0,0.5,0.0], ax=ax[2,1])
    colorbar.ax.tick_params(labelsize=8)
    ax.set_title('caudal', fontsize=10)

    fig.savefig('Kymographs.png', dpi=300)
```

7. Rather than plotting the kymograph, we can also represent the concentration of a given species across the whole segmenting tissue for all time points as a 3D surface. The following function plots and saves such a graph for the species with index specified by the argument `species_index` (here, from 1 to 5, and not from 0 to 4). We can also define a function that plots and saves the final concentration profile across the whole tissue, and a function that plots and saves the time evolution of the system in a given cell.

```
from mpl_toolkits.mplot3d import Axes3D

def plot_3d_surface(species_index, results, times):

    fig, ax = plt.subplots(subplot_kw={"projection":"3d"})

    x, t = np.meshgrid(ap_positions, times)
    ax.plot_surface(x, t, results[:,species_index-1,:].transpose(), color=
    species_colors[species_index-1])

    ax.set_title(species_names[species_index-1])
    ax.set_xlabel("AP position")
    ax.set_ylabel("Time")
    ax.set_zlabel("Concentration")

    fig.savefig('Conc_3d_surface.png', dpi=300)


def plot_final_conc_profile(results, morphogen_gradient):

    fig, ax = plt.subplots()

    ax.plot(ap_positions, morphogen_gradient[:,-1], color='tab:gray',
    label='caudal')
    for species_index in range(n_species):
        ax.plot(ap_positions, results[:,species_index,-1], species_colors[
    species_index], label=species_names[species_index])

    ax.set_xlabel('AP position')
    ax.set_ylabel('Concentration')
    ax.legend()

    fig.savefig('Final_conc.png', dpi=300)


def plot_time_course(cell_position, results, morphogen_gradient, times):

    fig, ax = plt.subplots()

    ax.plot(times, morphogen_gradient[cell_position,:], color='tab:gray',
    label='caudal')
    for species_index in range(n_species):
        ax.plot(times, results[cell_position,species_index,:],
    species_colors[species_index], label=species_names[species_index])

    ax.set_xlabel('Time')
    ax.set_ylabel('Concentration in the cell at '+str(int(100*ap_positions
    [cell_position]))+'% of the tissue')
    ax.legend()

    fig.savefig('Conc_time_course.png', dpi=300)
```

8. Lastly, we are ready to simulate and plot the desired behavior. The following lines initialize the simulations and use all previously defined functions to visualize results. Figure 2g represents the steady-state profile of this model for the "wild-type" network, with the correct relative position of various gap genes [17].

```
# Specify the value of all required parameters

integration_time = {
    "total_time" : 6.0,
    "time_step" : 0.02,
    "init_time" : 0.0
}

morphogen_parameters = {
    "peak" : 1.0,
    "shift" : 0.0,
    "flat_start" : 0.4,
    "n" : 4.0,
    "time_static_morphogen" : 2.0,
    "retracting_speed" : 0.2
}

rna_i = np.ones(n_species)


# Perform the integration

results, morphogen_gradient, times = integration(init_conc,
    integration_time, morphogen_parameters, rna_i)


# Generate the desired figures

plot_kymographs(results, morphogen_gradient)

species_index = 2
plot_3d_surface(species_index, results, times)

plot_final_conc_profile(results, morphogen_gradient)

cell_position = int(0.45*n_cells)
plot_time_course(cell_position, results, morphogen_gradient, times)
```

9. To model RNAi knockdown experiments, we run the simulations by putting to 0 the gene that we wish to remove. For instance, the following lines simulate and show the behavior of the embryo for *gt* RNAi. For another example where we modify the morphogen input, *see* **Note 14**.

```
rna_i = np.array([1,1,1,0,1])

results, morphogen_gradient, times = integration(init_conc,
    integration_time, morphogen_parameters, rna_i)

plot_kymographs(results, morphogen_gradient)
plot_final_conc_profile(results, morphogen_gradient)
```

4   Notes

1. We make several classical modelling approximations: we model protein concentration with a single continuous variable, we model self-regulation of the protein with a Hill function, and condense (fast) transcription and translation into a single term. We give a brief example of stochastic generalization of the formalism using the τ leaping algorithm [21] at the end of this section.

2. To keep bistability when varying parameters, we need to keep the same order of magnitude for all parameters referring to a number of proteins, i.e., the maximum production rate r, the level of P for half-activation P_0, and the maximum level of morphogen M_{max}. We invite the reader to multiply all these parameters by the same factor P^*. The same qualitative behavior should be observed, but at the new scale imposed by P^*. Note that the typical concentration scale of the system will then be multiplied by P^*, and thus the parameter Ω defined below in stochastic simulations needs to be divided by P^* to obtain the same level of noise. Furthermore, the maximum production rate r must be larger than the level of P for half-activation P_0 to observe bistability. The Hill coefficient n must be greater than 1 to allow for bistability. The Hill function becomes a step-function in the limit of very large Hill coefficients.

3. The Euler algorithm is potentially imprecise if the time-step $d\tau$ is too big, however, its simplicity makes it very convenient to explain basic concepts and for relatively simple simulations as described here. The "rule of thumb" to ensure a proper numerical convergence is to check that when we decrease the time step $d\tau$, the behavior of the integrated system does not change. To find a good initial value $d\tau$, a good strategy is to identify the presumptive fastest time-scale of the system, then take a much smaller value for $d\tau$. For instance, for Eq. 1, the time-scale is driven by the degradation rate of P, which is 1 in arbitrary units. We thus choose a much smaller time step $d\tau = 0.01$. In the following section we introduce a more efficient (but also less transparent) built-in numerical integrator.

4. The function that allows us to generate a figure with multiple subplots, `plt.subplots`, can be used in two ways. `fig, ax = plt.subplots()` creates a single figure, ax being the unique subplot. `fig, ax = plt.subplots(n,m)` generates a figure with $n \times m$ subplots on n rows and m columns. In that case, ax is an array of subplots, and `ax[i,j]` is used to plot the figure on row i and column j.

5. Since dP/dt is the derivative of V, we need to integrate it numerically with respect to P, which is done with the numerical package `scipy`. There is a little subtlety here since our numerical definition of dP/dt is done in the `dynamic` function above, which depends on both the morphogen and the protein. So to integrate with respect to P we need to fix the morphogen, and for this we use the so-called `lambda` operator to compactly redefine a new function g computing dP/dt as a function of P for a fixed value of the morphogen.

6. Equation 4 can be understood in the following way: Eq. 1 can be interpreted as the continuous version of a stochastic system

where protein P is produced with a Poisson rate $\rho(t) = M(t) + r\frac{P^n}{P^n + P_0^n}$ and degraded with a Poisson rate $\delta(t) = P(t)$. This means that during a small interval $[t, t + d\tau]$, there are typically $\rho(t)d\tau$ proteins produced (resp. $\delta(t)d\tau$ proteins degraded), and for a Poisson process we know that the variance is equal to the mean, so that variance is $\rho(t)d\tau$ as well for production (resp. $\delta(t)d\tau$ for degradation). The τ-leaping algorithm approximates those Poisson processes with Gaussian processes with same mean and variance (and thus is more accurate when those rates get bigger), so that

$$P(t + d\tau) \simeq P(t) + \mathcal{N}_\rho(\rho(t)d\tau, \rho(t)d\tau) - \mathcal{N}_\delta(\delta(t)d\tau, \delta(t)d\tau)$$

(6)

where $\mathcal{N}(a, b)$ is used to indicate a random Gaussian number of average a and variance b. The minus sign in front of $\mathcal{N}_\delta$ indicates that this process corresponds to a degradation. The last step is to write that for two independent Gaussian numbers $\mathcal{N}_x, \mathcal{N}_y$:

$$\mathcal{N}_x(a, b) - \mathcal{N}_y(c, d) = a - c + \mathcal{N}(0, b + d) = a - c + \sqrt{b + d}\mathcal{N}(0, 1)$$

(7)

which comes from the fact that the sum of two independent Gaussian numbers is the sum of their average plus a Gaussian number with average 0 and variance equal to the sum of variances, and finally rescaling the last Gaussian number to express it as a function of $\mathcal{N}(0, 1)$. Lastly, the reason why we can extract a typical concentration scale Ω comes from the fact that the differential equation 1 is invariant with respect to the rescaling $P \to \Omega P$, $M \to \Omega M$, $P_0 \to \Omega P_0$, $r \to \Omega r$. This means that we can define a typical concentration Ω for P, and rescale the differential equation using the previous rescaling without changing the continuous dynamics. With $r = 2$ as assumed in our notebook, Ω represents the half maximum number of proteins P. When computing the τ-leaping version of the equation, if we perform the same rescaling, there is a term $1/\sqrt{\Omega}$ left in front of the noise term, which comes from the fact that the noise level is tied to the number of molecules in the system. Intuitively, the smaller Ω, the bigger this term, consistent with the intuition that the effective noise "increases" as the number of molecules in the system decreases. *See* [22] for a study of this effect in a developmental context.

7. For plotting purposes, we define the array `ap_positions` consisting of the positions of the cells relative to the antero-posterior axis of the patterning tissue (normalized from 0 to 1). We find it more convenient to use the antero-posterior position of a cell rather than its index as the positional argument x since

it allows us to choose any number of cells to represent the whole segmenting tissue without affecting the resulting dynamics. This will be important later to define the position in corresponding functions. We also store the biochemical species' names (i.e., proteins) in the list `species_names`, and assign a different color to each species in the list `species_colors`.

8. The length of the `rna_i` vector needs to be equal to the number of species. Each of its elements must be equal to either 0 (meaning that we model this species' RNAi knockdown) or 1 (else).

9. To model a static morphogen gradient, we would simply need to set the time during which the morphogen gradient remains static, `time_static_morphogen`, to a value that is greater than the total simulation time.

10. Applying a condition to an array of a given size returns an array of the same size containing Boolean variables indicating whether or not the associated element of the original array satisfies the condition. We use this `numpy` functionality here to set the morphogen concentration to 0 in a specific region of the tissue.

11. This derivative is the function that will be passed in the argument of the `ode` function (from the `scipy` package) that will allow us to perform the integration to find the time course of the dynamics. `ode` accepts only functions that follow certain rules: `def derivatives(t, concentrations, param1, param2, ...)`. For the `ode` function to work correctly, the first argument of the function that we wish to integrate needs to be time, and the second argument needs to be a 1D array. The following arguments are optional and are seen as parameters. They need to be passed to `ode` using the `set_f_params` function.

12. When dealing with stiff equations for which the time-scale of the dynamics can vary greatly, it is better to use a stiff integrator like `lsoda`. Other (non-stiff) integrators include `dopri5`, which is based on the Runge–Kutta method. As mentioned in the previous note, the `ode` function requires the second argument of the function that we wish to integrate to be a single array. Therefore, we need to reshape the array for the initial concentrations `init_conc` accordingly. We also test if the integration was successful at each time step before keeping the results. If the integration was not successful, we stop the integration and print an error message. Finally, if needed, the current species concentrations and the current time can be accessed via `integrator.y` and `integrator.t`, respectively.

13. In the `imshow` function, we specify the interpolation scheme to get a smoother image. We also set the aspect to `auto`, which can improve the visual aspect of the kymograph, especially when we have a lot more time points than positions, or vice versa. We normalize the color scheme for the heatmaps to the same minimal and maximal values (`vmin` and `vmax`) for the five different species to allow better comparison. However, this means that we are not able to visualize variations in concentrations higher than `vmax`.

14. We can also model other RNAi perturbations that change the dynamics of the morphogen by modifying directly the parameters used to model the spatiotemporal distribution of the morphogen. For example, *pan* RNAi embryos exhibit a *cad* gradient that is reduced, stretched, and shifted anteriorly compared to wild-type embryos [17].

```
rna_i = np.array([1,1,1,1,1])

morphogen_parameters["peak"] = 0.5          # We reduce the gradient,
morphogen_parameters["n"] = 2.0             # stretch it,
morphogen_parameters["shift"] = -0.2/3.0    # and impose an anterior shift

results, morphogen_gradient, times = integration(init_conc, integration_time,
    morphogen_parameters, rna_i)

plot_final_conc_profile(results, morphogen_gradient)
```

Acknowledgements

We thank Ezzat El-Sherif for sharing the MATLAB code used in [17] and the referee for useful comments.

References

1. Turing AM (1952) The chemical basis of morphogenesis. Philos Trans R Soc Lond Ser B Biol Sci 237(641):37–72

2. Wolpert L (2006) Principles of development. Oxford University Press, Oxford

3. Cooke J, Zeeman EC (1976) A clock and wavefront model for control of the number of repeated structures during animal morphogenesis. J Theor Biol 58(2), 455–476

4. Meinhardt H (1982) Models of biological pattern formation. Academic, New York

5. Palmeirim I, Henrique D, Ish-Horowicz D, Pourquié O (1997) Avian hairy gene expression identifies a molecular clock linked to vertebrate segmentation and somitogenesis. Cell 91(5):639–648

6. Aulehla A, Wiegraebe W, Baubet V, Wahl MB (2008) Deng C, Taketo M, Lewandoski M, Pourquié O. A beta-catenin gradient links the clock and wavefront systems in mouse embryo segmentation. Nat Cell Biol 10(2):186–193

7. Hubaud A, Pourquié O (2013) Making the clock tick: right time, right pace. Dev Cell 24 (2):115–116

8. Lauschke VM, Tsiairis CD, François P, Aulehla A (2013) Scaling of embryonic patterning based on phase-gradient encoding. Nature 493(7430):101–105

9. Raspopovic J, Marcon L, Russo L, Sharpe J (2014) Modeling digits. Digit patterning is controlled by a Bmp-Sox9-Wnt Turing network modulated by morphogen gradients. Science 345(6196):566–570

10. Jaeger J, Surkova S, Blagov M, Janssens H, Kosman D, Kozlov KN, Manu, Myasnikova E, Vanario-Alonso CE, Samsonova M, Sharp DH, Reinitz J (2004) Dynamic control of positional information in the early Drosophila embryo. Nature 430(6997):368–371

11. Crombach A, Wotton KR, Jiménez-Guri E, Jaeger J (2016) Gap gene regulatory dynamics evolve along a genotype network. Mol Biol Evol 33:1293–1307

12. Balaskas N, Ribeiro A, Panovska J, Dessaud E, Sasai N, Page KM, Briscoe J, Ribes V (2012) Gene regulatory logic for reading the sonic hedgehog signaling gradient in the vertebrate neural tube. Cell 148(1–2):273–284

13. Wotton KR, Jiménez-Guri E, Crombach A, Janssens H, Alcaine-Colet A, Lemke S, Schmidt-Ott U, Jaeger J (2015) Quantitative system drift compensates for altered maternal inputs to the gap gene network of the scuttle fly Megaselia abdita. eLife 4:e04785

14. Corson F, Siggia ED (2017) Gene free methodology for cell fate dynamics during development. eLife 6:e30743

15. Rothschild JB, Tsimiklis P, Siggia ED, François P (2016) Predicting ancestral segmentation phenotypes from drosophila to anopheles using in silico evolution. PLoS Gen 12(5):e1006052–19

16. Peel A, Akam M (2003) Evolution of segmentation: rolling back the clock. Curr Biol 13(18):R708–10

17. Zhu X, Rudolf H, Healey L, François P, Brown SJ, Klingler M, El-Sherif E (2017) Speed regulation of genetic cascades allows for evolvability in the body plan specification of insects. Proc Natl Acad Sci USA 128(41):E8646–E8655

18. François P, Hakim V, Siggia ED (2007) Deriving structure from evolution: metazoan segmentation. Mol Syst Biol 3:9

19. Hobert O (2008) Regulatory logic of neuronal diversity: terminal selector genes and selector motifs. Proc Natl Acad Sci USA 105(51):20067–20071

20. Waddington CH (2014) The strategy of the genes. Routledge, London

21. Gillespie DT (2007) Stochastic simulation of chemical kinetics. Annu Rev Phys Chem 58:35–55

22. Perez-Carrasco R, Guerrero P, Briscoe J, Page KM (2016) Intrinsic noise profoundly alters the dynamics and steady state of morphogen-controlled bistable genetic switches. PLoS Comput Biol 12(10):e1005154

INDEX

Julien Dubrulle (ed.), *Morphogen Gradients: Methods and Protocols*, Methods in Molecular Biology, vol. 1863,
https://doi.org/10.1007/978-1-4939-8772-6, © Springer Science+Business Media, LLC, part of Springer Nature 2018

Printed in the United States
By Bookmasters